YEAST MOLECULAR BIOLOGY—RECOMBINANT DNA

YEAST MOLECULAR BIOLOGY—RECOMBINANT DNA

Recent Advances

Edited by

Michael S. Esposito

Lawrence Berkeley Laboratory
University of California
Berkeley, California

np NOYES PUBLICATIONS
Park Ridge, New Jersey, USA

Library of Congress Catalog Card Number 84-4096
ISBN: 0-8155-0987-1
Printed in the United States

Published in the United States of America by
Noyes Publications
Mill Road, Park Ridge, New Jersey 07656

10 9 8 7 6 5 4 3 2 1

Library of Congress Cataloging in Publication Data
Main entry under title:

Yeast molecular biology--recombinant DNA.

Proceedings of the First Berkeley Workshop on Recent Advances in Yeast Molecular Biology, sponsored by the Biology and Medicine Division, Lawrence Berkeley Laboratory, and the Dept. of Biophysics and Medical Physics, University of California, Berkeley, and held on May 20-22, 1982.

Includes bibliographies and index.

1. Yeast fungi--Genetics--Congresses. 2. Recombinant DNA--Congresses. 3. Molecular biology--Congresses. 4. Fungi--Genetics--Congresses. I. Esposito, Michael S. II. Berkeley Workshop on Recent Advances in Yeast Molecular Biology (1st : 1982) III. Lawrence Berkeley Laboratory. Biology and Medicine Division. IV. University of California, Berkeley. Dept. of Biophysics and Medical Physics.
QK617.5.Y47 1984 589.2′330487322 84-4096
ISBN 0-8155-0987-1

FOREWORD

Novel insights into recombinant DNA technology, as applied to yeast research, are described in this comprehensive volume based on the First Berkeley Workshop on Recent Advances in Yeast Molecular Biology, held in May 1982.

The workshop provided an excellent sounding board for researchers located throughout the U.S. The results of various genetic engineering techniques in recombinant DNA studies in yeasts were detailed and discussed. Since it was felt that the information presented at the workshop would be of interest and value to the private sector as well as to researchers in academia, and since a workshop is by its nature small, a number of participants were subsequently asked to summarize their presentations in writing. The 25 papers contained in this book are the result of that effort.

The papers describe experimental approaches using the newest technologies of DNA transformation, molecular cloning, and DNA sequence analysis. These techniques are employed to gain new information regarding chromosomal structure, gene regulation, DNA recombination and repair, and cell type control.

The information in the book is from *Recent Advances in Yeast Molecular Biology: Recombinant DNA,* edited by Michael S. Esposito of Lawrence Berkeley Laboratory, University of California, for the U.S. Department of Energy, September 1982.

The table of contents is organized in such a way as to serve as a subject index and provides easy access to the information contained in the book.

Advanced composition and production methods developed by Noyes Publications are employed to bring this durably bound book to you in a minimum of time. Special techniques are used to close the gap between "manuscript" and "completed book." In order to keep the price of the book to a reasonable level, it has been partially reproduced by photo-offset directly from the original report and the cost saving passed on to the reader. Due to this method of publishing, certain portions of the book may be less legible than desired.

EDITOR'S NOTE

Editor's Note: Each paper herein has been referenced, based on the papers from this first workshop being considered as the first volume of a series. The references appear following the authors' names at the beginning of each paper.

NOTICE

ACKNOWLEDGEMENTS

The First Berkeley Workshop on Recent Advances in Yeast Molecular Biology, sponsored by the Biology and Medicine Division of Lawrence Berkeley Laboratory and the Department of Biophysics and Medical Physics, was held on May 20-22, 1982. The meeting focused upon the novel insights regarding chromosomal structure, gene regulation, recombination, DNA repair, and cell type control, that have been obtained by experimental approaches incorporating the new technologies of yeast DNA transformation, molecular cloning, and DNA sequence analysis.

The workshop was instituted to provide yeast researchers having closely related interests an opportunity for informal discussion and exchange of scientific views. The organization of the workshop was initiated during a period of limited funding of scientific activities. We thank Prof. Edward L. Alpen, Director of Donner Laboratory, the Henry Miller Fund, and all of our colleagues, whose vital support made this meeting a reality.

Since workshops are by nature limited in attendance, we asked several participants to contribute manuscripts summarizing their presentations which could be shared with the scientific community. We are grateful to the authors for the care and rapidity with which their articles were prepared. The assembly of these proceedings was overseen by Kathleen Bjornstad, Loretta Lizama and Peggy Little, who are responsible for the timely appearance of this volume.

Berkeley Workshop Committee

Michael S. Esposito, Chairperson
Lawrence Berkeley Laboratory

Robert K. Mortimer
Dept. of Biophysics and Medical Physics, University of California, Berkeley, and Lawrence Berkeley Laboratory

Berkeley Workshop Committee (cont.)
Carlo Bruschi
Lawrence Berkeley Laboratory

David Schild
Lawrence Berkeley Laboratory

Ira Herskowitz
Dept. of Biochem. and Biophys. Genetics Div.,
University of California, San Francisco

CONTENTS AND SUBJECT INDEX

PART I
CHROMOSOME STRUCTURE, REPLICATION, AND ASSORTMENT

PART III
RECOMBINATION AND REPAIR

PART V
MATING TYPE: CONTROL OF SILENT CASSETTES AND MATING INTERCONVERSION

Part I

Chromosome Structure, Replication, and Assortment

1. IDENTIFICATION OF DNA SEQUENCES REQUIRED FOR MITOTIC STABILITY OF CENTROMERE PLASMIDS IN YEAST*

M. Fitzgerald-Hays and J. Carbon

Department of Biological Sciences
University of California
Santa Barbara, California 93106

Rec. Adv. Yeast Mol. Biol. (1982) 1: 1-12

Segments of Saccharomyces cerevisiae DNA that act as functional centromere units when introduced into yeast on autonomously replicating plasmids have been isolated from chromosomes III and XI. Portions of these DNA fragments have been subcloned into vector YRp7' to form plasmids pYe(CEN3)30 (627 bp fragment) and pYe(CEN11)5 (858 bp fragment), which exhibit the mitotic stabilization aspect of centromere function. Nucleotide sequence analysis of the small CEN DNA inserts has revealed several features of sequence organization (sequence elements I, II, III, IV) common to both CEN3 and CEN11 DNA. Deletion mutations were constructed which alter the CEN fragments and, in some cases, abolish mitotic stability. The deletion mutant plasmids that lack sequence elements I, II and III from the CEN region are extremely unstable during cell division and are present in elevated copy number in yeast cells. The DNA sequences removed by these deletions may confer mitotic stability by providing an attachment site for the spindle fiber to ensure proper segregation during cell division.

INTRODUCTION

The centromere regions of eukaryotic chromosomes play a fundamental role in chromosome movement to ensure that genetic information is stably maintained and properly transmitted during cell division. The DNA sequences within a functional centromere probably interact with nuclear proteins to provide a microtubule or membrane attachment site and allow stable chromosome segregation. Procedures used to isolate and characterize large segments of yeast DNA from the centromere region of chromosome III (CEN3) have been described previously (3). Centromere DNA from yeast chromosome XI (CEN11) was originally isolated on plasmid pYe(MET14)2 by taking advantage of the tight genetic linkage between the MET14 gene and the centromere of chromosome XI (12,5). Functional CEN11 sequences were localized by subcloning restriction fragments from pYe(MET14)2 and testing the resulting plasmids for mitotic stability in yeast (Figure 1). In the absence of a functional CEN fragment, autonomously replicating plasmids like YRp7' (TRP1 ars1), which replicate in yeast using an ars function, are very unstable during mitosis and are quickly lost under nonselective growth conditions (16,9). These unstable plasmids appear to attain an elevated copy number in a small portion of the population but segregate improperly during mitosis and continually generate progeny lacking plasmids. The addition of a functional centromere segment to an unstable plasmid allows it to be stably maintained and transmitted at low copy number to daughter cells during mitosis, presumably by offering an attachment site for the spindle fiber.

*M.F.-H. was supported by a Damon Runyon-Walter Winchell Cancer Fund postdoctoral fellowship. This research was funded by a grant (CA-11034) from the National Cancer Institute, National Institute of Health.

The smallest CEN fragments capable of mitotic stabilization are contained in plasmids pYe(CEN3)30 (627 bp) (4,6) and pYe(CEN11)5 (858 bp; Figure 1). These DNA fragments differ in primary nucleotide sequence but share several unique features of DNA sequence organization (6). The most obvious of these is an extremely [A+T]-rich core segment of about 90 bp (element II), which is flanked by two small regions (elements I and III) that are perfectly homologous in both CEN3 and CEN11 DNAs (Figure 2). A third homologous region, element IV, occurs in both centromere sequences. The most striking observation is that these three short regions of complete homology, elements I (14 bp), III (11 bp) and IV (10 bp), which surround the [A+T]-rich element II, are positioned in an almost identical spatial arrangement within the two centromere sequences as shown in Figure 2. In both CEN3 and CEN11 sequences, the most homologous region (71%) is a segment of about 130 bp including element I and extending 20 bp beyond element III.

To investigate the role these conserved centromere sequences might play in mitotic stabilization, we have constructed deletion mutations which alter the CEN DNA in pYe(CEN11)5. We find that DNA fragments conferring stable mitotic segregation in yeast include sequences extending from element I through element III; deletion of this region inactivates mitotic centromere function.

MATERIALS AND METHODS

Strains, Media and Enzymes

Saccharomyces cerevisiae strain J17(α *his2 ade1 trp1 met14 ura3*) was constructed as described in reference 5. *S. cerevisiae* strain J170B1 (*a ade1 trp1 his2 ura3*) is a canavanine-sensitive strain which demonstrates the canavanine response test useful for determining ploidy (14). It was constructed by crossing J17 with RH218(*a CUP1 gal2 mal suc trp1*) and analyzing the meiotic products for segregants with the appropriate genetic markers and the proper mutation response to canavanine resistance after irradiation with ultraviolet light. Media for bacterial and yeast growth have been described previously (9). Restriction endonucleases were purchased from New England Biolabs and used with buffers and reaction conditions as specified by the vendor. Transformation of yeast was performed by a modification described in references 8 and 9.

Construction of Bal31 Deletion Mutations

Full-length linear molecules of pYe(CEN11)5 were generated by cleavage with *Bam*HI and purified by extraction with phenol, chloroform:isoamyl alcohol (24:1), and ethanol precipitation. The *Bam*HI linears were treated with *Bal*31 (Bethesda Research Labs) in buffer containing 12 mM $CaCl_2$, 12 mM $MgCl_2$, 600 mM NaCl, 20 mM Tris-HCl (pH 8.1), and 1 mM EDTA. Two sets of reaction conditions were used to generate sets of nested deletion mutations; 2.5 μg linear DNA, 2.5 units *Bal*31, 50 μl final volume, 32°C for 30, 60, 90 and 180 seconds (13) or; 5 μg linear DNA, 2.5 units *Bal*31, 25 μl final volume, 20°C for 30 and 90 seconds (7). Reactions were stopped by the addition of EDTA to 50 mM and extraction with phenol, chloroform-isoamyl alcohol, and precipitation with ethanol. After ligation *in vitro*, DNA preparations were used to transform *E. coli* to ampicillin resistance. Plasmid DNA was isolated from single colonies and the deletion mutations were characterized by restriction analysis and

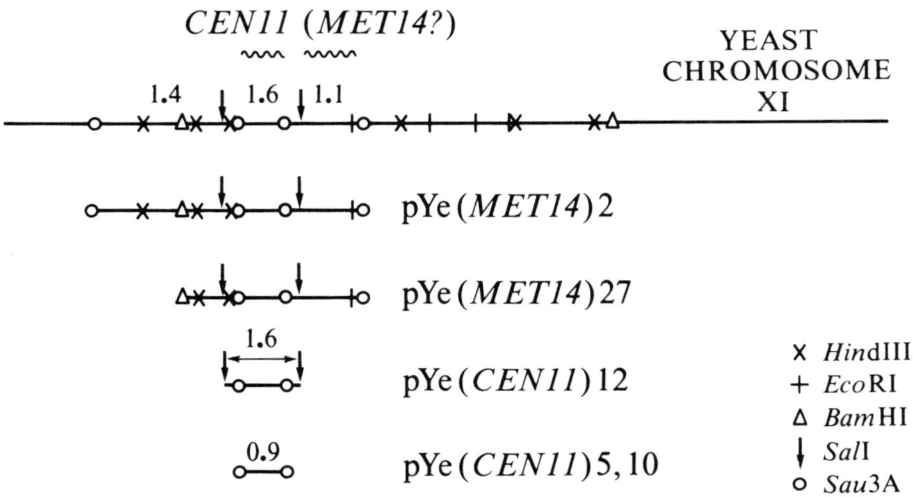

Figure 1. Subcloning Experiments to Localize CEN11.

The indicated segments of yeast DNA in pYe(MET14)2 (5.2 kb) and pYe(MET14)27 (3.4 kb) are cloned into the BamHI site of YRP7, the pYe(CEN11)12 fragment (1.6 kb) is in the SalI site of YRp7', and the 0.9 kb Sau3A fragment is cloned in two different orientations (pYe(CEN11)5 and pYe(CEN11)10) into the BamHI site of YRp7'. The reconstructed BamHI site is distal to the unique ClaI site (in pBR322) in pYe(CEN11)5 and proximal to it in pYe(CEN11)10. The only difference between YRp7 and YRp7' is the orientation of the 1.4 kb fragment of yeast DNA (TRP1 ars1) cloned into the EcoRI site of pBR322. Not all of the Sau3A sites are shown.

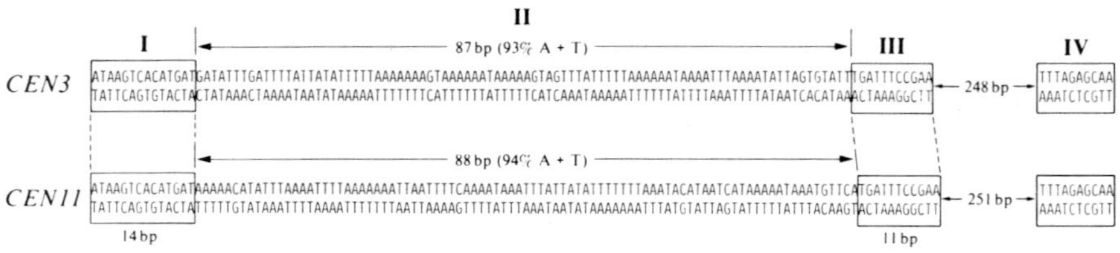

Figure 2. Spatial Arrangment of Sequence Elements Common to CEN3 and CEN11.

The regions of perfect sequence homology between CEN3 and CEN11, elements I (14 bp), III (11 bp) and IV (10 bp), are positioned in an almost identical spatial arrangement within the two centromere sequences. The very [A+T]-rich core (element II) lies between elements I and III in both DNA sequences. There is an overall homology of 71% in the region from element I through element III when CEN3 and CEN11 are compared. The complete DNA sequences compared for this analysis are the 858 bp *Sau*3A fragment in pYe(CEN11)5 and the 627 bp *Sau*3A fragment in pYe(CEN3)30 (6).

subsequent DNA sequence determination (11). Plasmid isolates containing the appropriate deletion mutations were tested for mitotic stability after transformation into strain J17.

Preparation of DNA for Southern Analysis

For copy number analysis, yeast strain J170B1 was transformed with pYe(CEN11)5 DNA or with various deletion mutant plasmid DNAs. Total yeast DNA (genomic and plasmid) was prepared according to Bloom and Carbon (1) from haploid J170B1 transformants grown overnight under non-selective conditions (plus tryptophan, 50 μg/ml). Ploidy was determined by the canavanine response test as described by Schild et al. (14). Total DNA preparations were cleaved to completion with *BglII*, electrophoresed in 1.4% agarose slab gels (90 mM Tris-borate pH 8.3, 2.5 mM EDTA) and transferred to nitrocellulose paper (15). Hybridization and nick translation procedures have been described (1).

RESULTS

Construction of Deletion Mutations in pYe(CEN11)5

When the mitotic stabilities of different *CEN*-containing plasmids were compared, no consistent, significant differences in stability that correlate with the size of the yeast centromere insert could be observed (6). In our earlier studies, the smallest segment of chromosome XI tested was the 858 bp *Sau3A* fragment contained in pYe(*CEN11*)5. Sequence analysis of the *CEN* insert in this mitotically stable plasmid revealed the presence of the [A+T]-rich core region (element II) flanked by the short sequence elements I, III and IV (Figure 2). To further delimit the centromere sequences required for mitotic stabilization, we now have constructed deletion mutations within the *CEN* insert in pYe(*CEN11*)5 and analyzed the behavior of the altered plasmids during cell division in yeast.

Plasmid pYe(*CEN11*)5 DNA was cut to completion with *BamHI* and the resulting full-length linear DNAs were treated with the double strand specific exonuclease, *Bal31* (7,13), ligated *in vitro*, and amplified by transformation into *E. coli*. Plasmid DNA was prepared from the progeny of single colonies and the extent of the deletion mutation was characterized by restriction enzyme analysis of each isolate. All the deletion mutations analyzed removed the unique *BamHI* site in pYe(*CEN11*)5, and the segments deleted ranged in size from less than 100 bp to greater than 1000 bp. Plasmids were selected with deletion mutations affecting a minimal amount of pBR322 vector DNA and extending an appropriate distance into *CEN* yeast sequences as shown in Figure 3. Nucleotide sequence analysis indicated that the deletion in plasmid *dl*31-28 removes sequences up to element I but does not alter the segment of DNA containing elements I through III or the remaining portions of the yeast DNA. The mutation introduced into plasmid *dl*31-48 deletes element I, the [A+T]-rich core and element III, but ends at the junction of element III with the flanking DNA. The deletion in *dl*31-6 removes these sequences plus an additional 100 bp of yeast DNA. Thus, the region of centromere DNA from element I through element III is present only in plasmids pYe(*CEN11*)5 and *dl*31-28, and has been deleted in both *dl*31-48 and *dl*31-6. None of the three deletion mutations described here alter the region of yeast DNA encompassing element IV, nor do they affect the *TRP1 ars1* segment required for autonomous replication and selection of the plasmids in yeast.

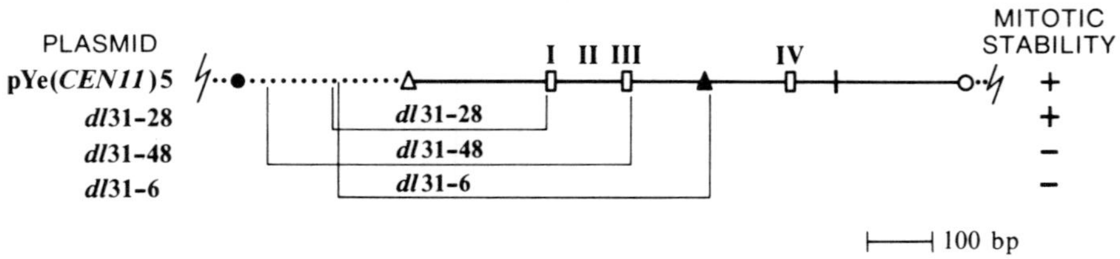

Figure 3. Diagram of the Deletion Mutations Constructed in pYe(CEN11)5.

The relevant region of plasmid pYe(CEN11)5 containing the 858 bp Sau3A CEN11 yeast DNA insert (solid lines) and flanking pBR322 DNA (dotted lines) is indicated. Segments of DNA removed by the deletion mutations constructed at the BamHI site in pYe(CEN11)5 are delineated by the brackets and designated by the plasmid names; dl31-28, dl31-48 and dl31-6. The mitotic stabilities of these plasmids were determined by the method described in Figure 4. Only selected restriction sites are shown; BamHI (△), Sau3A (○), SalI (●), HinfI (—|—) and HincII (▲). The endpoints of the deletion mutations are as follows: dl31-28, pBR322 nt 490 to CEN11 nt 217; dl31-48, pBR322 nt 584 to CEN11 nt 328; dl31-6, pBR322 nt 485 to CEN11 nt 462. Numbering is according to Sutcliffe (17) for pBR322 and Fitzgerald-Hayes et al. (6) for the CEN11 fragment.

Mitotic Stability and Copy Number in Yeast of pYe(CEN11)5 and the Deletion Mutant Plasmids

We have examined the effect of the deletion mutations introduced into pYe(CEN11)5 on the stable maintenance of these plasmids during mitotic cell division in yeast. Plasmid DNA from each mutant isolate was used to transform yeast strain J17 to $TRP1^+$. Plasmid mitotic stabilities were determined by assaying the percentage of $TRP1^+$ cells remaining in the population after several generations of non-selective growth. As shown in Figure 4, plasmids pYe(CEN11)5 and dl31-28 demonstrate the same relatively high level of mitotic stability. In contrast, both deletion mutant plasmids dl31-48 and dl31-6, which lack sequence elements I through III, show a dramatic decrease in stability and are lost very rapidly from the population. The mitotic behavior of these unstable deletion plasmids resembles that of the vector, YRp7', which segregates improperly during mitosis and thereby continuously produces progeny without plasmid. The presence of a DNA fragment including all or a portion of the region containing element I, the [A+T]-rich core, and element III appears to be required for stabilization and proper segregation of the plasmid during mitosis, presumably because this region contains sequences necessary for attachment to the mitotic apparatus.

The copy number maintained by pYe(CEN11)5 and the deletion mutant plasmids in haploid cells was determined using total yeast DNA (genomic and plasmid) prepared from J170B1 TRP^+ transformants grown overnight under non-selective conditions. Haploid J170B1 transformants were detected by assaying the frequency of mutation to canavanine-resistance (14). This avoids the testing of cell-fusion diploids (a/a) produced during transformation. Total DNA preparations were digested completely with BglII, which cleaves at a unique site within the TRP1 ars1 segment in each plasmid. After electrophoresis and transfer to nitrocellulose paper, the blot was hybridized with ^{32}P-labeled CEN11 Sau3A fragment (858 bp). The results of this analysis are shown in Figure 5. The larger molecular weight band in all the lanes (a-f) is the BglII fragment derived from the centromere region of chromosome XI. As expected, it is the only radioactive band in lane f, which contains BglII-cleaved total DNA from untransformed J170B1 cells. The lower molecular weight bands in lanes a through e of Figure 4 correspond to BglII linears of the various plasmids being tested. Comparison of the relative intensity of the plasmid band to the genomic band gives an estimate of the number of plasmid copies per haploid genome. Different autoradiographic exposures of this blot were subjected to tracing with a densitometer to aid in comparing relative intensities of the different bands. The copy number calculations on a per cell basis include a correction for the percentage of cells in the population that contained plasmid at the time the cells were harvested to prepare total DNA. The results of this analysis show that the copy number of pYe(MET14)27 (Figure 5, lane a) is approximately one plasmid per $TRP1^+$ cell. Plasmids pYe(CEN11)5 and dl31-28 (Figure 5, lanes b and c) are present in about one to two copies per $TRP1^+$ cell. A dramatically different result is obtained with the mitotically unstable plasmids dl31-48 and dl31-6 (Figure 5, lanes d and e), which have elevated copy numbers of about 50 and 25 copies per cell. These plasmids are present in only a small percentage (6% and 9%) of the total population. In addition, the copy number values for dl31-28, dl31-48 and dl31-6 are likely to be underestimates. The 858 bp probe used in the analysis is 75% homologous with the CEN insert in dl31-28, while the larger deletions in dl31-48 and dl31-6 contain even less yeast DNA complementary to the probe

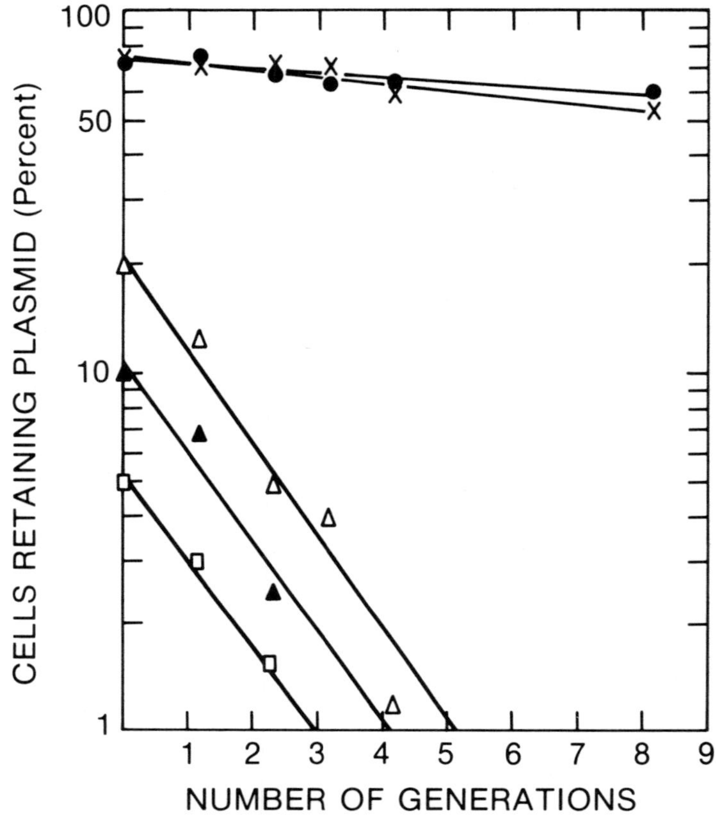

Figure 4. Mitotic Stability Analysis of pYe(CEN11)5 and Various Deletion Mutant Plasmids in Yeast.

The various plasmid DNAs were used to transform yeast strain J17. Single colony transformants containing pYe(CEN11)5 (●), dl31-28 (X), dl31-48 (△) and dl31-6 (□) were grown for two generations at 32°C in selective media, at which time tryptophan (50 μg/ml) was added and incubation continued for several generations. Aliquots were removed at appropriate intervals after a zero time point taken just before the addition of tryptophan, sonicated briefly, diluted and spread on YPD agar (2% peptone, 1% yeast extract, 2% glucose, 2% agar) supplemented with tryptophan. After growth for 2 days at 32°C, colonies were replica-plated onto YNB agar (Difco yeast nitrogen base, 2% glucose, 0.5% casamino acids, 20 μg/ml each adenine and uracil) in the presence or absence of tryptophan, to assay for the presence of plasmid. The percentage of Trp^+ and Trp^- colonies was scored after overnight incubation at 32°C.

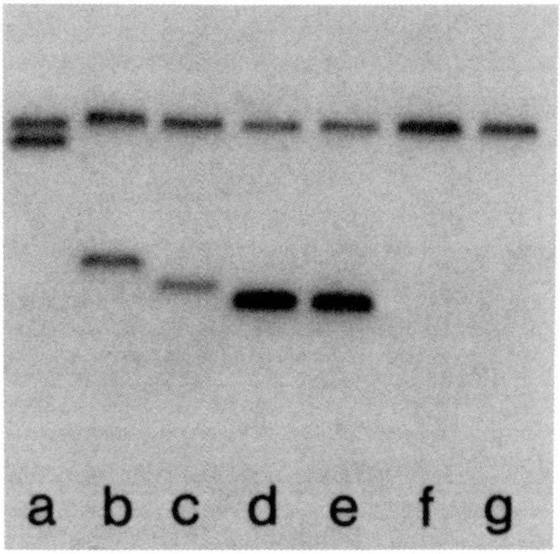

Figure 5. Copy Number Determinations for pYe(CEN11)5 and Deletion Mutant Plasmids in Yeast.

Haploid transformants of yeast strain J170B1 containing pYe(MET14)27 (a), pYe(CEN11)5 (b), dl31-28 (c), dl31-48 (d), dl31-6 (e) or YRp7' (g) were grown overnight under non-selective conditions and total DNA (genomic and plasmid) was prepared (1). A sample of DNA from untransformed J170B1 cells (f) was also prepared. At the time of harvest, the proportion of cells containing plasmid in each sample was determined from the percentage of Trp^+ cells in the populations (see legend to Figure 4). Total DNA was cut to completion with BglII, electrophoresed in 0.7% agarose, transferred to nitrocellulose and hybridized with ^{32}P-labeled 858 bp CEN11 Sau3A fragment. The upper bands present in lanes a through g represent the genomic BglII DNA fragment containing the centromere region from chromosome XI. The lower molecular weight bands in lanes a through e represent linearized plasmid DNA.

(62% and 46%, respectively). Thus, these unstable deletion plasmids resemble the vector YRp7' in displaying both an elevated copy number and an inability to segregate properly during mitosis.

DISCUSSION

We have cloned and analyzed small segments of yeast chromosomal DNA that function as centromere sequences in yeast, as assayed by mitotic stabilization of autonomously replicating arsl plasmids. Plasmids containing small fragments of DNA from the centromere regions of chromosomes III and IX, 858 bp (pYe(CEN11)5) and 627 bp (pYe(CEN3)30), demonstrate the same relative level of mitotic stability as do plasmids carrying larger DNA segments from these centromere regions (6). The rate of loss of these CEN plasmids during mitosis (1-3% per generation) is fairly consistent, although subtle differences in mitotic function might not be detected in the stability assays we have used.

Sequence analysis of the small centromere-containing DNA fragments in plasmids pYe(CEN11)5 and pYe(CEN3)30 reveals an [A+T]-rich region (element II) and three short sequence homologies (elements I, III and IV) common to both CEN3 and CEN11 DNAs. These features of sequence organization are positioned in an almost identical spatial arrangement within both centromere fragments (Figure 2) suggesting a role for these DNA sequences in centromere function. To determine the functional boundaries of the sequences required for mitotic stabilization, deletion mutations were constructed that alter the CEN insert in pYe(CEN11)5 (Figure 3). Mitotic analysis of the deletion mutant plasmids (Figure 4) demonstrated that the smallest CEN fragments capable of conferring mitotic centromere function (in pYe(CEN11)5 and dl31-28) contain DNA segments that include sequence elements I, II and III. Deletions that completely remove this region, in dl31-48 and dl31-6, result in unstable plasmid segregation during cell division. Recently, we have analyzed the effect of a deletion (dl31-1) in pYe(CEN11)5 that removes the first 5 bp of sequence element I, but does not alter the remainder of element I, or any DNA to the right of element I (as diagrammed in Figure 2). This deletion plasmid demonstrates the same level of mitotic stability as is observed with the parent plasmid, pYe(CEN11)5. Another deletion plasmid (dl31-35) lacks the region containing element IV, but does contain the entire region extending from element I through element III. This plasmid shows no reduction in mitotic stability as a result of removing sequence element IV (data not shown). Similar mutations constructed within the element I through element III region should allow further delineation of the sequences necessary for stable chromosome maintenance.

The CEN plasmids, pYe(MET14)27, pYe(CEN11)5 and dl31-28, are stably maintained through mitosis at low copy number, either one or two copies per cell (Figure 4). Thus, the active CEN element in some way controls the copy number of the minichromosome to a low value. This behavior is strikingly different from that of plasmids containing an ars but lacking a functional centromere. For example, plasmid YRp7' is mitotically quite unstable and always occurs at high copy number in a small proportion of the cells. This plasmid is apparently incapable of proper segregation and the replicated plasmid copies have a strong tendency to remain with the mother cell during cell division. While the centromere plasmid pYe(MET14)27, which contains a 3.4 kb DNA insert, is reproducibly present at one copy per plasmid-containing cell, the bias toward two copies per cell in the case of pYe(CEN11)5 and dl31-28 could reflect a slightly increased level of mitotic non-disjunction occurring during segregation of the plasmids with smaller CEN inserts. In contrast with the stable CEN plasmids, the unstable deletion mutant plasmids, dl31-48 and dl31-6, are present at an elevated copy number in a small proportion of the total population of cells, as is observed with the parent vector, YRp7'. This must reflect the inability of these plasmids to segregate, presumably because the mitotic attachment site has been deleted. The exact mechanism of the copy number control exerted by the CEN region is still unclear. However, these results suggest that in cases where the plasmids can segregate efficiently (pYe(MET14)27, pYe(CEN11)5, dl31-28), the low plasmid copy number per haploid genome is maintained. This may be due to a limitation in the number of extra spindle attachment points for extra centromeres, or it might reflect an active centromere-mediated copy number control mechanism.

In conclusion, we have identified small segments of yeast centromere DNA (about 130 bp or less) that are directly involved in maintaining the proper mitotic segregation of autonomously replicating minichromosomes in yeast. Proper segregation during mitotic cell division could result from the direct interaction of these DNA sequences in chromatin with specific DNA binding proteins to mediate attachment to the mitotic spindle fiber. Recent studies indicate that these DNA sequences are capable of binding specifically to certain nuclear proteins and are maintained in a unique chromatin conformation (1,2).

LITERATURE CITED

1. Bloom, K.S. and Carbon, J. 1982. Yeast centromere DNA is in a unique and highly ordered structure in chromsomes and small circular minichromosomes. Cell 29:305-317.

2. Bloom, K.S., Fitzgerald-Hayes, M. and Carbon, J. 1982. Structural analysis and sequence organization of yeast centromeres. Cold Spring Harbor Symp. Quant. Biol., in press.

3. Clarke, L. and Carbon, J. 1980. Isolation of a yeast centromere and construction of functional small circular minichromosomes. Nature 287: 504-509.

4. Clarke, L., Fitzgerald-Hayes, M., Buhler, J.-M. and Carbon, J. 1981. The structure and function of yeast centromeres. Stadler Symp. 13:9-24.

5. Fitzgerald-Hayes, M., Buhler, J.-M., Cooper, T. and Carbon, J. 1982. Isolation and subcloning analysis of functional centromere DNA (CEN11) from yeast chromosome XI. Molecular and Cellular Biology 2:82-87.

6. Fitzgerald-Hayes, M., Clarke, L. and Carbon, J. 1982. Nucleotide sequence comparisons and functional analysis of yeast centromere DNAs. Cell 29:235-244.

7. Guilfoyle, R. and Weinmann, R. 1981. The control regions for adenovirus VA RNA transcription. Proc. Natl. Acad. Sci. U.S.A. 78:3378-3382.

8. Hinnen, A., Hicks, J.R. and Fink, G.R. 1978. Transformation of yeast. Proc. Natl. Acad. Sci. U.S.A. 75:1929-1933.

9. Hsiao, C.-L. and Carbon, J. 1979. High frequency transformation of yeast plasmids containing the cloned yeast ARG4 gene. Proc. Natl. Acad. Sci. U.S.A. 76:3829-3833.

10. Hyman, B.C., Cramer, J.H. and Rownd, R.H. 1982. Properties of a *Saccharomyces cerevisiae* mtDNA segment conferring high-frequency yeast transformation. Proc. Natl. Acad. Sci. U.S.A. 79:1578-1582.

11. Maxam, A.M. and Gilbert, W. 1980. Sequencing end-labeled DNA with base specific cleavages. Methods in Enzymology 65:499-560.

12. Mortimer, R.K. and Hawthorne, D.C. 1966. Genetic mapping in *Saccharomyces*. Genetics 53:165-173.

13. Osborne, T.F., Schell, R.E., Burch-Jaffe, E., Berget, S.J. and Berk, A.J. 1981. Mapping a eukaryotic promoter: a DNA sequence required for *in vivo* expression of adenovirus pre-early function. Proc. Natl. Acad. Sci. U.S.A. 78:1381-1385.

14. Schild, D., Anathaswamy, H.N. and Mortimer, R.K. 1981. An endomitotic effect of a cell cycle mutation of *Saccharomyces cerevisiae*. Genetics 97:551-562.

15. Southern, E.M. 1975. Detection of specific sequences among DNA fragments separated by gel electrophoresis. J. Mol. Biol. 98:503-517.

16. Stinchcomb, D.T., Struhl, K. and Davis, R.W. 1979. Isolation and characterization of a yeast chromosomal replicator. Nature 282:39-43.

17. Sutcliffe, J.G. 1979. Complete nucleotide sequence of the *Escherichia coli* plasmid pBR322. Cold Spring Harbor Symp. Quant. Biol. 43:77-90.

2. CHROMATIN STRUCTURE OF YEAST CENTROMERES*

K.S. Bloom and J. Carbon

Department of Biological Sciences
University of California
Santa Barbara, California 93106

Rec. Adv. Yeast Mol. Biol. (1982) 1:13-27

The centromere sequences in the yeast chromosome are structurally distinct from the bulk of the nucleosomal chromatin. A discrete protected region of 220-250 bp of CEN sequence flanked by highly nuclease-sensitive sites was revealed by mapping the exact nuclease cleavage sites within the centromeric chromatin. On both sides of this protected region specific nuclease cutting sites exist at nucleosomal intervals (160 bp) for a total of 2.5-3 kb. The central protected region in the chromatin of both centromeres spans the 130 bp segment, extending from sequence element I through the AT-rich element II to about 20 bp beyond element III, that exhibits the highest degree of sequence homology (71%) between functional CEN3 and CEN11 DNAs. This unique chromatin conformation is maintained on CEN sequences introduced into yeast on autonomously replicating plasmids as well as on deletion mutant plasmids demonstrating centromere function in yeast. These results suggest that the same chromatin components are associated with centromere DNA sequences irrespective of their plasmid or chromosomal location to give rise to a region of structural differentiation that is required for centromere function *in vivo*.

INTRODUCTION

The centromere is an essential structural element of the eukaryotic chromosome. The presence of a centromere ensures the stable maintenance and proper segregation of chromosomes through mitosis and meiosis, presumably by providing an attachment point for spindle fibers or membrane sites involved in chromosome movement.

Functional centromere sequences have been isolated from chromosomes III (CEN3) and XI (CEN11) of the yeast, *Saccharomyces cerevisiae* (3,4). When DNA segments containing either CEN3 or CEN11 are introduced into yeast on autonomously replicating plasmids, genetic markers on these plasmids are stably maintained at low copy number through mitosis in the absence of selective pressure and segregate through meiosis in a manner similar to that expected for centromere-linked genes (3,4). Thus CEN DNA segments contain sequences required for stable mitotic maintenance and proper meiotic segregation of minichromosomes in yeast.

The DNA fragments containing functional CEN3 and CEN11 activity differ in overall nucleotide sequence but share several features of sequence organization (5,6). Both contain a region of about 90 bp (element II) that consists of greater than 90% A+T base pairs. Three regions of complete homology between CEN3 and CEN11, elements I (14 bp), III (11 bp) and IV (10 bp), surround

*K.S.B. was supported by a Jane Coffin Childs Postdoctoral Fellowship. This work was supported by NIH grant CA-11034 from the National Cancer Institute.

the A+T-rich core segment and are positioned in virtually identical spatial arrangements within the two centromere sequences (5,6). In both centromeres the most highly conserved region (71%) is a 130 bp segment from element I to about 20 bp beyond element III.

To determine specifically how the centromere DNA sequence is folded in the yeast chromosome and how this region of DNA gives rise to a functional centromeric unit, we have analyzed the structural organization of centromeric chromatin. Our results show that a specific region of 220-250 bp in the centromere DNA sequence is in a unique chromatin conformation, and the surrounding chromosomal DNA directs the nucleosomal subunits into a very ordered and specifically aligned array. The centromere DNA sequences are in this same conformation in the genome and on autonomously replicating plasmids in yeast.

MATERIALS AND METHODS

Plasmids and Yeast Transformation

Plasmids pYe(*CEN3*)41, pYe(*CDC10*)1 and pYe(*CEN3*)11 (3), pYe(*CEN11*)5 and pYe6 (6), and deletion mutant plasmids *dl*31-28, *dl*31-48 and *dl*31-6 (5) were described previously. Plasmid DNA was isolated as detailed by Tschumper and Carbon (16). Transformation of yeast was performed by a modification of the procedure of Hinnen et al. (8) as described by Hsiao and Carbon (9).

Preparation and Digestion of Nuclei

A modification of the procedures of Forte and Fangman (7) and Nelson and Fangman (12) as described by Bloom and Carbon (1) were used.

Preparation of DNA for Southern Analysis

DNA was extensively deproteinized, restriction digested and resolved by electrophoresis in agarose gels (1). Transfer of DNA from agarose gels to nitrocellulose paper was performed by the method of Southern (15). Nick translation and hybridization conditions have been described (1).

RESULTS

Chromatin Structure of Centromere DNA on Yeast Chromosomes

To understand how a specific DNA segment containing a functional centromere directs chromosomes through mitosis and meiosis, we have examined the chromatin structure of the centromere region as it exists in the yeast cell. Utilizing the mapping technique of Wu (17) we discovered the existence of a discrete nuclease-resistant core of 220-250 bp of *CEN* DNA in both chromosomes III and XI (1). The position of the nuclease cleavage sites within the centromere III and centromere XI region of chromatin DNA and naked, deproteinized DNA are shown in Figures 1 and 2, respectively. Following partial micrococcal nuclease or DNAase I cleavage of chromatin DNA from isolated yeast nuclei, DNA fragments were extracted and subsequently digested to completion with the restriction enzyme *Bam*HI (large arrow, Figures 1 and 2). The DNA fragments were separated electrophoretically, blotted to nitrocellulose and probed with the 627 bp *CEN3* fragment extending through the centromere from the *Bam*HI site on chromosome III (Figure 1) or the 800 bp fragment from chromosome XI

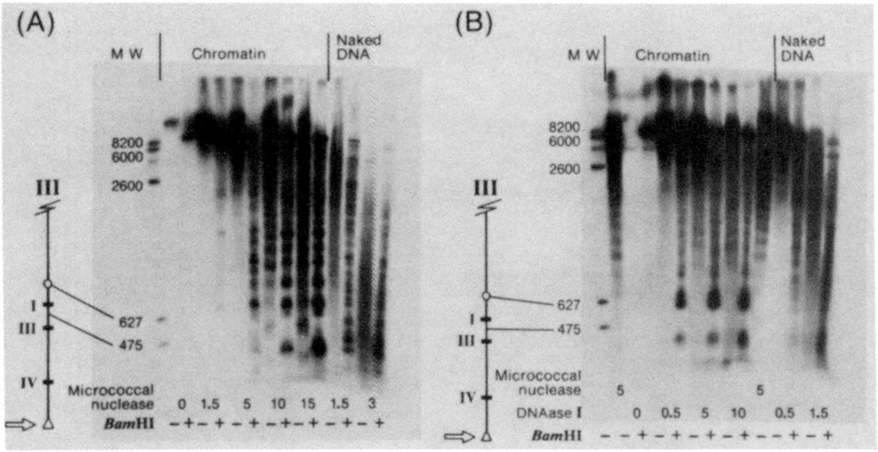

Figure 1. Mapping Nuclease Sensitive Sites on the Centromere Chromatin From Yeast Chromosome III.

Nuclei were prepared from yeast strain X2180a and were digested with micrococcal nuclease (50 units/ml) (A) or DNAase I (5 μg/ml) (B) for the times (min) indicated, as described in Materials and Methods. For the experiments with naked DNA, nuclei were prepared exactly as described for the chromatin digests, but immediately prior to nuclease cleavage, DNA was extensively deproteinized as described previously (1). The deproteinized DNA was subsequently digested under the same conditions as described for the chromatin digests with micrococcal nuclease (0.5 units/ml) (A) or DNAase I (50 ng/ml) (B) for the times (min) indicated. Following partial micrococcal nuclease of DNAase I cleavage, DNA samples were deproteinized, incubated in the presence (+) or absence (-) of BamHI and electrophoresed on 1.4% agarose gels. The DNAs were transferred to nitrocellulose filters and hybridized to the radiolabeled 627 bp CEN3 fragment. The identical pattern was obtained following hybridization to a radiolabeled 150 bp fragment extending toward the centromere from the BamHI site (data not shown). Molecular weight markers (M.W.) indicate yeast nuclear DNA fragments cut with BamHI, HindIII, BamHI/EcoRI, Sau3A and Sau3A/HinfI. These fragments contain regions complementary to the radiolabeled probe (3). To the left is shown a partial restriction site map of the centromere region of yeast chromosome III. Restriction enzyme sites are BamHI (△) and Sau3A (○). The large arrows indicate the BamHI site on yeast chromosome III that is immediately proximal to the 627 bp labeled DNA probe.

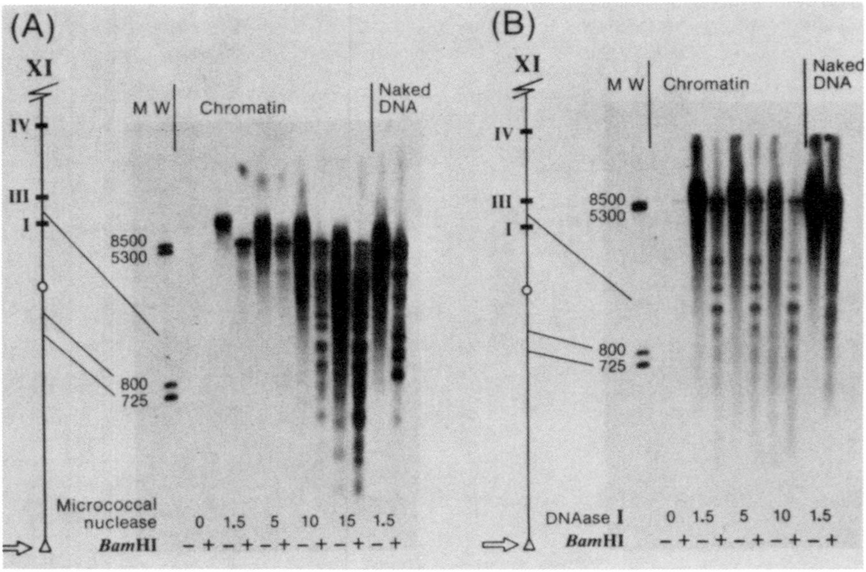

Figure 2. Mapping Nuclease Sensitive Sites on Centromere Chromatin From Yeast Chromosome XI.

Conditions for digestion of the chromatin or deproteinized DNA, cleavage with BamHI, gel electrophoresis and Southern transfer were as described in the legend to Figure 1 and Materials and Methods. After transfer to nitrocellulose, the filters were hybridized to the radiolabeled 800 bp fragment from chromosome XI, extending toward the centromere from the BamHI site (large arrows). Molecular weight markers (M.W.) indicate yeast DNA fragments containing regions complementary to the probe, cut with BamHI, EcoRI, BamHI/SalI, HindIII/SalI and HindIII (6). To the left is shown a partial restriction site map of the centromere region of yeast chromosome XI. Restriction enzyme sites are BamHI (△) and Sau3A (○).

extending toward the centromere from the BamHI site on chromosome XI (Figure 2). Only those fragments that contain the DNA in the same direction from the restriction site as the probe will be visualized following hybridization. The lengths of the hybridizing fragments provide a direct map of the points of nucleolytic cleavage within the chromatin DNA relative to the BamHI site. The appearance of a very distinct and ordered DNA fragmentation pattern (Figures 1 and 2, chromatin, restricted lanes) indicates that in all cells the same micrococcal nuclease and DNAase I cutting sites occur at regular intervals in the centromere region. The most striking feature of the centromere chromatin pattern is that a 220-250 bp region of centromere DNA from both chromosomes III and XI is completely protected from nucleolytic cleavage and is bounded by micrococcal nuclease and DNAase I cleavage sites. This discrete protected region can be visualized 350-600 bp in a centromere-proximal direction from the BamHI site on chromosome III (Figure 1) and 1200-1450 bp in a centromere-proximal direction from the BamHI site on chromosome XI (Figure 2). Specific micrococcal nuclease cleavage sites are also visualized following digestion of naked, deproteinized DNA (Figure 1A and 2A, naked DNA lanes). The cleavage pattern given by unprotected DNA is somewhat similar to that seen using whole chromatin, however specific sites in naked DNA are masked from nuclease action in chromatin. In particular, micrococcal nuclease-sensitive sites in naked, deproteinized DNA occurring 500 bp from BamHI site on chromosome III and 1350 bp from the BamHI site on chromosome XI are totally masked in the same region of chromatin DNA (Figure 1A and 2A, compare chromatin and naked DNA lanes). Thus the 220-250 bp protected region of DNA results in most part from the conformation of the DNA in chromatin, rather than from intrinsically nuclease-insensitive sites in naked DNA.

Mapping the chromatin structure in the centromere flanking region (1) revealed that the discrete 220-250 bp protected region of CEN DNA is flanked by micrococcal nuclease and DNAase I cleavage sites, and 2.5-3 kb of the surrounding chromatin is organized into a highly ordered and specifically aligned nucleosomal array.

Chromatin Structure of Centromere DNA on Minichromosomes

Since the unique structure of centromere chromatin is highly conserved between the two centromeres from chromosome III and XI in yeast (Figures 1 and 2), a similar organization might be expected to be associated with any yeast centromere sequence and the surrounding DNA, whether these sequences are present in chromosomes or on mitotically stable minichromosomes. To examine the structural organization of CEN DNA and the surrounding sequences on autonomously replicating plasmids in yeast, nuclei were prepared from yeast cells transformed with plasmids carrying yeast DNA from the centromere region of chromosome III (Figure 3). A time course of micrococcal nuclease cleavage of chromatin DNA in the isolated nuclei from transformed yeast cells is shown in Figure 4. To map only the centromere region cloned on the plasmid, the isolated DNA fragments were cleaved with HindIII, which cleaves the plasmid in the pBR322 vector DNA, and the hybridization probe was a 346 bp pBR322 DNA fragment (Figure 4) that extends toward the centromere from the HindIII site (large arrow, Figure 4). Only the centromere sequences that are contiguous to the pBR322 vector DNA should be visualized following hybridization and autoradiography.

The plasmids pYe(CEN3)41 and pYe(CDC10)1 are mitotically stable in yeast (3) and contain 1.6 and 7 kb, respectively, of yeast DNA from the centromere

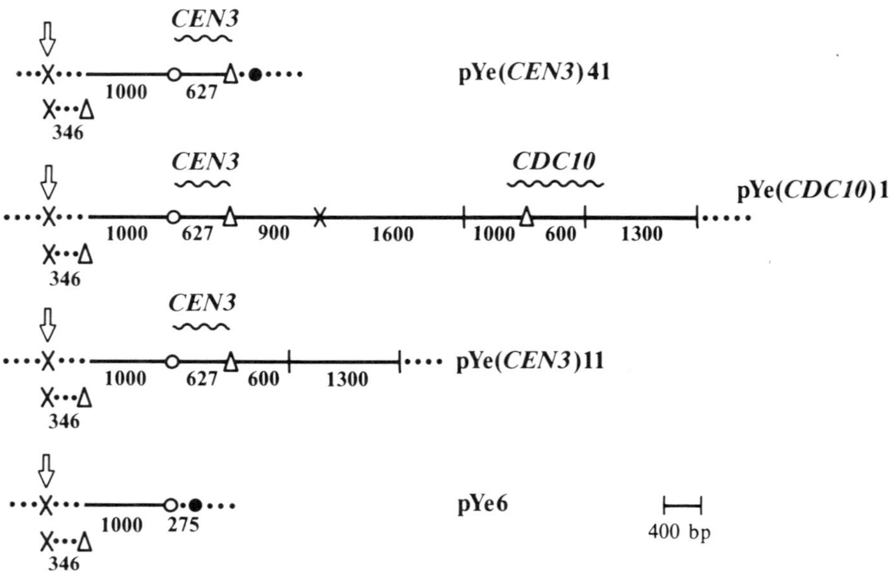

Figure 3. Comparative Restriction Maps of the Relevant Subcloned DNA Fragments from the Centromere Region of Yeast Chromosome III.

The position of the functional 627 bp CEN3 DNA fragment on the yeast DNA inserts is indicated by the wavy line above each restriction map. The solid lines indicate the yeast DNA inserts and the dotted lines denote vector (pBR322) sequences. The complete structures of pYe(CEN3)41, pYe(CDC10)1, and pYe(CEN3)11 are given in Clarke and Carbon (3). pYe6 is described in Fitzgerald-Hayes et al. (6). Only selected restriction sites are shown. The DNA fragments shown below each restriction map were used as hybridization probes; lengths are given in base pairs. The large arrows denote the restriction sites chosen for the nuclease mapping studies. Restriction enzyme cleavage sites are BamHI (△), Sau3A (○), SalI (●), HindIII (X) and EcoRI (|).

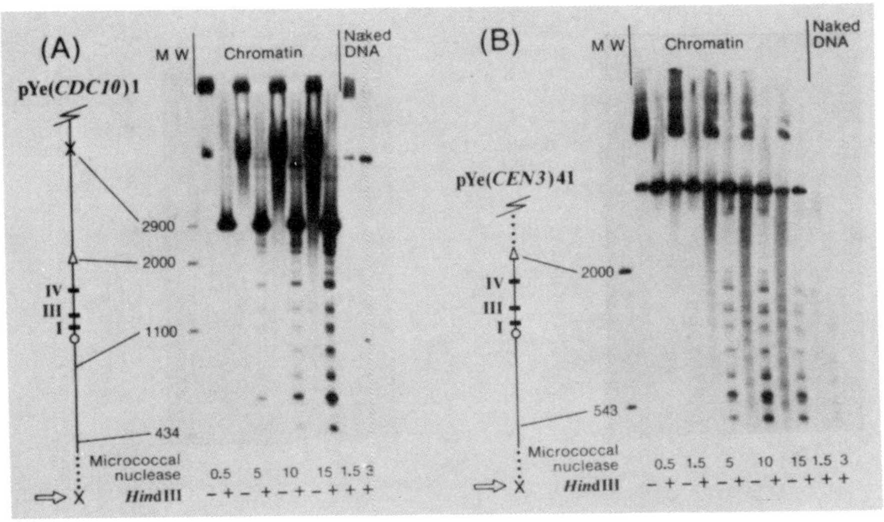

Figure 4. Mapping Nuclease Cleavage Sites on Plasmid pYe(CEN3)41, pYe(CDC10)1, pYe6 and pYe(CEN3)11 Chromatin From Transformed Yeast Cells.

Yeast cells (J17) were transformed with the appropriate plasmid DNA as described by Hsiao and Carbon (9). Restriction maps of the plasmid DNAs are shown in Figure 3. The relevant portion of each restriction enzyme map is shown to the left. Restriction enzyme sites are HindIII(X), BamHI (△), Sau3A (○) and EcoRI (|). Nuclei were prepared and digested for the times (min) indicated with micrococcal nuclease as described in the legend to Figure 1. The resulting DNA fragments were purified, digested with (+) or without (-) HindIII, and fractionated by electrophoresis. Samples were blotted and hybridized to the pBR322 bp DNA fragment shown in Figure 3. Molecular weight markers (M.W.) were prepared from DNA isolated from the appropriate yeast strain; for pYe(CDC10)1 (A), DNA was restricted with HindIII, HindIII/BamHI, HinfI and Sau3A; for pYe(CEN3)41 (B), DNA was restricted with BamHI/HindIII and Sau3A; for pYe6 (C), DNA was restricted with HindIII/BamHI, HinfI and Sau3A; and for pYe(CEN3)11 (D), DNA was restricted with HindIII/BamHI, HindIII/EcoRI and HinfI. The size of these DNA fragments serves to confirm the actual restriction map of the plasmid chromatin as it occurs in the transformed yeast cell. The large arrows indicate the HindIII site on the respective plasmid DNAs that is immediately proximal to the labeled 346 bp probe.

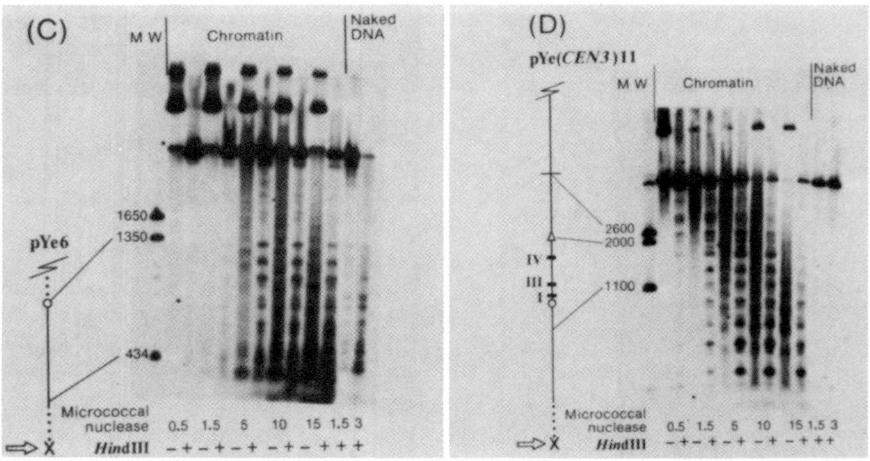

Figure 4 (continued).

region of chromosome III (Figure 3). Specific micrococcal nuclease cleavage sites at nucleosomal intervals are evident on the 1000 bp region of yeast DNA (400-1400 bp from the vector HindIII site, Figure 4A and B) that is contiguous with the CEN3 sequence in both plasmids (to the left of CEN3 in Figure 3). The CEN3 region, 1400-2000 bp from the vector HindIII site, occurs in a structure analogous to the chromosomal CEN3 region, and contains the 220-250 bp region of protected chromatin DNA (revealed by the 1400 and 1650 bp fragments). On pYe(CEN3)41 (Figure 4B), vector pBR322 DNA sequences juxtapose the BamHI site in the yeast DNA segment shown in Figure 3, whereas pYe(CDC10)1 contains an additional 5.4 kb of yeast DNA before pBR322 sequences are encountered. The yeast DNA contiguous with the BamHI site on pYe(CDC10)1 is cut into very ordered DNA fragments (Figure 4A), again very similar to the ordered pattern visualized on these sequences in chromosomal DNA (1). In marked contrast, the pBR322 DNA sequences immediately adjacent to the yeast DNA in the analogous region of pYe(CEN3)41 (Figure 3) are not cleaved in the chromatin at specific sites by micrococcal nuclease (Figure 4B, 2000-4000 bp region), but instead a uniform smear is visualized in the 2000-4000 bp region. It therefore appears that the nucleosomal array characteristic of the sequences flanking the centromere in yeast chromosome III is not simply being propagated from the centromere boundary through any DNA sequence. To examine if the nucleosomal ordering would be propagated through any yeast DNA sequence, the nuclease cleavage sites in chromatin were mapped on the mitotically stable plasmid pYe(CEN3)11 (3). A 3.5 kb BamHI yeast DNA fragment present in pYe(CDC10)1 was deleted in pYe(CEN3)11 (Figure 3). The yeast DNA sequences juxtaposed to the new position immediately adjacent to the CEN3 sequence in pYe(CEN3)11 (Figure 4D, 2000 bp from the HindIII site) are not cleaved at regular 160 bp intervals. Thus the highly ordered nucleosomal array is not propagated through pBR322 or yeast DNA sequences brought into new positions proximal to the CEN3 region on plasmid DNA, but instead is a characteristic property of the DNA sequence normally found flanking the centromere.

Since the sequences contiguous with the chromosomal CEN3 DNA are required for the higher order chromatin structure observed in this region of the chromosome, these sequences might be organized into highly ordered nucleosomal arrays regardless of their proximity to CEN3 DNA. To examine this possibility, the nuclease cleavage sites in chromatin were mapped on the plasmid pYe6. This plasmid carries 1000 bp of yeast DNA from the CEN3 flanking region without the 627 bp CEN3 Sau3A fragment (Figure 3) and is not stably maintained through mitosis (6). Micrococcal nuclease cleavage sites are evident at approximately nucleosomal intervals on this 1000 bp region of yeast DNA (400 to 1400 bp from the vector HindIII site, Figure 4C). Thus the sequences contiguous with the CEN3 DNA in yeast chromosome III are cleaved in the same specific manner by micrococcal nuclease when they are cloned into plasmids with or without the adjacent 627 bp CEN3 fragment (Figure 4). These results indicate that the nuclease-resistant centromere core as well as the flanking yeast DNA are in the same conformation whether present on the genome or on autonomously replicating plasmids in yeast.

Fine Structure Mapping of the Nuclease-Resistant Centromere Core

The 220-250 bp protected region in chromatin includes the region of DNA, extending from sequence element I to about 100 bp beyond sequence element III (Figure 1, 2 and 4), that is responsible for proper mitotic segregation of autonomously replicating plasmids in yeast (5). To relate the structure of these CEN sequences to their function we have examined the chromatin structure of CEN sequences on plasmids containing small DNA fragments capable of centromere function, and on deletion mutant plasmids with altered CEN sequences. If the nuclease-resistant core is a structural element necessary to the functional centromere in chromatin, then a similar distinct structure would be expected to be characteristic of any yeast centromere sequence that stabilizes minichromosomes through mitosis. Conversely, altered sequences that no longer demonstrate centromere function would not be expected to be in the same chromatin conformation as their functional counterparts. The plasmids used for these studies were constructed by introducing deletion mutations within the CEN insert in pYe(CEN11)5 (5). The resulting deletion mutant plasmids, dl31-28, dl31-48 and dl31-6 all have lost the unique BamHI site in pYe(CEN11)5 and varying amounts of pBR322 and yeast DNA as shown in Figure 5 (left). The structure of these sequences in chromatin was examined as described in Figure 4.

In pYe(CEN11)5 chromatin a 200-250 bp nuclease resistant region occurs between 800 and 1000 bp from the pBR322 HindIII site, and is bounded on both sides by micrococcal nuclease cleavage sites (Figure 5, wt-chromatin lanes). Micrococcal nuclease-sensitive sites occur in naked, deproteinized DNA 800 to 1000 bp from the HindIII site and are totally masked in this region of chromatin DNA (Figure 5, wt-naked DNA lanes). Thus the 250 bp protected region of DNA results in most part from the conformation of these sequences in chromatin. This region is the same 220-250 bp sequence of CEN11 DNA that is resistant to nuclease digestion in chromosome XI (Figure 2). Mapping the deletion mutant dl31-28, a mitotically stable plasmid (5), revealed a nuclease-resistant region between 800 and 1000 bp from the pBR322 HindIII site, that is bounded on only one side by a micrococcal nuclease cleavage site (Figure 5, 28-chromatin lanes). The resistant region is again the same 220-250 bp sequence of CEN11 DNA that is resistant to nuclease digestion in pYe(CEN11)5 and chromosome XI. The nuclease-sensitive site in pYe(CEN11)5 chromatin that bounds the protected region distal to the HindIII site is deleted in dl31-28. Since the pBR322 sequences adjacent to element I in dl31-28 DNA are not sensitive to cleavage in dl31-28 chromatin, the nuclease-sensitive site that bounds the protected region in pYe(CEN11)5 chromatin is not simply a consequence of the structure of the nuclease-resistant region in chromatin but may result from the interaction of specific yeast DNA sequences flanking the protected centromere core.

In deletion mutant dl31-48, a mitotically unstable plasmid (5), elements I through III are deleted. Mapping studies of dl31-48 revealed specific nuclease cleavage sites in the chromatin DNA 800 to 1000 bp from the HindIII site (Figure 5, 48-chromatin lanes). The same pBR322 DNA sequences in this region of dl31-48 are cleaved in chromatin and naked, deproteinized DNA (Figure 5, 48-naked DNA lanes). The intact yeast DNA in dl31-48, 850 bp from the HindIII site, is nuclease sensitive in chromatin and insensitive in naked, deproteinized DNA (compare Figure 5, 48-chromatin and 48-naked DNA lanes). In deletion mutant dl31-6, also mitotically unstable (5), elements I through III and the flanking 100 bp of yeast DNA proximal to the HindIII site are deleted (Figure 5, left). Specific nuclease cleavage sites occur in the 800-1000 bp region from the HindIII site and are the same in both chromatin or naked, deproteinized DNA (Figure 5, compare 6-chromatin and 6-naked DNA lanes). Thus, the nuclease-resistant region is eliminated when the sequence

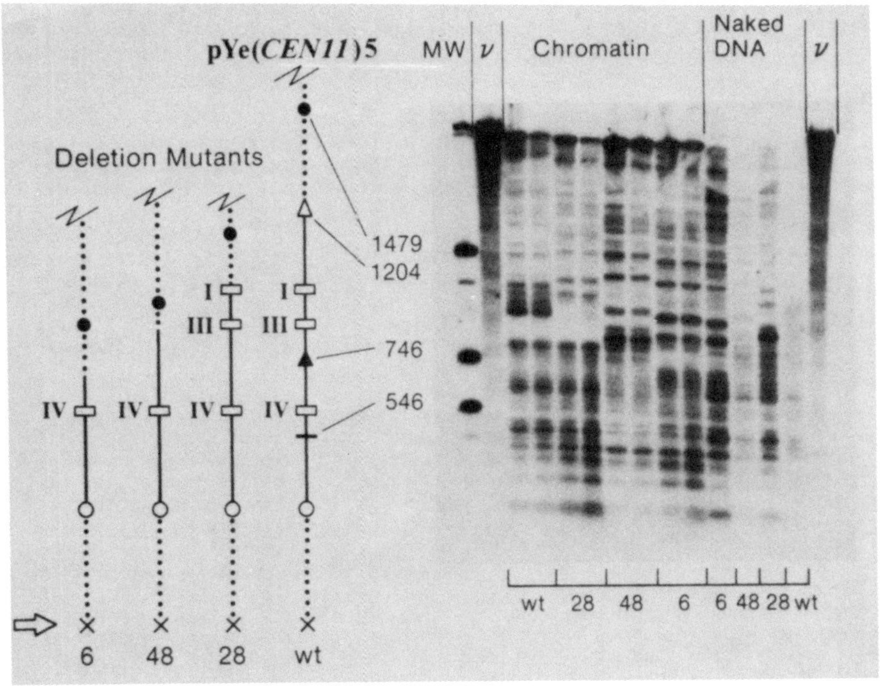

Figure 5. Mapping Nuclease Sensitive Sites on pYe(CEN11)5 and Deletion Mutant Plasmids From Transformed Yeast Cells.

The relevant portions of each restriction site map for pYe(CEN11)5 (wt), dl31-28 (28), dl31-48 (48) and dl31-6 (6) are shown to the left. Restriction enzyme sites are HindIII (X), Sau3A (○), HinfI (|), HincII (▲), BamHI (△) and SalI (●). The solid lines represent yeast DNA inserts and the dotted lines denote vector (pBR322) sequences. Nuclei were prepared from the appropriate yeast transformants as described in the legend to Figure 1. Micrococcal nuclease digestion of chromatin and naked, deproteinized DNA was performed as described in Figure 1 for 5 min. The resulting DNA fragments were purified and incubated in the presence (chromatin and naked DNA lanes) or absence (ν) of HindIII. ν DNA was prepared from pYe(CEN11)5 transformants. The samples were electrophoresed on a 1.7% agarose gel, blotted and hybridized with the 346 bp pBR322 fragment extending from the HindIII site (large arrow) to the Sau3A site in a centromere proximal direction. Molecular weight markers (M.W.) were prepared from DNA isolated from pYe(CEN11)5 transformants and restricted with HindIII/HinfI, HindIII/HincII, HindIII/BamHI and HindIII/SalI. The size of these DNA fragments serves to confirm the actual restriction map of the plasmid chromatin as it occurs in the yeast cell.

containing elements I-III is deleted, and coincides directly with the loss of ability of these fragments to confer mitotic stability to autonomously replicating plasmids in yeast.

DISCUSSION

To determine how the sequences common to both CEN regions might mediate centromere function, we have examined the chromatin structure of these sequences in the yeast cell. A highly nuclease-resistant centromere core was found that includes the region extending from sequence element I to about 100 bp beyond sequence element III (Figures 1, 2 and 6). This unique chromatin structure was found to be associated with the smallest CEN sequences capable of stabilizing autonomously replicating plasmids in yeast (Figure 5, wt and dl31-28) but was not evident on the plasmids with altered CEN sequences that were mitotically unstable (Figure 5, dl31-48 and dl31-6). Thus a region of structural differentiation has been defined that encompasses the sequences required for centromere function.

A highly ordered and specifically aligned array of nucleosomes spans 2.5-3 kb of DNA surrounding the central nuclease-resistant core. Mapping the nuclease cleavage sites within plasmid chromatin allowed us to determine that the ordered distribution of nucleosomes in the CEN flanking region does not result as a consequence of propagation from the centromere boundary, but rather these sequences must contain specific recognition signals for nucleosome alignment. Since the centromere flanking sequences do not stabilize autonomously replicating plasmids in yeast (6), it is possible that the highly ordered nucleosomal arrays surrounding the centromere core play a significant role in other centromere functions, such as the ability to segregate properly through meiosis in a typical Mendelian pattern and to maintain a controlled copy number. The highly ordered array of nucleosomes adjacent to the centromere in yeast may resemble the chromatin structure associated with the highly repetitive satellite DNAs occurring in the centromeric heterochromatin of higher eukaryotes (2). Although the biological significance of these regions of DNA remains to be elucidated, there could be an evolutionary relationship between the centromere regions in yeast and satellite sequences in higher eukaryotes.

The centromere is defined in cytological terms as the primary constriction of the chromosome where the spindle fibers attach. In yeast, a single microtubule attaches directly to each chromatin fiber without any visible structural differentiation (13). Similar direct attachments of microtubules with the chromatin fiber also seem to occur in other organisms (10). It is possible that the nuclease-resistant centromere core we have observed in yeast chromatin might serve as the microtubule attachment site, a structurally primitive kinetochore. As shown in Figure 6, the structural parameters that have been identified for the individual components of the mitotic apparatus are consistent with this view. The structural description of the centromere core as revealed by chromatin mapping studies (Figure 1, 2, 4 and 5) indicate that 220-250 bp of DNA are in a unique chromatin structure. The most extensively characterized unit of chromatin, the nucleosome core particle, consists of 160 bp of DNA wrapped around histone proteins to give rise to a roughly cylindrical particle with a diameter of about 11 nm and a height of approximately 5 nm (11). We can therefore estimate that the centromere core may be folded into a chromatin particle 15-20 nm in diameter. Microtubules, which are the major component of the mitotic apparatus, are about 20 nm in diameter (Figure

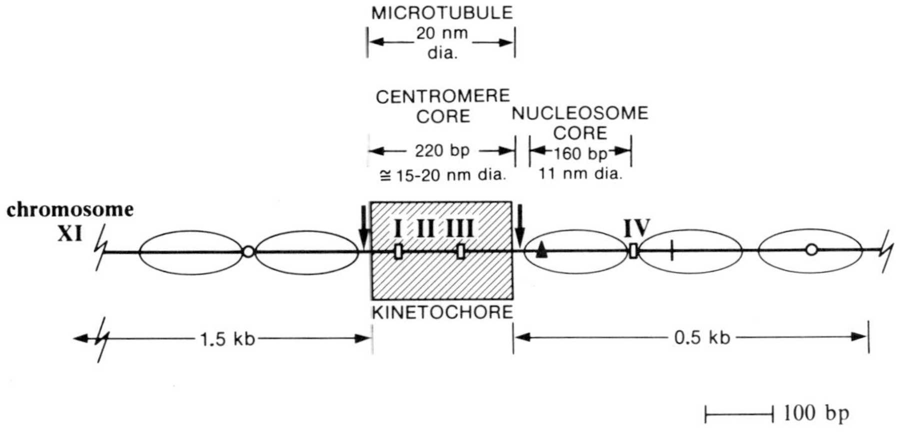

Figure 6. Schematic Representation of the Centromere Region of Chromosome XI.

A map of the chromosomal region surrounding CEN11 shows the 858 bp CEN11 fragment and the relative positions of sequence elements I, III and IV (5,6) that are homologous between CEN3 and CEN11. The very [A+T]-rich region (element II) lies between elements I and III. Restriction enzyme sites are Sau3A (○), HincII (▲) and HinfI (┼). Micrococcal nuclease (↓) cleavage sites that bound the nuclease-resistant centromere core (shaded box) are indicated. The DNA fiber is presented in linear form to visualize the position of the 220-250 bp nuclease-resistant core and the nucleosome cores relative to the restriction map of chromosomal DNA. In chromatin the nucleosomal DNA is wrapped around histone proteins to form a roughly cylindrical particle with a diameter of 11 nm as indicated and a height of approximately 5 nm (11). By extrapolation, we estimate the centromere core particle to be at least 15-20 nm in diameter, however the conformation of DNA within the 220-250 bp nuclease-resistant core remains to be elucidated. The arrows below the restriction map indicate how far the ordered nucleosomal arrays extend from the nuclease-resistant centromere core (1).

6) (13) and have been visualized in electron micrographs to make direct contact with the chromatin fiber (13). Thus, despite the lack of visual chromosome condensation, mitosis in yeast may be very similar to mitosis in higher eukaryotes. Recent evidence in support of this view revealed that the highly differentiated mammalian kinetochore actually consists of a series of coiled DNA fibers in close apposition, that are resolved only upon high resolution electron microscopy (14). Since a yeast chromosome is on the average 100 times smaller than a mammalian chromosome, a single coiled DNA fiber that serves as a microtubule attachment site may comprise the entire yeast kinetochore (see Figure 6).

LITERATURE CITED

1. Bloom, K.S. and Carbon, J. 1982. Yeast centromere DNA is in a unique and highly ordered structure in chromosomes and small circular minichromosomes. Cell, 29:305-317.

2. Brutlag, D.L. 1980. Molecular arrangement and evolution of heterochromatic DNA. Ann. Rev. Genetics 14:121-144.

3. Clarke, L. and Carbon, J. 1980. Isolation of a yeast centromere and construction of functional small circular minichromosomes. Nature 287: 504-509.

4. Fitzgerald-Hayes, M., Buhler, J-M., Cooper, T.G. and Carbon, J. 1982. Isolation and subcloning analysis of functional centromere DNA (CEN11) from *Saccharomyces cerevisiae* chromosome XI. Molecular and Cellular Biology 2:82-87.

5. Fitzgerald-Hayes, M. and Carbon, J. 1982. Identification of DNA sequences required for mitotic stability of centromere plasmids in yeast. In: C. Bruschi, M. Esposito, R. Mortimer, and D. Schild (ed.), Proceedings of Berkeley Workshop on Recent Advances in Yeast Molecular Biology: Recombinant DNA. Dept. of Energy Publication, Berkeley, CA.

6. Fitzgerald-Hayes, M., Clarke, L. and Carbon, J. 1982. Nucleotide sequence comparisons and functional analysis of yeast centromere DNAs. Cell 29:235-244.

7. Forte, M.A. and Fangman, W.L. 1976. Naturally occurring cross-links in yeast chromosomal DNA. Cell 8:425-431.

8. Hinnen, A., Hicks, J.B. and Fink, G.B. 1978. Transformation of yeast. Proc. Natl. Acad. Sci. USA 75:1929-1933.

9. Hsiao, C.-L. and Carbon, J. 1979. High frequency transformation of yeast plamids containing the cloned yeast ARG4 gene. Proc. Natl. Acad. Sci. USA 76:3829-3833.

10. Kubai, D. 1975. The evolution of the mitotic spindle. Int. Rev. Cytol. 43:167-227.

11. McGhee, J.D. and Felsenfeld, G. 1980. Nucleosome structure. Ann. Rev. Biochem. 49:1115-1156.

12. Nelson, R.G. and Fangman, W.L. 1979. Nucleosome organization of the yeast 2-μm DNA plasmid: A eukaryotic minichromosome. Proc. Natl. Acad. Sci. USA 76:6515-6519.

13. Peterson, J.B. and Ris, H. 1976. Electron-microscopic study of the spindle and chromosome movement in the yeast *Saccharomyces cerevisiae*. J. Cell Sci. 22:219-242.

14. Ris, H. and Witt, P.L. 1980. Structure of the mammalian kinetochore. Chromosoma 82:153-170.

15. Southern, E.M. 1975. Detection of specific sequences among DNA fragments separated by gel electrophoresis. J. Mol. Biol. 98:503-517.

16. Tschumper, G. and Carbon, J. 1980. Sequence of a yeast DNA fragment containing a chromosomal replicator and the *TRP1* gene. Gene 10:157-166.

17. Wu, C. 1981. The 5' ends of *Drosophila* heat shock genes in chromatin are hypersensitive to DNase I. Nature 286:854-860.

3. THE FIDELITY OF MITOTIC CHROMOSOME REPRODUCTION IN S. cerevisiae*

Leland H. Hartwell, Susan K. Dutcher, John S. Wood and Barbara Garvik

Department of Genetics, SK-50
University of Washington
Seattle, Washington 98195

Rec. Adv. Yeast Mol. Biol. (1982) 1:28-38

We have found conditions that permit the addition of a chromosome to a nucleus of S. cerevisiae (kar1 mediated chromosome transfer) and conditions that permit removal of a chromosome (MBC induced chromosome loss). These methods were utilized to construct disomes and monosomes. Experiments demonstrated that the loss of chromosome VII from chromosome VII disomic haploid cells or chromosome V from diploid cells could be efficiently selected. A fluctuation analysis demonstrated that the frequency of mitotic chromosome loss in either of these cases is about one in 5×10^4 divisions. We conclude that this is an accurate estimate of the fidelity with which mitotic yeast cells replicate and segregate their chromosomes.

kar1 Mediated Chromosome Addition

We (6) and Nillson-Tillgren et al. (12) have found that chromosomes are transferred from one nucleus to another during abortive nuclear fusion in heterokaryons having one parent with the kar1-1 mutation (3). Haploid progeny, cytoductants, were selected in crosses of the following type: MATa KAR1 cyh2 x y M^+ N^+ [rho] x MATα kar1-1 X^+ Y^+ m n [rho^+], where cyh2 is a recessive nuclear mutation conferring resistance to cycloheximide, x, y, m, and n are nuclear mutations conferring auxotrophy for various nutrients, [rho] and [rho^+] signify the absence and presence of mitochondrial DNA, respectively. Cytoductants were selected by plating conjugation mixtures onto medium containing cycloheximide with glycerol as the carbon source. Neither of the two parents nor the true diploids arising from the cross grow on this medium. Cytoductants in this cross that have the genotype of the KAR1 parent bearing the cyh2 allele and have received functional mitochondria from the cytoplasm of the kar1-1 parent will grow on this medium. The unselected nuclear markers, x, y, m, and n were scored in the cytoductants. Most cytoductants retained the nuclear genotype (x y M^+ N^+) of the KAR1 parent, but a small proportion of the cytoductants had acquired the X^+, Y^+, m, or n markers of the kar1-1 parent (exceptional cytoductants).

Of the 60,800 KAR1 cytoductants examined from the selective medium monitoring 10 of the 17 chromosomes from four independent crosses, 349 exceptional cytoductants were found. Transfers involving 9 (chrom. I, II, III, IV, V, VI, XI, XII and XVI) chromosomes were found. Only chromosome VII failed to be transferred in these experiments and subsequent experiments (see below) demonstrate transfer of this chromosome as well.

*Support was obtained from NIH Grant GM17709 to L.H. and NIGMS Training Grant GM07735.

Coincident acquisition of linked markers from the donor nucleus by the recipient nucleus indicates that whole chromosomes are being transferred between nuclei. The loci his7, tyr1, and lys2 map to a region about 120 centimorgans long on the right arm of chromosome II. In a cross in which the two parents had the genotypes his7 tyr1 lys2 and HIS7 TYR1 LYS2, 20 of 21 of the exceptional cytoductants that became prototrophic for lysine were also prototrophic for histidine and tyrosine. Loci, ade1 and cys1, located on opposite arms of chromosome I, were also found to be acquired together in exceptional cytoductants. In the cross MATa kar1-1 ade1 CYS1 cyh2 [rho] x - MATa ADE1 cys1 [rho$^+$], 15 of 16 ADE$^+$ exceptional cytoductants were also cys1.

Cytoductants were also selected in crosses of the reciprocal type: MATa KAR1 x y M$^+$ N$^+$ [rho$^+$] x MATα kar1-1 cyh2 X$^+$ Y$^+$ m n [rho]. Cytoductants from these crosses, selected to have the nucleus of the kar1 parent, were tested for exceptional nuclear markers coming from the KAR1 parent. Of the 18,260 cytoductants examined, 83 exceptional cytoductants were observed. Again, acquisition of linked markers was almost always coincident. Of the 31 exceptional cytoductants that became prototrophic for leucine (LEU2) or histidine (HIS4), 25 were coincident for the acquisition of the mating type locus. Acquisition of markers on different chromosomes showed only slight coincidence. The frequency of exceptional KAR1 cytoductants was the same as that of exceptional kar1 cytoductants. Thus, chromosomal markers are transferred at the same frequency and with the same coincidence to kar1 and KAR1 exceptional cytoductants.

Cytoductants were also selected in crosses in which no kar1 parent was involved. Cytoductants are produced at a low frequency (1 cytoductant per 1,000 diploids) in crosses of the type: MATa KAR1 CYH$^+$ x y M$^+$ N$^+$ (rho$^+$] x MATα KAR1 cyhs2 X$^+$ Y$^+$ m n [rho]. No exceptional cytoductants were found when 5 of the 17 chromosomes were monitored among 8,750 cytoductants. Therefore, transfer of chromosomal markers is reduced at least 50-fold in KAR1 x KAR1 zygotes compared to KAR1 x kar1 zygotes.

The frequency with which various nuclear markers were transferred varied over a 24-fold range, from 0.7 to 19.5 exceptional cytoductants per 10,000 cytoductants examined. The frequency of transfer for a particular nuclear marker was inversely correlated with the genetic size of the chromosome upon which it was located. Chromosomal I, the shortest chromosome, based on known genetic loci, with a genetic length of approximately 100 centimorgans, was transferred most frequently. Chromosome IV, one of the longest chromosomes, was transferred infrequently. Chromosomes of intermediate size were transferred with intermediate frequencies. However, the frequencies of transfer of two chromosomes, VI and XII, were anomalous with respect to their known genetic map lengths. The discrepancy for chromosome XII could be accounted for since the physical length of chromosome XII is greater than predicted by its genetic map length because it contains the ribosomal DNA cistrons (16), which show little or no recombination and thus have no genetic length.

Disomes for the recipient nucleus would be expected to be the primary product of chromosome transfers. The fact that exceptional cytoductants are usually haploid suggests that such disomes are usually transient. However chromosomes III transfers frequent lead to stable chromosome III disomes (6, 12) suggesting that the failure to recover disomes for most chromosomes is merely a consequence of the fact that most disomes are at a growth disadvantage compared to the haploid progeny that they invariable produce by chromosome loss or nondisjunction.

We set out to determine whether disomes were frequently produced as the primary event for other chromosomes as well. In order to test for this possibility it was necessary to be able to select for exceptional cytoductants without selecting for disomy and to have a convenient method for distinguishing disomes from haploids. Exceptional cytoductants for a particular chromosome are rare (10^{-3} to 10^{-4} of the cytoductants) and the mating mixture will contain unmated haploids, normal cytoductants, and diploid cells at much greater frequencies than the exceptional cytoductants. Therefore we used recipient haploid cells that were multiply resistant for recessive drug resistance alleles (permitting selection against the donor haploid or cytoductant and the diploid) and were auxotrophic for a marker on the itinerant chromosome homologue (permitting selection for the prototrophic allele and hence the chromosome from the donor strain, fig. 1). Furthermore, the recipient haploid nucleus contained a recessive drug resistance allele on the itinerant chromosome homologue. Thus if the product of the transfer is a disome the exceptional cytoductant should be drug-sensitive and prototrophic for the markers on the itinerant chromosome but should produce drug-resistant progeny at a frequency of about 10^{-4} due to mitotic recombination. The drug-resistant mitotic recombinants are easily detected by replica plating to drug containing plates where the disome will display a drug sensitive patch with papillations to drug resistance. If the primary product of the transfer is a haploid where the donor chromosome replaces the recipient chromosome the exceptional cytoductant will also be prototrophic and drug sensitive but will not produce drug resistant mitotic recombinants at detectable frequencies.

The rare clones that grew up were picked and streaked onto nonselective medium for marker tests by replica plating or in the case of itinerant chromosome V first cloned onto medium lacking tryptophan since in this case there was a noticeable background of contaminating cells on the selective plate (Table 1).

The frequency of selected clones was about 1×10^{-4} that of diploids produced in each cross but this number strongly underestimates the actual frequency of exceptional cytoductants since many generations intervened between mating and selection during which time the disomes will be at a growth disadvantage relative to the diploids. The recipient nucleus in each case contained some recessive markers on chromosomes that were not selected and these markers should display the recessive allele in all cases of chromosome addition (disomes) or chromosome replacement (haploids). In each case some clones were found that were not recessive for these unselected markers and such clones were subtracted from the number of selected clones to give the number of potential disomes. Clones that received dominant alleles from the donor strain for unselected chromosomes may be cases of coincident transfer of more than one chromosome as observed previously (6).

Clones were considered disomes if they had the genotype of potential disomes for unselected markers (thus assuring that they were haploid and contained the selected dominant nutritional marker from the itinerant chromosome of the donor) and the resistance allele from the itinerant homologue of the recipient. These clones must have contained two copies of both homologues

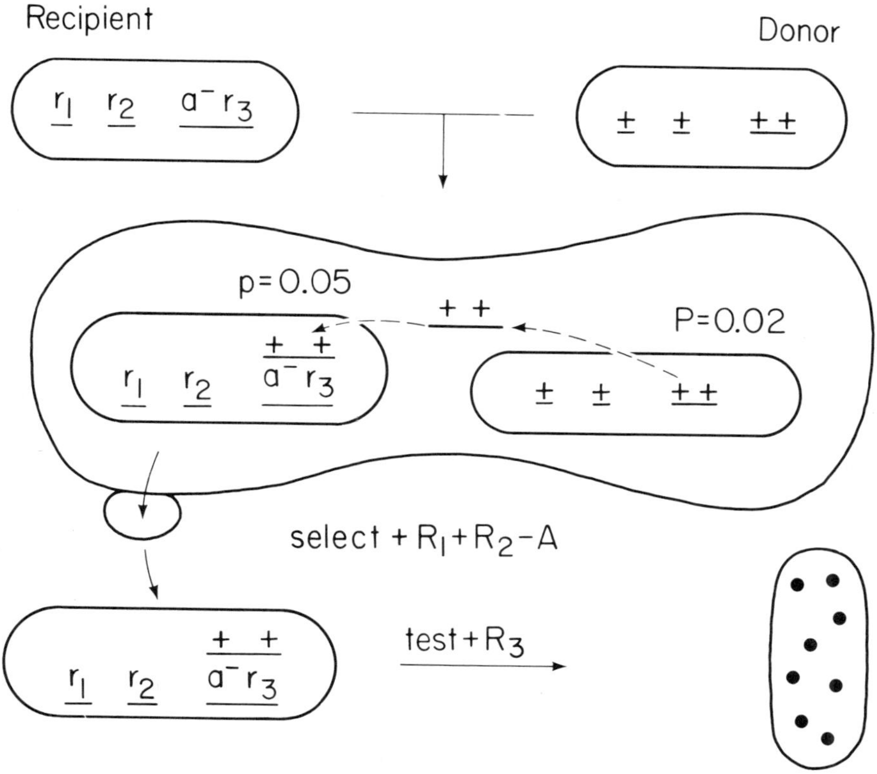

Fig. 1. Cells containing the recipient nucleus with three recessive resistence alleles, r1, r2 and r3 fuse with cells containing the donor nucleus during conjugation and since one of the parents is karl a transient heterokaryon results. With a probability of about 0.02 the donor nucleus loses an itinerant chromosome and with a probability of about 0.05 a recipient nucleus captures the itinerant chromosome (see 6). Exceptional cytoductants are selected that receive the A+ allele from the donor and the r1 and r2 alleles from the recipient. Exceptional cytoductants are then examined for the r3 allele which will be absent if chromosome replacement has occurred and present in the heterozygous form (evident as a papillations of growth on medium containing inhibitor 3) if chromosome addition has occurred.

Table 1

Parent Recipient	Strains kar1-1 Donor	Itinerant Chromosome	Markers on Itinerant Chromosome: Auxotrophy	Markers on Itinerant Chromosome: Drug Resistence	Selected Markers	Number of Selected Clones Examined	Number of Potential Disomes	Number of Actual Disomes
3400-10-3	JCK5-25	V	trp2	can1	cyh2, sap3, TRP2	32	27	5
"	"	VII	ade5	cyh2	can1, sap3, ADE5	64	53	53
"	"	XVI				96	83	70
3450-25a	3449-9-1	XV	ade2	tcm1	can1, cyh2, ADE2	189	136	98
3450-46	3449-5-4	XV	ade2	tcm1	can1, cyh2, ADE2			

of the itinerant chromosome although they were selected to contain only the itinerant chromosome donated by the karl-1 strain. For chromosome VII, XV and XVI the majority of the rare clones arose from disomes. In each of these three cases the difference between potential disomes and the actual disomes are clones that contain only the itinerant chromosome from the donor but did not have the resistance allele from the recipient. For chromosome V only a minority of the clones retained alleles from both parents and surprisingly most of the difference between potential and actual disomes were missing the TRP2 allele from the donor even though this marker had been selected. Nevertheless, it is likely in this case as well that the majority of the primary events were disomes since the clones initially without tryptophan. The potential chromosome V disomes grow extremely slowly (see below) and strong selection exists for chromosome loss. We conclude that in the majority of the cases where a chromosome transfer occurs the resulting cell is a disome capable of propagating both homologous chromosomes for many generations.

MBC Induced Chromosome Loss

Methyl-2-yl-benzimidazole carbamate (MBC) is the active component of the widely used fungicide Benomyl. In both fungal and mammalian cells MBC has been found to cause metaphase arrest (4, 8, 19). This block in mitosis observed cytologically has been found functionally to be a stage-specific block in early nuclear division (13; Wood and Hartwell, in press). The primary genetic effect of MBC has been reported to be chromosome nondisjunction, observed as an increased frequency of mitotic segregation for heterozygous markers (1, 7, 11).

The effects observed with MBC are all apparently due to the disruption of microtubules. Davidse and Flach found that MBC bound to a fungal protein closely resembling mammalian tublin (5). Conclusive evidence that this protein was tubulin has been presented by Sheir-Neiss, *et al.* (18); they demonstrated that Benomyl resistant strains of Aspergillus had β-tubulins with altered electrophoretic charge or peptide fingerprints. Resistant or supersensitive mutants also had altered binding coefficients for Benomyl to tubulin. Cytological evidence suggests that this is the case; MBC arrested cells had few or no spindle microtubles (17; B. Byers, personal communication).

The ability of MBC to induce chromosome loss in *S. cerevisiae* was tested in strains heterozygous for a recessive drug marker and heterozygous for a recessive auxotrophic marker on the opposite arm of the same chromosome, with both of the recessive markers linked in coupling. The frequency of drug resistant cells was assayed before and after MBC treatment, and then 50 to 100 independent, drug resistant clones were tested for the appearance of the auxotrophic marker located on the opposite arm of the chromosome from the drug resistance allele. This design distinguished between mitotic recombination (one chromosome arm altered) and chromosome loss (both chromosome arms altered) events.

A haploid strain disomic for chromosome III was treated with MBC. Prior to MBC treatment the frequency of cryptopleurine resistant (cry^R) clones was approximately 1.2×10^{-5} and 74% of these were due to mitotic recombination. After 24 to 48 hours in MBC, the frequency of resistant clones was increased more than 100 fold to $4\text{-}5 \times 10^{-3}$ and 90% of these had uncovered markers on both arms of chromosome III.

Since the frequency of these presumptive chromosome III loss colonies increased more than 100 fold while viability declined less than 10 fold, MBC treatment could not be simply selecting for pre-existing or spontaneous chromosome loss cells, unless such cells were resistant to division arrest by MBC. This latter possibility was tested by pooling 10 cry^R chromosome loss colonies into one culture and 10 cry^R mitotic recombination colonies into another, and treating each of these pooled cultures with MBC. Neither culture was resistant to the inhibitory effect of MBC: the viable cell counts after 48 hours in MBC were 84% and 125%, respectively, of their starting values. Thus the increase in chromosome loss frequency caused by MBC must have been due to the induction of new events.

To see if loss occurred for other chromosomes as well, a similar experiment was performed with a diploid strain heterozygous for both the recessive drug resistance markers to canavanine on chromosome V and cycloheximide on chromosome VII as well as for auxotrophic markers on the opposite arms of these same chromosomes. At the start of the experiment canavanine resistant (can^R) clones occurred at a frequency of approximately 2×10^{-4} and 62% of these were the result of mitotic recombination. After 13 hours of MBC treatment the frequency of can^R was increased 150 fold to approximately 3×10^{-2} and 92% of these were due to chromosome loss. In the same experiments cells were plated on cycloheximide medium to determine the frequency of cycloheximide resistant (cyh^R) cells. The initial frequency of cyh^R clones was about 4×10^{-5} and essentially all of these (98%) were caused by mitotic recombination. After 13 hours in MBC the frequency had increased 20 fold to 8×10^{-4} cyh^R and 66% of these were the result of chromosome loss. Although MBC did not induce intragenic recombination we found a small and irreproducible induction of intergenic recombination.

Chromosome loss was examined in a strain which was heterozygous for auxotrophic markers on 13 chromosomes to examine the relative frequencies of loss of each of these chromosomes after MBC treatment. We found that all chromosomes were lost at approximately equal frequencies.

The Fidelity of Chromosome Replication and Segregation

One of our interests is the fidelity of chromosome replication and segregation. We wished to monitor the frequency with which a cell fails to donate a specific chromosome to one of its progeny. In order to make this measurement it is necessary to be able to propagate the parent cell and progeny derived from it by chromosome loss and to demonstrate unequivocally by meiotic analysis that the parent and progeny differ in their chromosome complement. Furthermore, since chromosome loss events are likely to be rare it is necessary to be able to select for the progeny that have lost a chromosome. Finally, in order to determine the frequency of loss one must determine the phenotypic lag necessary for the genetic event that produces a chromosome loss to be manifest in a cell that can reproduce under the selective conditions. These conditions have not previously been met although in a number of studies estimates of the spontaneous frequency of chromosome loss in *S. cerevisiae* have been made but none of these were designed to quantitatively measure this value. All monitored the yield of chromosome loss events after growth of a culture rather than rate of chromosome loss. Parry and Zimmerman (15) selected purported monosomes for chromosome VII on cycloheximide containing medium from *cyh2*/+ heterozygous diploid strains and found that the frequency of monosomes was between 0.86 and $289/10^6$ cells in various cultures (14, 15), and were unable to confirm the monosomy of the products by meiotic analysis. Campbell et al. (2) selected mating cells from chromosome III Mata/α disomes and determined the proportion that were due to haploids; they estimate the spontaneous yield of chromosome III loss events at about 10^{-4}/cell. Liras et al. (10) examined the appearance of auxotrophs in diploids marked heterozygously on both chromosome arms for several chromosomes and estimated the spontaneous yield of monosomic cells at less than 10^{-5}/cell.

Phenotypic Expression

When heterozygous cells lose a dominant allele for antibiotic resistance they become genotypically resistant. However, it might take several divisions to dilute out the sensitive gene product and hence there might be a phenotypic lag before such cells would be detected by plating on antibiotic containing medium. Conversely, heterozygous cells might be able to complete one or more divisions on antibiotic medium thus producing resistant offspring on the selective plate. We wished to determine whether recessive resistance markers showed phenotypic lag or leakiness so that the observed frequency of resistant cells from a heterozygous parent could be properly interpreted. karl generated heterokaryons provide a convenient method for accomplishing this goal since the heterokaryotic zygote can be genotypically heterozygous if one haploid parent carries the sensitive allele and the other parent carries the resistant allele. Further, the progeny produced by end buds of the zygote are haploid and hence half are genotypically resistant but newly arisen from a sensitive heterokaryon. We have demonstrated that heterokaryons with one sensitive and one resistant nucleus do not divide when placed onto cycloheximide containing medium but that first buds containing the cyh2 allele do divide and produce colonies (Dutcher, in press). A similar result is observed with canl/CANl heterokaryons (Hartwell, unpublished). Thus the cyh2 and canl alleles are neither leaky nor show phenotypic lag.

Our experience with disomic strains generated by karl mediated chromosome transfer demonstrated that chromosome VII disomes were able to divide mitotically only slightly slower than the haploids they produce by chromosome loss. Meiotic analysis of the chromosome VII disomic strains showed that they contained two copies of chromosome VII while the haploid cells produced by chromosome loss contained only one copy of chromosome VII (unpublished).

The frequency of chromosome VII loss from disomic haploids was determined in chromosome VII disomes heterozygous for the cyh2 allele on one chromosome arm in coupling to an auxotrophic marker on the other arm. Individual clones were picked and plated onto cycloheximide medium. The cyh2 colonies that were prototrophic for the auxotrophic marker were assumed due to recombination and those that were auxotrophic were assumed due to chromosome loss. The frequency of recombination and loss was determined by the method of the median (9). One copy of chromosome VII was lost once in about 2×10^5 cell divisions.

Although the fidelity of chromosome replication and segregation seems reasonably high in these disomic strains it is possible that the fidelity is much lower than in a euploid cell either because of the abnormal dosage of chromosome VII genes or the abnormal number of centromeres per spindle pole. Consequently we wished to determine the frequency of chromosome loss from a euploid cell. We needed to find therefore a monosome that could be efficiently selected from a diploid cell. MBC induces the loss of all chromosomes with nearly equal frequency when the monosomes are detected nonselectively. However when we attempted to select monosomes for chromosomes III, V, or VII the frequencies of recovered monosomes differed by orders of magnitude. Chromosome V monosomes were about 30 times as frequent as chromosome VII monosomes. For chromosomes V and VII we know that this discrepancy is not due to phenotypic expression since *can1* and *cyh2* do not show a phenotypic lag. Furthermore, since we selected chromosome V monosomes at nearly the same frequency as they were detected without selection after MBC treatment we conclude that chromosome V monosomes can be efficiently selected on canavanine containing medium while chromosome VII monosomes cannot be selected on cycloheximide containing medium.

A diploid *Mat*α/α strain heterozygous for *can1* on one arm of chromosome V and, in coupling, *hom3* on the other was cloned onto nonselective medium. Individual clones were picked and plated onto canavanine containing medium and the frequency of auxotrophic (chromosome loss) and prototrophic (recombinant) colonies were determined. Ten of the purported monosomes were crossed to a *Mat*a/a strain and monosomy was confirmed by dissection of the resulting tetraploid was dissected. The frequency of chromosome V loss from this euploid diploid cell was found to be essentially the same as that observed for the frequency of chromosome VII loss from a disomic cell. We conclude that the normal fidelity of chromosome replication and segregation in *S. cerevisiae* is about 1 failure in 5 x 10^4 cell divisions.

Literature Cited

1. Bignami, M., Aulicino, F., Velcich, A., Carere, A. and Morpugo, G. 1977. Mutagenic and recombinogenic action of pesticides in Aspergillus nidulans. Mut. Res. 46:395-402.

2. Campbell, D.A., Fogel, S. and Lusnak, K. 1975. Mitotic chromosome loss in a disomic haploid of Saccharomyces cerevisiae. Genetics 79:383-396.

3. Conde, J. and G. Fink. 1976. A mutant of Saccharomyces cerevisiae defective for nuclear fusion. Proc. Natl. Acad. Sci. USA 73:3651-3655.

4. Davidse, L.C. 1975. Antimitotic activity of methyl benzimidazole-2-yl-carbamate in fungi and its binding to cellular protein, p. 483. In: M. Borgers and M. DeBrabander (eds.), Microtubules and Microtubule Inhibitors. North Holland Publishing Co., Amsterdam.

5. Davidse, L.C. and Flach, W. 1977. Differential binding of methyl benzimidazole-2-yl-carbamate to fungal tubulin as a mechanism of resistance to this antimitotic agent in mutant strains of Aspergillus nidulans. J. Cell Biol. 72:174-193.

6. Dutcher, S.K. 1981. Internuclear transfer of genetic information in karl-1/KAR1 Heterokaryons in Saccharomyces cerevisiae. Molec. Cell Biol. 1:245-253.

7. Hastie, A.C. 1970. Benlate induced instability of Aspergillus diploids. Nature 226:771.

8. Howard, R.J. and Aist, J.R. 1977. Effects of MBC on hyphal tip organization, growth, and mitosis of Fusarium acuminatum, and their antagonism by D2O. Protoplasma 92:195-210.

9. Lea, D.E. and Coulson, C.A. 1949. The distribution of the numbers of mutants in bacterial populations. J. Genetics 49:264-285.

10. Liras, P., McCuskre, J., Mascioli, S. and Haber, J.E. 1978. Characterization of a mutation in yeast causing nonrandom chromosome loss during mitosis. Genetics 88:651-671.

11. Morris, N.R. and Oakley, C.E. 1979. Evidence that p-fluorophenyl-alanine has a direct effect on tubulin in Aspergillus nidulans. J. Gen. Microbiol. 114:449-454.

12. Nilsson-Tillgren, T., Litske, J.G., Holmberg, S. and Kielland-Brandt, M.C. 1980. Transfer of chromosome III during kar mediated cytoduction in yeast. Carlsberg Res. Commun. 45:113-117.

13. Orr, E. and Rosenberger, R. 1976. Determination of the execution points of mutations in the nuclear replication cycle of Aspergillus nidulans. J. Bact. 126:903-906.

14. Parry, J.M., Sharp, D., Tippins, R.S. and Parry, E.M. 1979. Radiation-induced mitotic and meiotic aneuploidy in the yeast Saccharomyces cerevisiae. Mutation Res. 61:37-55.

15. Parry, J.M. and Zimmerman, F.K. 1976. The detection of monosomic colonies produced by mitotic chromosome non-disjunction in the yeast Saccharomyces cerevisiae. Mutation Res. 36:49-66.

16. Petes, T. 1979. Yeast ribosomal DNA genes are located on chromosome XII. Proc. Natl. Acad. Sci. USA 76:410-414.

17. Quinlan, R.A., Pogson, C.I. and Gull, K. 1980. The influence of the microtubule inhibitor methyl benzimidazole-2-yl-carbamate (MBC) on nuclear division and the cell cycle in Saccharomyces cerevisiae. J. Cell Sci. 46:341-352.

18. Sheir-Neiss, G., Lai, M. and Morris, N.R. 1978. Identification of a gene for β-tublin in Aspergillus nidulans. Cell 15:639-647.

19. Styles, J.A. and Garner, R. 1974. Benzimidazolecarbamate methyl ester - evaluation of its effects in vivo and in vitro. Mut. Res. 26:177-187.

4. REGULATION OF YEAST HISTONE GENE EXPRESSION*

L.M. Hereford+ and M.A. Osley+

Rosenstiel Basic Medical Sciences Research Center
Brandeis University
Waltham, Massachusetts 02254

Rec. Adv. Yeast Mol. Biol. (1982) 1:39-51

Yeast histone mRNA levels are tightly regulated during cell division by distinct transcriptional and post-transcriptional mechanisms. Transcriptional regulation is evidenced by the restriction of histone mRNA synthesis to an interval between late G_1 and early S. Using <u>cdc</u> mutants it can be shown that activation and termination are temporally separable and dependent upon continued progression of cells through the cell division cycle: transcription is activated in late G_1 but is not terminated until cells enter S phase. Sequences responsible for this phenomenon appear to reside in or near functional origins of replication located at the 3' ends of each of the two H2B genes. This observation suggests that histone gene expression is closely tied to the replicative state of the DNA, and may be regulated by mechanisms responsible for the periodicity of DNA replication.

INTRODUCTION

The eucaryotic cell division cycle is comprised of a precise series of temporally ordered events which result in the production of two progeny cells. A striking example of this orderliness is the DNA replication process, where, at a defined point within the cell cycle, DNA synthesis is initiated. Genetic analysis of cell division in yeast has clearly established that this event is dependent upon the sequential function of a number of gene products and, further, that the completion of S is necessary for coninued progression through cell division (for a review see reference 19). While this process in and of itself is remarkable, it is also the case that during the S phase newly replicated DNA must be assembled into functional chromosomes. By analogy with phage morphogenesis, this is most likely an exquisitely orchestrated series of events dependent upon mechanisms which ensure maintenance of correct stoichiometry among the various chromosomal proteins.

---

* Support was obtained from National Institutes of Health grant GM 27334.

+ Present address: Sidney Farber Cancer Institute and the Department of Microbiology and Molecular Genetics, Harvard Medical School, Boston, Massachusetts 02115

To date, the most dramatic example of such mechanisms is the regulation of histones, the most ubiquitous chromosomal proteins. In a variety of organisms, including yeast (28), histone synthesis is tightly coupled to DNA replication as evidenced by the observation that newly synthesized histones are found only in S phase cells, and, further, that inhibition of DNA replication results in a parallel inhibition of histone synthesis (for a review, see reference 24). While these phenomena are well documented, the mechanisms underlying them have been the subject of some controversy. Extremes in interpretation range from no regulation at all (16) to post-transcriptional regulation (27), or to combined post-transcriptional and transcriptional regulation (2,3,5,8,13,14). In spite of the controversies, we were sufficiently encouraged by what appeared to be overwhelming evidence for some type of regulation to reinvestigate the problem in yeast, where we felt that the obvious biochemical and genetic tractability, combined with the ease of synchrony and the availability of cdc mutants, would allow us to make more definitive conclusions regarding regulatory mechanisms. Our approach was to clone the histone genes and then utilize the cloned sequences as probes to investigate the behavior of histone mRNA during cell division. In this paper we will briefly review our findings on the structure of the histone genes and then describe, in some detail, the regulatory mechanisms which govern their expression.

RESULTS

Histone Gene Structure

Presumably as a reflection of its small genome size, the histone genes of yeast do not comprise a highly repetitive gene family. As we showed some years ago, there are only two copies of H2A and H2B genes per haploid genome and they are arranged into two divergently transcribed H2A-H2B pairs (21). Mitchell Smith (personal communication) has identified the same pairwise arrangement for the H3 and H4 genes, which again are present in only two copies per haploid genome. All 4 histone pairs are genetically unlinked to one another (Smith and Hereford, unpublished observations). One H2A-H2B pair (TRT-2,3) is tightly linked to the centromere of chromosome 2. The location of the other pair (TRT-1), which is not centromere linked, is unknown.

The relevant structural features of TRT-1 and TRT-2,3 are summarized in Figure 1.

Our initial analysis suggested that TRT-1 and TRT-2,3 share limited homology and that this homology is restricted to the H2A and H2B coding sequences, that is, both the spacer DNA between the genes as well as the 3' flanking regions are single copy DNA. The plasmids also contain additional, non histone genes. Genes 1 and 2 are genes identified by both r-looping and hybrid arrest translation; 3 and 4 are genes which code for minor transcripts

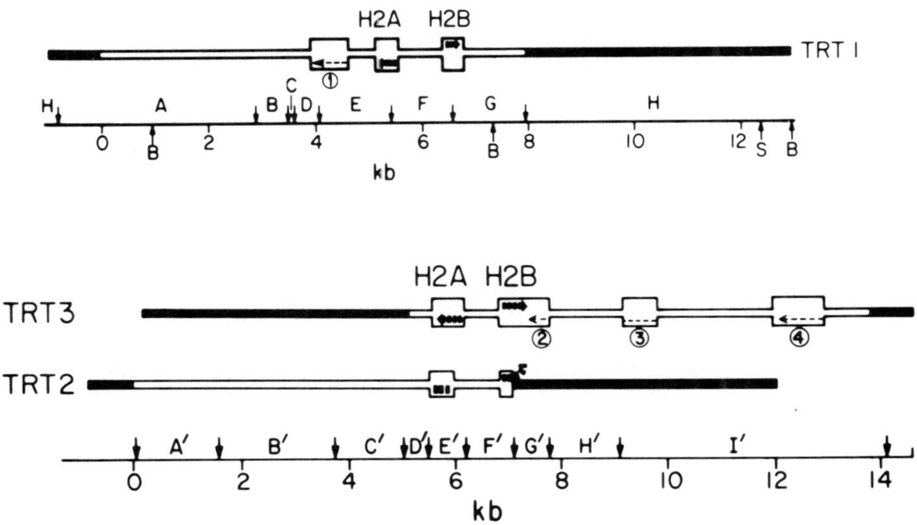

Figure 1. Organization of yeast H2A and H2B. Dark lines are vector sequences. Letters above the lines are Hind III fragments. Figure is taken from reference 20 and is reprinted with permission of Cell.

transcripts which have been identified only by r-loops (21).

The limited homology between TRT-1 and TRT-2,3 has been directly confirmed by sequence analysis of both of the H2B and H2A genes. The H2B genes show 12.6% sequence divergence and differ by 4 amino acids (33), while the H2A genes show a 6% sequence divergence and differ by only 2 amino acids (7). In contrast to the high degree of conservation of coding sequences, the 5' flanking regions show no obvious homology either between homologous genes or gene pairs (7,33).

The finding that there are two copies of the H2A and H2B genes raised the question of whether both copies were expressed. By Northern blot analysis, we established that both H2B genes were expressed and that the levels of expression were approximately equal (22). This observation has been corroborated by the elegant experiments of Rykowski *et al*. (3) which demonstrate, by frame shift mutations, that cells can exist with either copy of the H2B genes.

In the case of the H2A genes, we were unable to detect unique transcripts and therefore did not know whether only one gene was expressed or whether both were expressed but produced transcripts of identical size. This issue has been recently resolved by Kolodrubetz *et al*. (26), who, using the same mutational approach described for the H2B genes, have shown that both the H2A genes are functional.

Histone Gene Regulation

Transcriptional Regulation. We have found that the synthesis of H2A and H2B mRNA is restricted to the S-phase of the cell cycle and is the consequence of two levels of regulation - transcriptional and post-transcriptional. Preliminary evidence that transcriptional control is an important component of histone synthesis originated from measurements of the steady state levels of histone transcripts throughout synchronous cell division (22). This analysis showed that the accumulation of histone mRNA is coordinate and periodic. Since the peak accumulation of these transcripts appeared to precede the point of maximal DNA synthesis, we argued that the pattern was consistent with a model of periodic transcription of histone genes.

A more direct approach to the question of periodic transcription was undertaken in a series of experiments in which the *rate* of synthesis of histone mRNA was measured (20). Two important conclusions were produced by these investigations. The first was that synthesis of histone mRNA is indeed periodic. The second was somewhat unexpected and was that significant synthesis of histone mRNA begins well before the onset of DNA replication. The latter observation eliminated a simple model of transcriptional activation in which the synthesis of histone mRNA is triggered by active chromosome replication. To dissect the steps required for histone gene transcription, the synthesis of histone mRNA was examined in synchronous cultures of two _cdc_ mutants. Cells arrested late in G_1 by the _cdc7_ mutation synthesize histone mRNA at rates comparable to those observed in wild type cells. However, in these cells, histone mRNA synthesis is not periodic but continues at maximal rates for as long as cells are maintained at the restrictive temperature. Periodicity is restored in cells arrested in early S by the _cdc8_ mutation. Taken together, these results suggest that the periodicity of histone mRNA synthesis is not an intrinsic property which is simply entrained by some event late in G_1. Rather, activation and cessation of synthesis are temporally separable events which are dependent upon continued progression of cells through the cell division cycle.

Post-Transcriptional Regulation

Three lines of evidence demonstrate that post-transcriptional controls, mediated via changes in histone mRNA stability, are also important in expression of histone genes. First, when DNA synthesis is inhibited, the levels of histone mRNA dramatically fall (22). This phenomenon cannot be accounted for simply by inhibition of transcription followed by normal turnover of existing transcripts, since, under these conditions, the mRNA decays with a half-life on the order of 3-5 minutes, in contrast with the normal histone mRNA half-life which is about 15 minutes (29).

Second, a direct demonstration that histone mRNA stability can be altered arises from the consequences of increasing the copy number of histone genes (29). Duplication of one H2A-H2B gene pair results in a two-fold elevation in the rate of synthesis of mRNA encoded by these genes. The extra RNA, however, is degraded two times faster than normal, resulting in a concentration of histone mRNA corresponding to only one gene copy.

Finally, a more detailed examination of the pattern of accumulation of histone transcripts vis-`a-vis the rate of chromosome replication (20) reveals that histone mRNA is most abundant at the time in the cell cycle when the rate of DNA synthesis is maximal. This response can be interpreted to mean that histone transcripts synthesized _before_ this period are differentially stabilized as a function of the rate of replication: when the replication rate is high, histone transcripts are stable; when the replication rate is reduced, these transcripts are degraded.

Model of mRNA Regulation

A general model incorporating our results on transcriptional and post-transcriptional control of histone gene expression is shown in Figure 2.

Histone mRNA synthesis is initiated in G_1 at a point which is

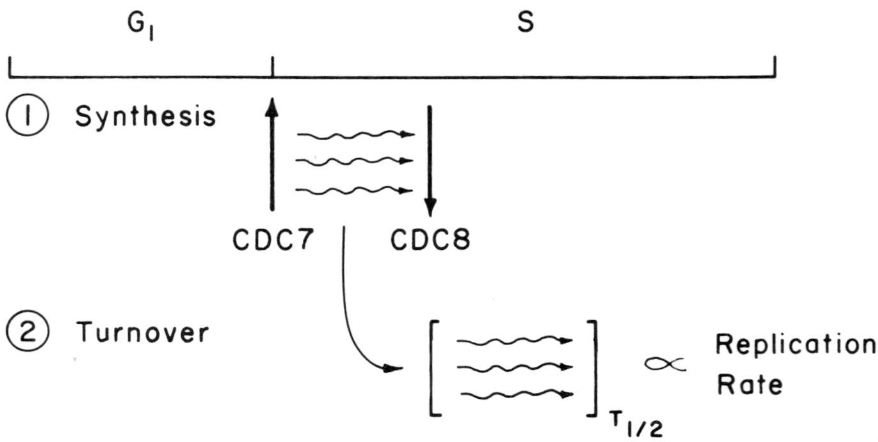

Figure 2. Model of Histone Gene Regulation. Figure is taken from reference 21 and is reprinted with permission of Cell.

either prior to or concomitant with the CDC7 function. Termination of transcription occurs early in S. The histone mRNA produced within this interval is then regulated post-transcriptionally by changes in its half life. These changes, which we believe are a consequence of the DNA replication rate, result in the nearly perfect coincidence of histone mRNA accumulation and this rate. We assume that the two modes of regulation have complementary functions. The transcriptional regulation ensures that histone mRNA is produced at the proper time during cell division, while the post-transcriptional regulation ensures that it is produced in the proper amount.

As regards the mechanisms responsible for these two levels of control, we have postulated that the post-transcriptional regulation may be similar to the autogenous regulation exhibited by certain phage proteins such as gene 32 (15) or reg A (25) as well as the *E. coli* ribosomal proteins (4,11,12,35) in which these proteins modulate expression by specifically binding to their cognate mRNAs. We propose that changes in half-life may be a reflection of the translational efficiency of the mRNA, which in turn is governed by the level of free histones. Thus histone mRNA would be maximally translated and thereby stabilized during periods of peak DNA synthesis. While we have no evidence that this is, in fact, the mechanism, the appeal lies in the similarity between histones and other proteins which exhibit autogenous regulation: namely, that all of them are proteins with high affinities for nucleic acid.

Any mechanism postulated to explain the transcriptional regulation must take into account the observation that activation and cessation of histone mRNA synthesis are temporally separable and dependent upon continued progression of cells through the cell division cycle. The simplest way to achieve this regulation, without evoking complex cycles of transcription factors, is to couple transcription to some cell cycle event. The most obvious would be the DNA replication cycle itself. One could imagine that mechanisms which govern the periodic activation of origins of replication also serve to activate histone gene transcription. Thus, alterations in chromatin structure which lead to the activation of an origin could also activate transcription. Replication through the region would then restore the chromatin to its prereplicative state and would serve to terminate transcription. The appeal of this model is its simplicity. Moreover, it is also consistent with the observed behavior of histone gene transcription: namely, that histone mRNA is activated in G_1, but is not terminated until cells have entered S.

Correlation between ARS Sequences and Histone Gene Transcription

One way to envision this model is simply to have histone genes positioned near an origin of replication. We were therefore struck by the observation that both TRT-1 and TRT-2,3 appeared to contain functional origins of replication (ARS sequences (32)) as assayed by yeast transformation. Further analysis revealed that they were located at the 3' ends of each of the H2B genes. The fact that ARS sequences occur on the average about every 40 kb in the yeast genome (1,6) argued that it was unlikely that this arrangement would occur randomly and therefore suggested that these sequences might indeed play a role in histone gene expression.

To assess the role of the origin sequences in histone gene expression, we took advantage of the recent observations of Rose *et al.* (30), and Guarente and Ptashne (18), that gene fusions to beta-galactosidase can be utilized as a rapid and convenient assay of yeast promoter activity. Thus, fusions were created between the histone promoters, either with or without the origin sequences, and the levels of beta-galactosidase produced after transformation of these plasmids into yeast were determined.

The features of the TRT-1 H2A-H2B gene pair relevant to the plasmid constructions are shown in Figure 3.

Transcription of the histone genes is divergent and initiates within the 800 nucleotides of spacer DNA which separates them. A 1.1 kb Hind III fragment encompasses the entire spacer region as well as a portion of the 5' coding sequences of both the H2A and H2B genes. A second Hind III fragment of 1.3 kb immediately adjacent to the 1.1 kb region includes the 3' end of the H2B gene and flanking sequences as well as the origin of chromosome replication in an 800 bp region which is bounded by one of the Hind III sites and an internal Bam H1 site. Insertion of the 1.1 kb Hind III fragment into the 5' end of *lacZ* in either orientation produces a fusion in the correct reading frame. Although both hybrid genes are expressed in yeast, our analysis has been restricted to the H2A-lacZ fusion (pfusA1-1. Figure 3B).

Two additional plasmids were constructed by inserting the origin-containing 1.3 kb Hind III fragment of *TRT-1* (Figure 3A) into the distal Hind III site on the pfusA1-1 fusion plasmid. Insertion of this fragment in the correct orientation (pfusA1-2, Figure 3C) reconstitutes an intact H2B gene, while insertion in the opposite orientation (pfusA1-3, Figure 1C) results in the separation of the 5' and 3' H2B coding regions by the origin sequences.

To determine whether the origin sequences play any role in the expression or regulation of the H2A gene, we took advantage of the fact that integration of plasmid sequences introduced by transformation occurs by homologous recombination (23). The plasmids which we constructed, therefore, integrated either at the *TRT-1* locus or at loci homologous to the selectable marker present on the plasmid. The selectable marker which we used is the yeast *LEU2+* gene (Figure 3B,C). Since the DNA fragment containing this gene also contains a portion of a repetitive Ty sequence (H. Klein, personal communication), integration occurred either at the *leu2* locus or at random Ty sites within the genome.

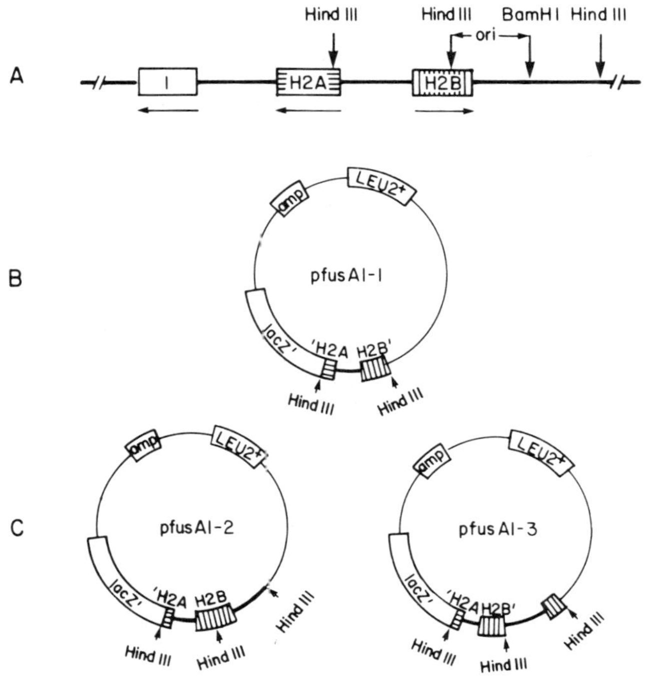

Figure 3. Construction of H2A fusion plasmids.

A. Organization of genes at TRT-1 locus showing H2A and H2B and nonhistone protein 1 gene (21). The arrows represent the direction of transcription. The two Hind III fragments used to construct the fusion plasmids are indicated. The 1.1 kb Hind III fragment includes 40 nucleotides of H2A coding sequence, 236 nucleotides of H2B coding sequence and about 800 nucleotides of spacer DNA separating the two genes. The 1.3 kb Hind III fragment contains the 3' end of the H2B gene and its flanking DNA and also includes a functional origin of chromosome replication (ori) which has been localized to a 800 bp Hind III-BamHl fragment.

B. pfusAl-1 was created by inserting the 1.1 kb Hind III fragment shown in A. into the unique Hind III site at the 5' end of the lacI-lacZ fragment carried on plasmid pG200 (17). This plasmid also contains the yeast $LEU2^+$ gene to provide a selectable marker for maintenance of the plasmid in yeast.

C. pfusAl-2 results from the insertion of the 1.3 kb Hind III fragment shown in A. into pfusAl-1 in the correct orientation. pfusAl-3 has this same fragment in the reverse orientation.

As a consequence of integration at these different genomic locations, the sequences flanking pfusA1-1 vary greatly. When pfusA1-1 is integrated at the TRT-1 locus, it displaces the resident H2A gene and is thus properly positioned with respect to the origin sequences. At leu2 or any other chromosomal location, however, the only histone gene sequences present are those 5' sequences in which transcription is initiated. If the latter sequences are all that are required for expression and regulation, we would expect the levels of beta-galactosidase to be independent of the chromosomal location of the fusion gene.

Leu^+ transformants of pfusA1-1 were initially screened on X-Gal indicator plates. Only two phenotypic classes, which differed markedly in their degree of beta-galactosidase expression, were obtained. In all cases, transformants producing high levels of enzyme contained the fusion integrated at the TRT-1 locus, as determined by Southern blot analysis. This difference was confirmed by enzyme assays which showed 4-fold differences in beta-galactosidase levels (data not shown).

To determine whether the enhancement of beta-galactosidase expression reflected a role for the origin sequences in regulation, we assayed the levels of beta-galactosidase in synchronous cell cultures. We found that in the absence of these sequences beta-galactosidase is synthesized at low constitutive levels during cell division, whereas, in their presence, a dramatic step in enzyme synthesis was observed. This step occurs exactly at the point at which histone mRNA is maximally accumulated (data not shown). These results, which have been confirmed at the RNA level by Northern gel analysis, lead us to the conclusion that sequences in or near the origin sequences are necessary for the transcriptional regulation of the histone genes. As such, they lend support to our notion that this transcriptional regulation may be inextricably linked to the periodic activation of DNA replication sites.

DISCUSSION

As we have demonstrated, histone synthesis is tightly regulated during cell division by concerted transcriptional and post-transcriptional mechanisms. Our preliminary results suggest that the first of these mechanisms may be mediated by the periodicity of the DNA replication cycle: that is, that histone transcription is regulated by the periodic activation of origins of replication. As we have said, we suggest that the coupling of replication and transcription is a consequence of alterations in the state of the histone gene chromatin. Thus, we envision that activation of DNA replication origins results in changes in chromatin structure which allow activation of transcription. Termination of transcription can then be readily achieved simply by replication through the histone gene region and restoration of the chromatin to its G_1 state. Formally this model is similar to those in which restructuring of chromatin during DNA synthesis ultimately changes the developmental fate of a particular cell (34). The only difference is that the changes we propose are oscillatory rather than permanent.

At present we do not know whether these interpretations are correct or whether, in fact, the mechanisms are far more complex than our experiments suggest. In any case we feel that we have identified a potentially novel form of regulation. Even though this type of regulation appears to be novel, we feel that it need not be limited to histones. We therefore suggest that there are other proteins which may be similarly regulated and such proteins would be those which are required during S phase either for structural or enzymatic functions.

LITERATURE CITED

1. Beach, D., Piper, M., and Shall, S. 1980. Isolation of chromosomal origins of replication in yeast. Nature **284**:185-187.
2. Borun, T.W., Gabrielli, F., Ajiro, K., Zweidler, A., and Baglioni, C. 1975. Further evidence of transcriptional and translational control of histone messenger RNA during the HeLa cell cycle. Cell **4**:59-67.
3. Borun, T.W., Scarff, M.D., and Robbins, E. 1967. Rapidly labelled polyribosome-associated RNA having the properties of histone messenger. Proc. Natl. Acad. Sci. USA **58**:1977-1983.
4. Brot, N., Caldwell, P., and Weissbach, H. 1980. Autogenous control of *Escherichia coli* ribosomal protein L10 synthesis *in vitro*. Proc. Natl. Acad. Sci. USA **77**:2592-2595.
5. Butler, W.B. and Mueller, G.C. 1973. Control of histone synthesis in HeLa cells. Biochim. et Biophys. Acta **294**:481-486.
6. Chan, C. and Tye, B.K. 1980. Autonomously replicating sequences in *Saccharomyces cerevisiae*. Proc. Natl. Acad. Sci. USA **77**:6329-6333.
7. Choe, J., Kolodrubetz, D., and Grunstein, M. 1982. The two yeast histone H2A genes encode very similar protein subtypes. Proc. Natl. Acad. Sci. USA **79**:1484-1497.
8. Detke, S., Lichtler, A., Philips, I., Stein, J., and Stein, G. 1979. Reassessment of histone gene expression during cell cycle in human cells using homologous H4 histone cDNA. Proc. Natl. Acad. Sci. USA **76**:4995-4999.
9. Elgin, S.C.R. 1981. DNAse I-hypersensitive sites in chromatin. Cell **27**:413-415.
10. Elliot, S.C. and McLaughlin, C.S. 1978. Rate of macromolecular synthesis through the cell cycle of the yeast *Saccharomyces cerevisiae*. Proc. Natl. Acad. Sci.USA **75**:4384-4388.
11. Fallon, A.M., Jinks, C.S., Strycharz, G.D., and Nomura, M. 1979. Regulation of ribosomal protein synthesis in *E. coli*: by selective mRNA inactivation. Proc. Natl. Acad. Sci. USA **76**:3411-3415.

12. Fallon, A.M., Jinks, C.S. and Nomura, M. 1979. Expression of ribosomal protein genes cloned in a hybrid plasmid of Escherichia coli: Gene dosage effects on synthesis of ribosomal protein messenger ribonucleic acid. J. Bacteriol. **138**:383-396.
13. Gallwitz, D. and Brendl, M. 1972. Synthesis of histones in a rabbit reticulocyte cell-free system directed by a polysomal RNA fraction from synchronized HeLa cells. Biochem. Biophys. Res. Commun. **47**:1106-1111.
14. Gallwitz, D. and Miller, G.C. 1969. Histone synthesis in vitro on HeLa cell microsomes. J. Biol. Chem. **244**:5947-5952.
15. Gold, L., O'Farrell, P.Z., and Russel, M. 1976. Regulation of gene 32 expression during bacteriophage T4 infection of Escherichia coli. J. Biol. Chem. **251**:7251-7262.
16. Groppi, C.E. and Coffino, P. 1980. G_1 and S phase cells synthesize histones at equivalent rates. Cell **21**:1195-240.
17. Guarente, L., Lauer, G., Roberts, T.M. and Ptashne, M. 1980. Improved methods for maximizing expression of a cloned gene: a bacterium that synthesizes rabbit beta-globin. Cell **20**:543-553.
18. Guarente, L. and Ptashne, M. 1981. Fusion of Escherichia coli lacZ to the cytochrome c gene of Saccharomyces cerevisiae. Proc. Natl. Acad. Sci. USA **78**:2199-2203.
19. Hartwell, L.H. and Pringle, J.R. The Saccharomyces cerivisiae Cell Cycle. In: J. Strathern, E. Jones, and J. Broach (ed.), The Molecular Biology of the Yeast Saccharomyces. Cold Spring Harbor Laboratory, Cold Spring Harbor, N.Y., in press.
20. Hereford, L.M., Bromley, S., and Osley, M.A. Periodic transcription of yeast histone genes. Cell, in press.
21. Hereford, L., Fahrner, K., Woolford, J. Jr., Rosbash, M., and Kaback, D.B. 1979. Isolation of yeast histone genes H2A and H2B. Cell **18**:1261-1971.
22. Hereford, L.M., Osley, M.A., Ludwig, J.R., and McLaughlin, C.S. 1981. Cell-cycle regulation of yeast histone mRNA. Cell **24**:367-375.
23. Hinnen, A., Hicks, J., and Fink, G.R. 1978. Transformation of yeast. Proc. Natl. Acad. Sci. USA.**75**:1929-1933.
24. Kedes, L.H. 1979. Histone genes and histone messengers. Ann. Rev. Biochem. **48**:837-870.
25. Karam, J., Gold, L., Singer, B.S., and Dawson, M. 1981. Translational regulation: Identification of the site on bacteriophage T4 rIIB mRNA recognized by the reg A gene function. Proc. Natl. Acad. Sci. USA **78**:4469-4673.
26. Kolodrubetz, D., Choe, J., Rykowski, and Grunstein, M. 1981. Abstracts from the meeting on the Molecular Biology of Yeast, p. 105. Cold Spring Harbor Laboratory, Cold Spring Harbor, N.Y.
27. Melli, M., Spinelli, G., and Arnold, E. 1977. Synthesis of histone messenger RNA of HeLa cells during the cell cycle. Cell **12**:167-174.
28. Moll, R. and Wintersberger, E. 1976. Synthesis of yeast

histones in the cell cycle. Proc. Natl. Acad. Sci. USA **73**:1863-1867.

29. Osley, M.A. and Hereford, L.M. 1981. Yeast histone genes show dosage compensation. Cell **24**:377-384.
30. Rose, M., Casadaban, M., and Botstein, D. 1981. Yeast genes fused to beta-galactosidase in *Escherichia coli* can be expressed normally in yeast. Proc. Natl. Acad. Sci. USA **78**:2460-2464.
31. Rykowski, M.C., Wallis, J.W., Choe, J., and Grunstein, M. 1981. Histone H2B subtypes are dispensable during the yeast cell cycle. Cell **25**:477-487.
32. Struhl, K., Stinchcomb, D.T., Scherer, S., and Davis, R.W. 1978. High frequency transformation of yeast: autonomous replication of hybrid DNA molecules. Proc. Nat. Acad. Sci. USA **76**:1035-1039.
33. Wallis, J.W., Hereford, L., and Grunstein, M. 1980. Histone H2B genes of yeast encode two different proteins. Cell **77**:799-806.
34. Weintraub, H., Flint, S.J., Leffak, I.M., Groudine, M., and Grainger, R.M. 1977. The generation and propagation of variegated chromosomal structures. Cold Spring Harbor Symposium on Quantative Biology **17**:401-416.
35. Yates, J.L., Arfsten, A.E., and Nomura, M. 1980. *In vitro* expression of *Escherichia coli* ribosomal protein genes: autogenous inhibition of translation. Proc. Natl. Acad. Sci. USA **77**:1837-1841.

5. MAPPING AUTONOMOUSLY REPLICATING SEGMENTS ON A CIRCULAR DERIVATIVE OF CHROMOSOME III[1]

Carol S. Newlon, Rodney J. Devenish[2] and Lolya R. Lipschitz

Department of Zoology, University of Iowa
Iowa City, Iowa 52242

Rec. Adv. Yeast Mol. Biol. (1982) 1:52-68

A method for purifying a 190 kilobase (kb) ring chromosome derived from chromosome III of Saccharomyces cerevisiae has been developed. The extracts, which contain 70-85% of the expected yield of ring chromosome, also contain other circular DNAs, including the 2μm plasmid, 3-5 copies per cell of circular ribosomal DNA, and less than 1 copy per cell of a circular molecule which hybridizes to Ty1-17. The electrophoretic properties of the ring chromosome are anomalous: nicked circles fail to enter an 0.7% agarose gel, and closed circles do not enter such a gel in the presence of ethidium bromide.

Replication origins used in vivo are being mapped on the ring chromosome by electron microscopic techniques. In addition, a library has been constructed which consists of BamHI fragments of the ring chromosome inserted into the integrating URA3 vector YIp5. We have identified at least three BamHI fragments from the ring which contain an autonomously replicating segment (ARS). One of these ARS-containing fragments is within the 125kb of the chromosome for which we have a restriction map and is located approximately 20 kb distal to the HIS4 locus.

INTRODUCTION

DNA replication initiates at many sites spaced along eukaryotic chromosomal DNAs at intervals of 50 - 330 kb (reviewed in 16). The precision with which origins of replication are spatially and temporally ordered on eukaryotic chromosomes is largely unknown. Saccharomyces cerevisiae offers an ideal system for studies of eukaryotic DNA replication for two reasons. First, although its chromosomal DNAs replicate like those of higher eukaryotes (26), they are at least 100 times smaller than a typical eukaryotic chromosome, and the smallest yeast chromosomal DNA's are only the size of T4 DNA (28). Their small size makes it feasible to isolate intact chromosomal

[1]Support was obtained from NIH research grant GM 21510 and NIH RCDA GM 00265 to C.S.N.

[2]Present address: Department of Biochemistry, Monash University, Clayton, Victoria, 3168, Australia

DNAs (25, 29). Second, the yeast transformation system makes possible genetic manipulations which cannot yet be done in other systems.

Two lines of evidence suggest that replication origins on yeast chromosomal DNA may be at defined sequences. We have used electron microscopy to examine the replication of the smallest yeast chromosomal DNA's and have concluded that replication origins are located non-randomly at intervals of about 36 kb (25). Others have isolated DNA sequences from yeast which, when inserted into plasmids, confer on those plasmids the ability to replicate autonomously in yeast (1, 7, 19, 20, 32). These autonomously replicating segments (ARSs) seem likely to be origins of replication, since prokaryotic sequences which by a variety of criteria are replication origins allow autonomous replication of prokaryotic plasmids. In addition, the frequency of occurrence of ARSs in yeast chromosomal DNA, one per 40 kb (1, 7), is consistent with the spacing of replication origins, and ARS-containing plasmids replicate under the same cell cycle control as chromosomal DNA (40). However, it remains to be demonstrated that these sequences function as initiation sites on chromosomes or plasmids *in vivo*.

A number of studies in both higher and lower eukaryotes suggest that DNA replication is temporally ordered (reviewed in 16). However, the resolution of these studies was at best approximately one-fifth of S-phase, a time much longer than the time required for an individual replication unit to complete DNA synthesis. Evidence from both Drosophila and Triturus demonstrates that the spacing between replication origins varies with developmental stage (3, 4). These results mean that either not every origin is used in every S phase or that the temporal order of activation of origins can vary.

The identification of a circular derivative of chromosome III (34, 13) has provided an opportunity to further define the location and temporal order of activation of replication origins. This ring chromosome contains approximately 190 kb of DNA. Based on the average spacing between origins (25), this chromosome should have five or six replication origins. Because it is circular, it can be uniquely identified by electron microscopy. We are taking two approaches in studying the replication of this chromosome. First, we are using electron microscopy to map replication origins and define their temporal order of activation on the molecule as it replicates *in vivo*. Second, we are identifying and mapping ARSs on the ring chromosome. Ultimately we will be able to determine whether ARSs function as replication origins *in vivo* and whether there is a perfect correlation between ARSs and origins. In this paper we report progress in identifying ARSs.

MATERIALS AND METHODS

Strains and Media

Yeast strains DC021/DC022 #62 and DC021/DC022 #20, a [cir°] ring chromosome strain and its wildtype counterpart were isolated by J. Strathern as described for the [cir$^+$] strains XG1 #24 and XG1 #1 (34). Detailed genotypes are given in reference 13. A364A (18) was from L. Hartwell and YNN27 (33) was obtained from D. Stinchcomb. E. coli strain JA226 (13) was used as the transformation recipient. Yeast strains were grown in Y-minimal medium (26) supplemented with 2% glucose and when necessary with amino acids, purines, and pyrimidines at 50 mg/l. Bacteria were grown in L broth.

Preparation of Genomic DNA

Total DNA was prepared from yeast by the method of Fitzgerald-Hayes and Bloom (per. comm.). Spheroplasts were suspended in 0.1M Tris-Cl, 0.05M Na EDTA, pH 8.0 and lysed by the addition of 10% sodium dodecyl sulfate to a final concentration 0f 0.5%. The lysate was made 0.5M in NaCl by the addition of a 5M solution and extracted with 0.75 volume phenol: $CHCl_3$: isoamyl alcohol (50:24:1) and then with 0.5 volume $CHCl_3$: isoamyl alcohol (24:1). DNA was precipitated with ethanol and the DNA pellets redissolved in STE (10mM NaCl, 10mM Tris-Cl, 1mM EDTA, pH 8.0) and treated with 5μg/ml pancreatic RNase at 37° for 1 hr. SDS and salt were added as above and the organic extractions repeated before DNA was precipitated with ethanol.

Preparation of Covalently Closed Circular (ccc) DNA

The method is a modification of the procedure of Casse et al. for isolating Ti plasmids from Agrobacterium (6), and has been described in detail (13). Briefly, spheroplasts were lysed at pH 12.45 which denatures linear but not covalently closed circular (ccc) DNA. The lysate was then taken through a gentle high salt-phenol extraction to remove single stranded DNA and protein. The ccc DNA was recovered from the aqueous phase by ethanol precipitation. A standard gel track contained lysate from about 4×10^8 cells.

Cloning of Ring Chromosome DNA

ccc DNA prepared from the [cir°] or [cir$^+$] ring chromosome strain was centrifuged to equilibrium in CsCl gradients, precipitated with ethanol and redissolved. It was cut to completion with BamHI, for which no sites exist in 2μm DNA (17) or in ribosomal DNA (2), and then ligated into BamHI-cut YIp5 vector which had been treated with bacterial alkaline phosphatase to prevent recircularization of the vector. The ligation mix was transformed into E. coli JA226 and approximately 500 tetracycline-sensitive, ampicillin-resistant colonies, containing plasmids with inserts, were picked from each starting yeast strain. The plasmid-containing strains were stored in 96-well microtiter dishes at -80° in the presence of 8% dimethylsulfoxide.

Gels, Transfers, Nick Translations, and Hybridizations

Electrophoresis of DNA was in 0.7% agarose gels using a Tris-borate-EDTA buffer. Transfer of DNA to nitrocellulose filters, nicktranslation of DNA probes, and hybridizations were by standard procedures as described previously (13).

RESULTS

Isolation of the Ring Chromosome

The ring chromosome results from an intrachromosomal recombination between homologous sequences at the mating type locus (MAT) on the right arm of chromosome III and the silent cassette, HML, on the left arm as depicted in Figure 1 (34). Early efforts to isolate DNA from this ring chromosome bearing strain yielded only about 1% of the expected amount of ring chromosome DNA. To isolate enough ring chromosome DNA for cloning, we have developed a method of extracting ccc DNA which yields up to 85% of the theoretical yield of the ring chromosome (1 molecule per diploid, or 0.6% of the total DNA). Tracks a and b of Figure 2 show an ethidium bromide-stained agarose gel in which ccc DNA extracts from a [cir^+] ring chromosome strain and an isogenic wildtype diploid have been electrophoresed. A slowly migrating DNA species is present in the ring strain that is absent from the wildtype diploid extract. The autoradiograms in the same figure demonstrate that this DNA, labeled "ring", is derived from chromosome III. DNA from pairs of tracks corresponding to the stained tracks shown was transferred to nitrocellulose and hybridized to various ^{32}P-labeled probes. The ring band hybridizes to the chromosome III probe (MATα, tracks c and d), but not to a probe from chromosome V (CYC7, tracks e and f). In addition to MATα, probes from the HIS4 and LEU2 loci on chromosome III hybridize to the ring band. However, as expected, a probe from the region of chromosome III distal to HML fails to hybridize to the ring band (13). Probes from chromosomes IV, X, and XV also fail to hybridize to the ring band. The chromosome III probe also hybridizes to a second band which migrates at the same rate as linear fragments of chromosomal DNA. Since this band contains little or no fragmented DNA from other linear chromosomes, as evidenced by its failure to hybridize to non-chromosome III probes, presumably it results from broken ring chromosome.

Other DNA species in the extract. We have extended the hybridization analysis to identify other circular DNA species in the extract. The 2µm plasmid monomer is clearly visible in the EtBr stained tracks. Hybridization with a 2µm probe (Figure 2, tracks g and h) reveals multimers of 2µm DNA up to at least pentamers. In experiments not shown, we have quantitated the yield of 2µm DNA and estimate that the monomer band contains 75-90% of the theoretical yield of 2µm DNA (13). The extracts also contain circular copies of ribosomal DNA (Figure 2, tracks i and j). The monomer species is often weakly visible in the stained gel. We estimate that we recover 3-4 copies per cell of this ribosomal DNA circle (13).

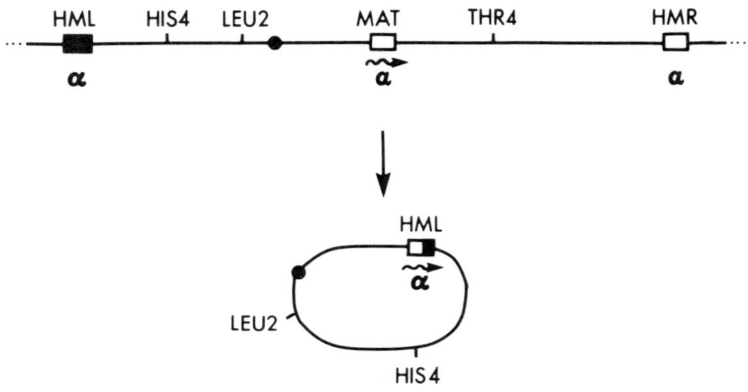

Figure 1. Generation of the ring chromosome. Homologous recombination between the regions of homology at MATa and HMLα yields the ring chromosome and an acentric fragment (not shown). The acentric fragment is presumably lost during mitotic growth. The recombination between HMLα and MATa yields a hybrid locus which expresses α-information derived from HML.

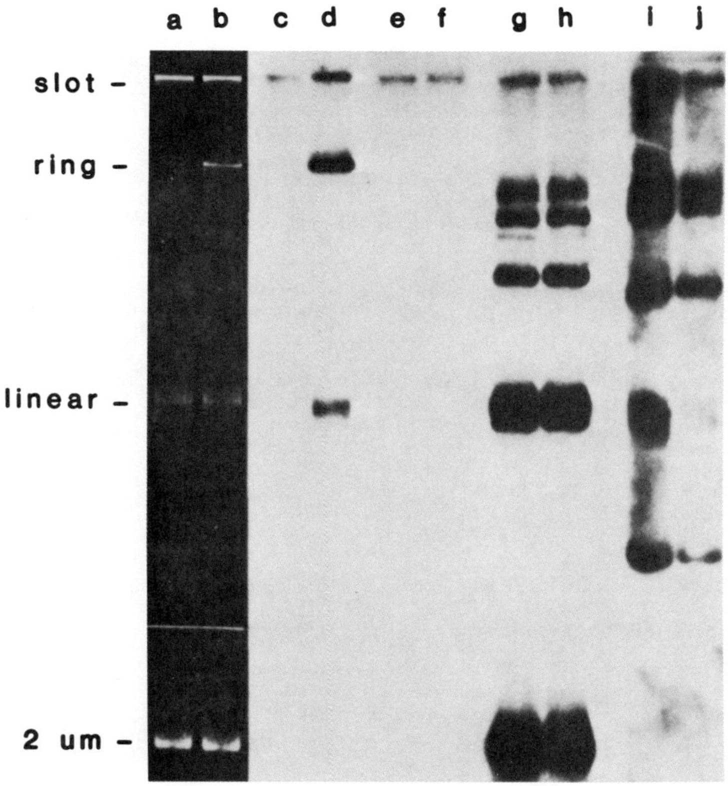

Figure 2. Identification of DNA species in ccc DNA extracts. ccc DNA extracts from XG1 #1, a $[cir^+]$ wildtype diploid, and XG1 #24, its isogenic ring chromosome strain were electrophoresed in an 0.7% agarose gel (20 mA, 17 hrs). The gel was stained with EtBr and photographed, and DNA from pairs of tracks was transferred to nitrocellulose after partial depurination (31). (a) and (b) show the EtBr-stained gel containing extract from XG1 #1 (a) and XG1 #24 (b). DNA from pairs of tracks equivalent to (a) and (b) was hybridized to ^{32}P-labeled DNA fragments as follows and hybridization was detected by autoradiography. (c) and (d) the EcoRI-HindIII yeast fragment from MATα on chromosome III (35); (e) and (f) the PstI-EcoRI yeast fragment from pYeCYC7, isolated from chromosome V (24); (g) and (h) the 2μm DNA monomer excised from plasmid 82-6B with PstI (17); (i) and (j) the SmaI fragment containing a single rDNA repeat from pBD4 (2).

In view of recent reports of circular copia elements in Drosophila (14), we tested ccc DNA extracts of yeast for hybridization to a Ty1-17 probe. In Figure 3 it can be seen that Ty1-17 hybridizes to three DNA species in extracts from [cir$^{\circ}$] ring strains (tracks e and f): the ring chromosome, linear DNA, and a third, faster migrating species. The hybridization to the ring chromosome is expected because there are at least two Ty elements in the ring, one just distal to LEU2 (Ty1-17, ref. 21) and one near PGK (Ty1-161, ref. 21). The faster migrating species is also present in the [cir^{+}] ring strain (track g) and in an unrelated haploid, YNN27 (track h). Because pBR322 sequences in the probe hybridize to 2μm DNA (34), the position of this Ty species can be easily compared with the 2μm monomer in tracks g and h. It migrates more slowly than 2μm DNA (6.3 kb, ref. 17), but faster than the ribosomal DNA monomer circle (8.5 kb, ref. 12). It is therefore larger than the 5.6 kb Ty1 element described by Cameron et al. (5), and it is close to the size of the 6.8 kb Ty1-161 element near PGK (21). Although we have found Ty-related DNA migrating at this position in extracts from four different strains, there is variability in the amount. Extracts from the [cir$^{\circ}$] ring strain consistently show less hybridization than those from the [cir^{+}] ring strain, and there is variability among extracts from the same strain (compare tracks e and f). A rough estimate of the number of copies of the circular Ty can be made by comparing the intensity of hybridization to the ring chromosome and the small circle. Assuming that there are two Ty's on the ring chromosome, the number of copies of Ty circles varies from approximately 0.2 per cell to approximately 2 per cell.

Electrophoretic properties of the ring chromosome. A somewhat surprising finding of our analysis of the ccc DNA extract is that we have found only a single ring chromosome DNA species. The expectation, based on previous experience with bacterial plasmids, was that we would find two forms, supercoils and nicked circles, which would migrate at different rates. Figure 4 shows data which suggest that the ring chromosome band identified by hybridization contains supercoiled molecules and that nicked circles do not enter the gel. Thermal hydrolysis of the extract (75° for 90'), a treatment which nicks supercoiled DNA (15), results in loss of the ring chromosome band with an increase in the intensity of hybridization to DNA at the slot. Treatment of the extract at 75° for a much longer time (48 hr) results in the conversion of 2μm supercoils to nicked circles (data not shown).

In addition, the migration of the ring is altered in gels run in the presence of EtBr (Fig. 4). Under these conditions, the ring chromosome also fails to enter or barely enters the gel. This behavior contrasts with that of small ccc DNA's which migrate faster in the presence of EtBr. Although the reason for the anomalous migration of the ring chromosome is not understood, we suggest that it may be related to the stiffness of highly twisted supercoils (38). As EtBr intercalates in the molecule, the DNA forms a compact, stiff, highly branched structure which is impeded from entering the gel. The difference between this molecule and smaller plasmids would be the size of the compact structure.

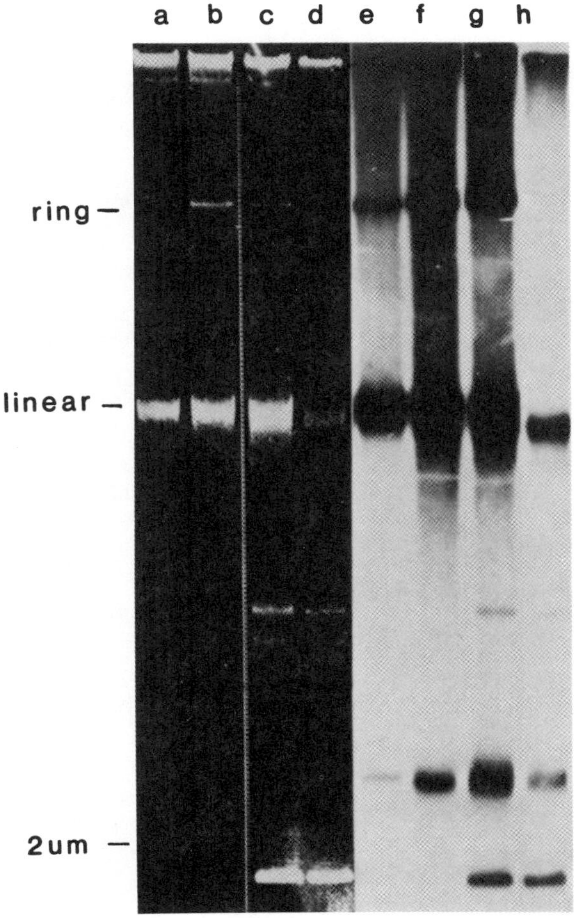

Figure 3. Hybridization of ccc DNA extracts with a Ty probe. ccc DNA extracts from the following strains were electrophoresed in an 0.5% agarose gel (20mA, 16 hr) and DNA was transferred to nitrocellulose as described in the legend to Figure 2. The filters were hybridized to a ^{32}P-labeled Tyl-17 probe consisting of the Ty-containing XhoI fragment excised from λgt KG17 (21). (a) and (b) two different ccc DNA extracts from DC021/DC022 #62; (c) ccc DNA extract from XG1 #24; (d) ccc DNA extract from YNN27 transformed with pLC544 (20); (e) through (h) autoradiograms of (a)-(d) after hybridization to Tyl-17.

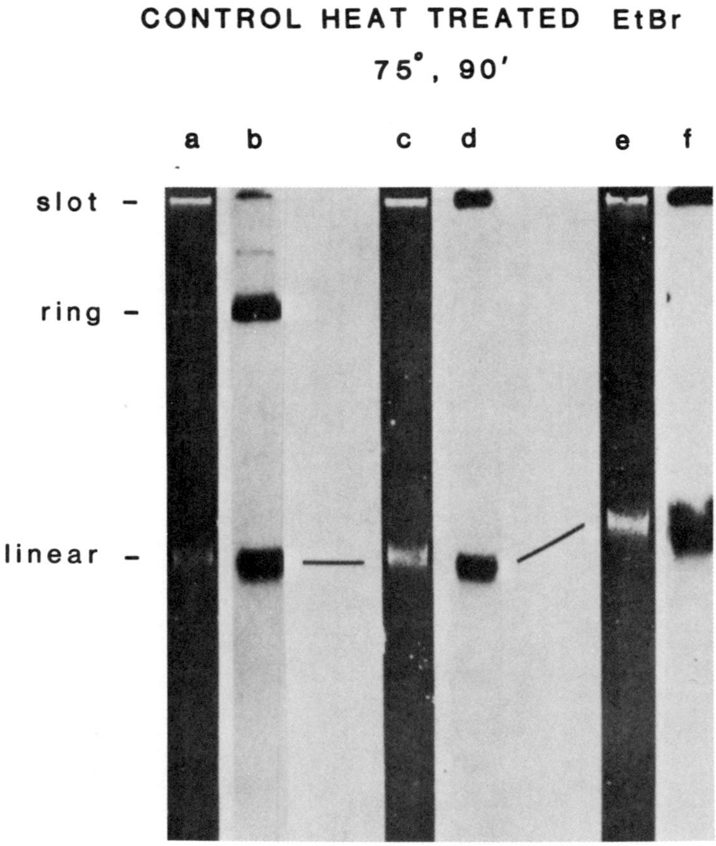

Figure 4. Electrophoretic properties of the ring chromosome. Aliquots of ccc DNA extract from DC021/DC022 #62 (in 10 mM Tris-Cl, 1mM EDTA, pH 8.0) were treated as follows and electrophoresed in 0.7% agarose gels (20 mA, 16 hr). DNA was then transferred to nitrocellulose and hybridized with ^{32}P-labeled 3.4 and 3.6 kb EcoRI fragments from pYe(LEU2)10 (29) to detect chromosome III sequences. (a) control, no treatment; (b) autoradiogram of (a); (c) extract treated 75° for 90 minutes before electrophoresis; (d) autoradiogram of (c); (e) extract electrophoresed on a gel containing 1μg/ml EtBr in the gel and running buffers; (f) autoradiogram of (e).

Cloning and Mapping of the Ring Chromosome

To identify ARSs on the ring chromosome, our strategy has been to clone restriction endonuclease fragments of the ring in a plasmid incapable of autonomous replication and to screen the resulting hybrid plasmids for those which have acquired the ability to replicate autonomously. To this end we have used the URA3 integrating vector YIp5 (36) and ligated BamHI digested ccc DNA extracts from [cir$^+$] and [cir°] ring strains into its single BamHI site in the TET^R gene. BamHI was used because it is expected to cut the ring chromosome at intervals of, on average, 9.3 kb, and because it does not cleave the major known contaminants in the ccc DNA extract, 2μm DNA and ribosomal DNA. Therefore, the ring chromosome should contain approximately 20 BamHI fragments. Assuming that the efficiency of cloning each fragment is the same, and that the ring chromosome is the only DNA in the extract cut by BamHI, then a bank of 97 plasmids with inserts has a 99% probability of having all the fragments (8). We isolated 500 hybrid plasmids each from both [cir°] and [cir$^+$] ring extracts, identifying those with inserts by their tetracycline-sensitive, ampicillin-resistant phenotype.

To precisely map ARSs on the ring chromosome, we need to generate a restriction map of the molecule. To facilitate mapping, we have screened a lambda-Charon 4A library of a partial EcoRI digest of DNA from yeast strain A364A for chromosome III fragments which overlap with the BamHI fragments isolated by direct cloning. The library was constructed by Dr. John Woolford. In order to identify phage with chromosome III inserts, we screened recombinant plaques with a ^{32}P-labeled ring chromosome probe. For this purpose, we extracted the ring chromosome band from agarose gels using the sodium perchlorate method of (39). The recovered DNA was nick translated and used as a hybridization probe. One hundred forty-two such phages have been isolated with an average yeast DNA insert of approximately 13 kb.

Initially using cloned fragments containing the HIS4, LEU2, CEN3, PGK, and MAT loci to screen the λ and YIp5 libraries, and then walking from those regions by overlap hybridization techniques, we have identified and mapped with BamHI, EcoRI, and XhoI, 127 kb of the anticipated 190 kb molecule.

Identification of ARSs on the Ring Chromosome

ARS-containing plasmids transform yeast with efficiencies 10^2- to 10^3-fold higher than non-replicating plasmids (36). When a mixture of autonomously replicating and integrating plasmids is used to transform yeast, most of the transformants carry autonomously replicating plasmids (1, 7). Therefore we pooled the YIp5 library in groups containing approximately 300 independent transformants each and isolated plasmid DNA from each pool. Plasmid DNA from one such pool was used to transform a ura3$^-$ yeast strain to URA3$^+$ and 124 transformants were picked for further analysis. Of these, 122 of 124 showed some degree of mitotic instability, as expected for transformants carrying autonomously replicating plasmids (Table 1). On the basis of the stability of the URA^+ phenotype, the transformants appear to fall into two classes. In the majority class (79%), the URA^+ phenotype was very unstable, with less than 10% of the cells from a colony grown under selective conditions able to form a URA^+ colony. Thirteen percent were more than 80% stable in the same assay.

Table 1. Mitotic Stability of $URA3^+$ Yeast Transformants[1]

Stability class %	100	91-99	81-90	71-80	61-70	51-60	41-50	31-40	21-30	11-20	1-10	<1
Number of transformants	2	9	7	2	1	1	0	0	2	2	36	62

[1] Independent transformants were picked and streaked on selective (-ura) plates. URA^+ colonies were then picked, diluted, and plated for single colonies on non-selective plates. The percent of URA^+ single colonies was then determined by replica plating to -ura plates. Between 100 and 300 colonies were scored for each transformant.

We have begun to characterize the plasmids in these transformants. ccc DNA extracts have been prepared from fourteen transformants and the plasmid DNA from five has been recovered in E. coli and mapped. Four of the five have the same 6.1 kb BamHI fragment which has been recovered in both orientations. The prototype plasmid is called YRp1-2. The fifth has an 8.4 kb BamHI fragment and is called YRp4-1.

The ARS-containing fragments of YRp4-1 and YRp1-2 have been nick translated and used to probe blots of gels containing yeast genomic DNA digested with BamHI and EcoRI as well as blots of gels with the ring chromosome. Such an analysis of the ARS4-1 fragment is shown in Figure 5. The following conclusions can be drawn. 1) The BamHI fragment in YRp4-1 is present in a single copy in the genome of the three strains tested and shows no restriction site polymorphism. 2) It hybridizes to three EcoRI fragments, as expected from its restriction map and shows no restriction site polymorphism. 3) It shows no homology detectable by hybridization to the ARS1-2 BamHI fragment. 4) It is from the ring chromosome as demonstrated by its strong hybridization to the ring chromosome band in a ccc DNA extract. The ARS1-2 fragment is also uniquely represented in the genome and shares homology with the ring chromosome. It is from a region of the chromosome we have mapped and lies about 20 kb distal to the HIS4 locus.

By hybridizing the ARS fragments to blots of gels containing ccc DNA extracts of the first fifteen transformants, we have determined that seven are homologous to ARS1-2, four are homologous to ARS4-1, and four show no homology to either, but clearly carry plasmids. Therefore, there are at least three different ARSs in this collection.

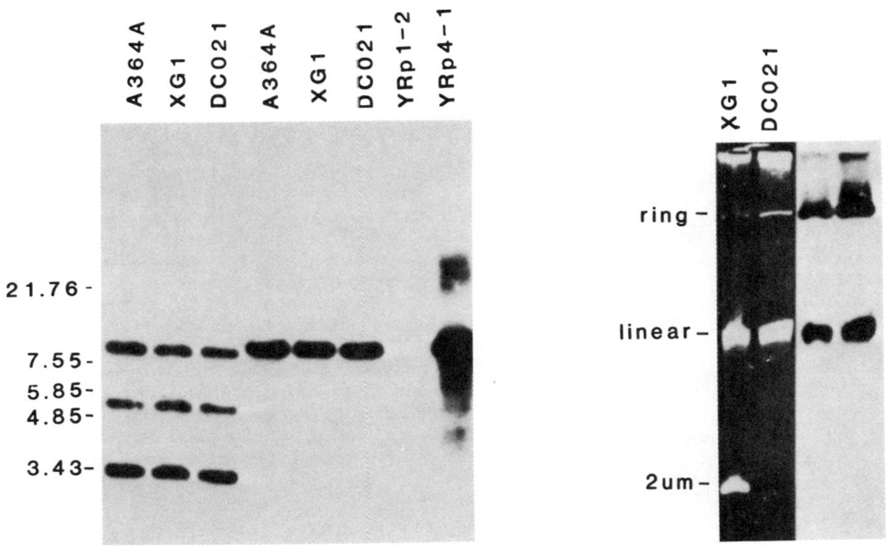

Figure 5. Characterization of *BamHI* fragment containing *ARS4-1*. Left panel: Two micrograms of total genomic DNA from yeast strains A364A, XG1 #24, and DC021/DC022 #62 were digested to completion with *EcoRI* (left three tracks) or *BamHI* (next three tracks) and electrophoresed with *BamHI*-digested YRp1-2 and YRp4-1 on an 0.7% agarose gel. The numbers on the left are sizes in kb of molecular weight standards. DNA was transferred to nitrocellulose and hybridized with ^{32}P-labeled *ARS*-containing *BamHI* fragment from YRp4-1. Right panel: ccc DNA extracts were electrophoresed in an 0.7% agarose gel, transferred to nitrocellulose and hybridized with the ^{32}P-probe described above. The left tracks show the stained gel and the right tracks the autoradiogram.

DISCUSSION

The method we have developed for purifying the ring chromosome is generally applicable to preparing other circular DNAs from yeast. In this paper we demonstrate the presence of three other endogenous circular DNA species in the extract: 2μm plasmid, ribosomal DNA, and Ty DNA. We have also used it as a means of isolating autonomously replicating plasmids.

Our finding of circular ribosomal DNA confirms the work of others (10, 22, 23), and our estimate of the number of copies per cell of this species agrees well with the early estimate of Clark-Walker and Miklos (11). This is the first report of circular Ty elements in yeast. We do not have direct evidence for the circularity of the fast-migrating DNA species we have found, but it is the most likely interpretation of our results. The significance of the circular ribosomal DNA and Ty DNA is unknown. They could arise from homologous recombination events, in the case of ribosomal DNA between adjacent copies in the tandem array and in the case of Ty between the 330 base pair direct repeats (delta sequences) that border the element. Alternatively, the circular copies could replicate autonomously. The available evidence supports the recombination model for Ty elements. Roeder and Fink (30) have found that the most frequent excision event involving the Ty912 element is a recombination between the delta sequences flanking the element, leaving a single delta in the genome and presumably releasing a circular Ty. However, this event is still quite rare, and to account for circular Ty elements at the frequency with which we find them requires postulating other Ty elements which excise much more frequently than Ty912. In the case of ribosomal DNA, there is evidence that at least some non-coding spacer regions may contain an ARS (37), but there is no direct evidence that the extrachromosomal circular copies do replicate autonomously.

The ARS-containing fragments we have found are similar to those described by others (19, 20, 32). They are mitotically unstable, and the two we have studied are unique in the genome. There are two possible explanations for the two apparent stability classes (Table 1). First, the more stable class could result from the presence on some plasmids of stabilizing sequences, for example, centromeric sequences (9). Second, the more stable class could arise from the integration of an unstable plasmid early during selective growth of the colony analyzed. We have found the same plasmid, YRp1-2, in transformants from both stability classes, supporting the second explanation. We cannot eliminate the possibility that some of the more stable transformants carry plasmids with stabilizing sequences.

We are more than half way to our goal of identifying and mapping ARSs on the ring chromosome. The ultimate success of this project depends on the completeness of the libraries we have prepared. The YIp5 library of BamHI fragments appears to be complete. We have now sought seven fragments, ranging in size from 8 kb to more than 20 kb, by direct screening and have found multiple representatives of six and a single copy of the seventh. Some sequences, for example LEU2 and CEN3, are clearly missing from the phages we selected. As an alternative source for fragments overlapping the BamHI library, we have constructed a SalI library of the ring chromosome by direct cloning.

We will use these cloned ARSs from a single chromosome to study the temporal order of replication of chromosome III, to determine whether ARSs function as origins *in vivo*, and to assess the effect of deleting one or more ARSs on the replication and stability of chromosome III.

LITERATURE CITED

1. Beach, D., Piper, M., and Shall, S. 1980. Isolation of chromosomal origins of replication in yeast. Nature (London) 284:185-187.

2. Bell, G.I., DeGennaro, L.J., Gelfand, D.H., Bishop, R.J., Valenzuela, P., and Rutter, W.J. 1977. Ribosomal genes of *Saccharomyces cerevisiae*. 1. Physical map of the repeating unit and location of the regions coding for 5S, 5.8S, 18S, and 25S ribosomal RNAs. J. Biol. Chem. 252:8118-8125.

3. Blumenthal, A.B., Kriegstein, H.J., and Hogness, D.S. 1974. The units of DNA replication in *Drosophila melanogaster* chromosomes. Cold Spring Harbor Symp. Quant. Biol. 38:205-223.

4. Callan, H.G. 1972. Replication of DNA in the chromosomes of eukaryotes. Proc. Roy. Soc. London B 181:19-41.

5. Cameron, J.R., Loh, E.Y., and Davis, R.W. 1979. Evidence for transposition of dispersed repetitive DNA families in yeast. Cell 16: 739-751.

6. Casse, F., Boucher, C., Julliot, J.S., Michel, M., and Dénarié, J. 1979. Identification and characterization of large plasmids in *Rhizobium meliloti* using agarose gel electrophoresis. J. Gen. Microbiol. 113: 229-242.

7. Chan, C.M., and Tye B.-K. 1980. Autonomously replicating sequences in *Saccharomyces cerevisiae*. Proc. Natl. Acad. Sci. U.S.A. 77:6329-6333.

8. Clarke, L., and Carbon, J. 1976. A colony bank containing synthetic Col E1 hybrid plasmids representative of the entire *E. coli* genome. Cell 9:91-99.

9. Clarke, L., and Carbon, J. 1980. Isolation of a yeast centromere and construction of functional small chromosomes. Nature 287:504-509.

10. Clark-Walker, G.D., and Azad, A.A. 1980. Hybridizible sequences between cytoplasmic ribosomal RNAs and 3-micron circular DNAs of *Saccharomyces cerevisiae* and *Torulopsis glabrata*. Nucl. Acids Res. 8:1009-1022.

11. Clark-Walker, G.D., and Miklos, G.L.G. 1974. Localization and quantification of circular DNA in yeast. Eur. J. Biochem. 41:359-365.

12. Cramer, J.H., Farrelly, F.W., and Rownd, R.H. 1976. Restriction endonuclease analysis of ribosomal DNA from Saccharomyces cerevisiae. Molec. Gen. Genet. 148:233-241.

13. Devenish, R.J., and Newlon, C.S. 1982. Isolation and characterization of yeast ring chromosome III by a method applicable to other circular DNAs. Gene, in press.

14. Flavell, A.J., and Ish-Horowicz, D. 1981. Extrachromosomal circular copies of the eukaryotic transposable element copia in cultured Drosophila cells. Nature (London) 292:591-595.

15. Freifelder, D., and Dewitt, R. 1977. Thermal hydrolysis as a means of opening supercoiled circles of DNA. Gene 1:385-387.

16. Hand, R. 1978. Eukaryotic DNA: Organization of the genome for replication. Cell 15:317-325.

17. Hartley, J.L. and Donelson, J.E. 1980. Nucleotide sequence of the yeast plasmid. Nature 286:860-865.

18. Hartwell, L.H. 1967. Macromolecular synthesis in temperature-sensitive mutants of yeast. J. Bacteriol. 93:1662-1670.

19. Hsiao, C.-L., and Carbon, J. 1979. High-frequency transformation of yeast by plasmids containing the cloned yeast ARG4 gene. Proc. Natl. Acad. Sci. U.S.A. 76:3829-3833.

20. Kingsman, A.J., Clarke, L., Mortimer, R.K., and Carbon, J. 1979. Replication in Saccharomyces cerevisiae of plasmid pBR313 carrying DNA from the yeast trp1 region. Gene 7:141-152.

21. Kingsman, A.J., Gimlich, R.L., Clarke, L., Chinault, A.C., and Carbon, J. 1981. Sequence variation in dispersed repetitive sequences in Saccharomyces cerevisiae. J. Mol. Biol. 145:619-632.

22. Larionov, V.L., Grishin, A.V., and Smirnov, M.N. 1980. 3μm DNA - an extrachromosomal ribosomal DNA in the yeast Saccharomyces cerevisiae. Gene 12:41-49.

23. Meyerink, I.H., Klootwijk, J., and Planta, R.J. 1979. Extrachromosomal circular ribosomal DNA in Saccharomyces carlsbergensis. Nucl. Acids Res. 7:69-76.

24. Montgomery, D.L., Leung, D.W., Faye, G., Shalit, P., Smith, M., and Hall, B.D. 1980. Isolation and sequence of the gene for iso-2-cytochrome c in Saccharomyces cerevisiae. Proc. Natl. Acad. Sci. U.S.A. 77: 541-545.

25. Newlon, C.S., and Burke, W.G. 1980. Replication of small chromosomal DNAs in yeast, p.399-409. In: B. Alberts and C.F. Fox (ed.), Mechanistic Studies of DNA Replication and Recombination, ICN-UCLA Symposia on Molecular and Cellular Biology, Vol. XIX. Academic Press, New York, N.Y.

26. Newlon, C.S., Petes, T.D., Hereford, L.M., and Fangman, W.L. 1974. Replication of yeast chromosomal DNA. Nature 247:32-35.

27. Petes, T.D., Byers, B., and Fangman, W.L. 1973. Size and structure of yeast chromosomal DNA. Proc. Natl. Acad. Sci. U.S.A. 70:3072-3076.

28. Petes, T.D., and Fangman, W.L. 1972. Sedimentation properties of yeast chromosomal DNA. Proc. Natl. Acad. Sci. U.S.A. 69:1188-1191.

29. Ratzkin, B., and Carbon, J. 1977. Functional expression of cloned yeast DNA in *Escherichia coli*. Proc. Natl. Acad. Sci. U.S.A. 74:487-491.

30. Roeder, G.S., and Fink, G.R. 1980. DNA rearrangement associated with a transposable element in yeast. Cell 21:239-249.

31. Smith, G.E., and Summers, M.D. 1980. The bidirectional transfer of DNA and RNA to nitrocellulose or diazobenzyloxymethyl-paper. Anal. Biochem. 109:123-129.

32. Stinchcomb, D.T., Struhl, K., and Davis, R.W. 1979. Isolation and characterization of a yeast chromosomal replicator. Nature (London) 282:39-43.

33. Stinchcomb, D.T., Thomas, M., Kelly, J., Selker, E., and Davis, R.W. 1980. Eukaryotic DNA segments capable of autonomous replication in yeast. Proc. Natl. Acad. Sci. U.S.A. 77:4559-4563.

34. Strathern, J.N., Newlon, C.S., Herskowitz, I., and Hicks, J.B. 1979. Isolation of a circular derivative of yeast chromosome III: Implications for the mechanism of mating type interconversion. Cell 18:309-319.

35. Strathern, J.N., Spatola, E., McGill, C., and Hicks, J.B. 1980. Structure and organization of transposable mating type cassettes in *Saccharomyces* yeasts. Proc. Natl. Acad. Sci. U.S.A. 77:2839-2843.

36. Struhl, K., Stinchcomb, D.T., Scherer, S., and Davis, R.W. 1979. High-frequency transformation of yeast: Autonomous replication of hybrid DNA molecules. Proc. Natl. Acad. Sci. U.S.A. 76:1035-1039.

37. Szostak, J.W., and Wu, R. 1980. Unequal crossing over in the ribosomal DNA of *Saccharomyces cerevisiae*. Nature (London) 284:426-430.

38. Upholt, W.B., Gray, H.B., and Vinograd, J. 1971. Sedimentation velocity behavior of closed circular SV40 DNA as a function of superhelix density, ionic strength, counterion and temperature. J. Mol. Biol. 62:21-38.

39. Yang, R.C.-A., Lis, J., and Wu, R. 1979. Elution of DNA from agarose gels after electrophoresis, p. 176-182. In: L. Grossman and K. Moldave (ed.), Methods in Enzymology, Vol. 68. Academic Press, New York, N.Y.

40. Zakian, V.A., and Scott, J.F. 1982. Construction, replication, and chromatin structure of TRP1 RI circle: a multiple copy synthetic plasmid derived from yeast chromosomal DNA. Mol. Cell Biol. 2:221-232.

6. RECOMBINATION IN YEAST RIBOSOMAL DNA*

T.D. Petes, S. Smolik-Utlaut, and M. McMahon

Department of Microbiology, University of Chicago
920 E. 58th Street, Chicago, Illinois 60637

Rec. Adv. Yeast Mol. Biol. (1982) 1:69-75

We describe meiotic and mitotic recombination events within the repeated yeast ribosomal RNA (rRNA) genes. During meiosis, non-sister strand recombination is suppressed within the ribosomal DNA (rDNA) but unequal sister strand exchange occurs frequently. During mitosis, both non-sister and sister strand recombination events occur. These exchanges may be either reciprocal or non-reciprocal (gene conversion) events.

INTRODUCTION

We have analyzed recombination events within the repeated yeast genes that encode ribosomal RNA. Before describing these genetic studies, we will summarize the results obtained from physical analysis of these repeated genes.

Each haploid strain has about 100 copies of the rRNA genes (11). Each gene has about 9 kilobases (kb) of DNA and encodes 4 species of rRNA, the 5S, 5.8S, 18S and 25S species (summarized in ref. 7). The 5S coding sequences are separated from the coding sequences of the other rRNA species by approximately 2 kb of non-transcribed spacer DNA. All 100 copies of the chromosomal rRNA genes are located in a single tandem array which is on chromosome XII (5,17).

Within a single haploid strain, the rRNA genes are usually very similar in their restriction pattern, indicating a high degree of sequence conservation between different repeats. Most laboratory strains contain rRNA genes that have seven EcoR1 sites; this type of repeat has been called "form I" rDNA. A strain in which each rRNA gene has only six EcoR1 sites ("form II" rDNA) has recently been identified (9). Restriction and sequence analysis of form I and form II genes (McMahon, Smolik-Utlaut and Petes, unpublished data) have shown several changes in addition to the EcoR1 site difference. For example, in 300 base pairs of non-transcribed spacer DNA surrounding the 5S rRNA coding sequences, there are seven differences between form I and form II. Three of these changes are transversions in A-T base pairs and four are single base insertions of T into the noncoding strand of the form II gene. Despite these changes, the two forms of rDNA are more than 95% homologous.

The genetic studies described below were done using two different approaches. First, we used the naturally-occuring form I-form II polymorphism as a genetic marker. Second, we used recombinant DNA procedures and the yeast transformation technique to construct strains that had single-copy selectable yeast genes inserted within the rDNA. In the first section below, we examine meiotic recombination within the rRNA gene cluster; in the second section, we describe the mitotic properties of recombination in the rDNA.

* Support was obtained from N.I.H. (GM24110 and R.C.D.A. 1 KO4 AG00077) and the N.C.I. DHHS (CA19265).

MEIOTIC RECOMBINATION

Non-Sister Strand Recombination

In order to examine meiotic recombination between non-sister tandem arrays of rRNA genes, we crossed a haploid strain with form I rDNA to a haploid that had form II (8). When the resulting diploid underwent meiosis, most of the tetrads (12 of 14) segregated two form I spores to two form II spores. This pattern indicates that meiotic recombination between non-sister tandem arrays is infrequent since a non-sister exchange should produce one form I spore, one form II spore and two mixed form I-form II spores. In a similar but more extensive study (5), 3 of 59 tetrads had the segregation pattern expected for meiotic exchange between non-sister clusters. From the estimated number of recombination events per meiosis (100 per cell) and the fraction of nuclear DNA that is rDNA (about 5%), we calculated that non-sister meiotic recombination within the rDNA is suppressed by about two orders of magnitude relative to non-rDNA sequences (5,8).

In the experiments described above, we could not exclude the possibility that the observed suppression of recombination was a result of the sequence heterogeneity between form I and form II. In later experiments (17), we constructed yeast strains in which the single-copy _ura3_ gene was inserted distal to the rRNA gene cluster. We then measured recombination between _ura3_ and _pep3_ (a marker centromere-proximal to the rDNA) in a diploid strain that had two form I gene clusters. We found that meiotic recombination was suppressed by at least a factor of 15 in the rRNA gene cluster. Thus, the observed suppression is not a result of the form I or form II heterogeneity. Before discussing possible explanations for this suppression, we will describe meiotic sister strand recombination within the rDNA.

Sister Strand Recombination

Since sister chromatids are identical in DNA sequence, the form I-form II heterogeneity could not be used to analyze sister strand exchange within the rRNA gene cluster. In order to monitor this type of recombination, we constructed strains in which the rRNA gene cluster contained a selectable single-copy gene. These strains were constructed by transforming a yeast strain that was mutant at the _leu2_ locus with a recombinant plasmid (pTP9) that had a wild type _leu2_ gene in addition to a fragment of yeast rDNA (6). Since Hinnen _et al_ (2) had demonstrated that integrative recombination between a transforming plasmid and the yeast chromosome occurs by sequence homology, we expected that most Leu^+ transformants would contain the _leu2_ gene integrated within the rDNA. Both genetic and physical tests showed that this expectation was justified (6).

In one experiment, a form I haploid containing _leu2_ integrated into the rDNA was mated to a form II haploid that also had a _leu2_ gene in the rDNA (6). This diploid strain was put through meiosis. Approximately 88% of tetrads segregated $4^+:0^-$ for the leucine requirement, 10% segregated $3^+:1^-$ and 2% segregated $2^+:2^-$. When the tetrads that segregated $3^+:1^-$ were analyzed, we found that the form I-form II heterogeneity segregated 2 form I spores: 2 form II spores. By quantitative Southern analysis, we showed that the loss of a _leu2_ gene from one chromatid (generating a leu^- spore) was correlated with the duplication of the _leu2_ gene in the sister chromatid; spores having the same form of rDNA were considered to contain chromosomes derived from sister chromatids. This result demonstrates that unequal sister strand recombination occurs in at least 10% of the meiotic cells. Since leu⁻ spores are not formed unless the exchange event occurs between the misaligned leu2 insertions, it is likely that the frequency of unequal sister strand exchange is actually considerably higher than 0.10. This high frequency of unequal sister strand recombination indicates that the rRNA gene cluster has a high degree of polymorphism within a cell population in the number of repeats per cluster.

The rRNA gene cluster, therefore, shows the properties of high levels of meiotic sister strand recombination and suppressed non-sister exchange. One possibility is that these two phenomena may be causally related. Our favored hypothesis to explain the causal relationship requires two assumptions. First, we assume that sister strand interactions occur more frequently for repeated sequences than single-copy sequences. This assumption seems reasonable since, in the absence of structures to align sequences for recombination, the frequency of recombination is likely to be related to the gene dosage. The second assumption is that sister strand recombination precedes and inhibits non-sister exchange. If these two assumptions are correct, it is likely that all tandem arrays of repeated genes will have suppressed non-sister recombination and high levels of sister strand exchange. To determine whether these unusual recombinational properties are specific to meiosis in yeast, we examined mitotic recombination within the rDNA.

MITOTIC RECOMBINATION

We examined two types of mitotic recombination in our studies (12). First, we analyzed recombination between a transforming plasmid containing yeast rDNA and the chromosomal rDNA sequences. Second, we studied recombination between two chromosomal rRNA gene clusters.

Plasmid-Chromosome Recombination

Since the integration of a transforming plasmid into a yeast chromosome is a mitotic recombination event, we examined this type of recombination as a model system for investigating the specificity of mitotic exchange. We constructed a diploid yeast strain (SSU10) that was homozygous for mutations of the leu2 and ura3 loci and was heterozygous for the form I-form II heterogeneity. We transformed SSU10 with a recombinant plasmid (pSS2) that contained an intact form II rRNA gene as well as the wild type leu2 and ura3 genes. 19 Leu$^+$ Ura$^+$ transformant were isolated and the transforming genes were mapped. In 16 of 19 transformants, the plasmid containing form II rDNA integrated into the form II rDNA; in 3 of 19, the integration was into a form I cluster. In a second series of experiments, SSU10 was transformed with a recombinant plasmid (pSS41) that contained an intact form I rRNA gene as well as selectable single-copy genes. In 33 of 44 transformants, the recombinant plasmid integrated into the form I cluster. Both of these results show significant ($p < .01$) deviation from random integration. Thus, the integrative recombination system can distinguish between the perfect homology of a form I -form I or form II-form II interactions and the very good homology of a form I-form II interaction.

In order to find out whether the preferential integration described above was the result of DNA sequences localized in a small section of the rRNA gene, we did other transformation experiments in which fragments of the intact repeat were used (12). For example, we constructed recombinant plasmids containing Bgl II fragments of the rRNA gene and a selectable marker. Since Bgl II cuts both form I and form II genes into two equal-sized fragments, all four recombinant plasmids constructed with the Bgl II halves were identical in size. We found that none of these four plasmids was capable of directing preferential integration into a homologous cluster. We also found that if a recombinant plasmid contained a complete form II rRNA gene in which the continuity of the repeat was interrupted by an insertion of non-ribosomal DNA, then the plasmid was incapable of directing preferential integration. We believe, therefore, that preferential integration of a form I plasmid into a form I cluster or a form II plasmid into a form II cluster requires a long region (greater than 4.5 kilobases) of uninterrupted sequence homology.

Mitotic Recombination Between rRNA Gene Clusters on Homologous Chromosomes

Before discussing the experimental system that we used to study mitotic recombination between chromosomal rRNA gene clusters, we will briefly describe the classical approach for the detection of mitotic recombination in yeast. If a diploid cell is heterozygous at a genetic locus, mitotic recombination can result in the formation of homozygous cells. In yeast, mitotic recombination is usually examined using diploids that are heterozygous for one or more auxotrophic mutations. Thus, the appearance of homozygotes is monitored by allowing single diploid cells to divide mitotically, forming colonies on a rich nutrient agar. These colonies are then replica-plated onto plates containing media that is deficient for specific nutrients. If mitotic recombination occurs immediately upon plating the single diploid cells on the rich media, then a sectored colony can be formed on the supplement-deficient media.

Two different types of mitotic recombination can yield a sectored colony (4), a reciprocal mitotic exchange between the heterozygous genes and the centromere or a non-reciprocal (gene conversion) event. These two classes of events can be distinguished by meiotic analysis of the prototrophic half of sectored colonies. If the prototrophic sector is homozygous for the wild type allele of the gene, then the sectoring is likely to be the result of reciprocal recombination. If the prototrophic sector is heterozygous, then the sectoring may be the result of a mitotic gene conversion (4).

In order to examine mitotic recombination within the rDNA, we constructed a number of different diploid strains (SSU12-15) that were heterozygous for several genetic markers on chromosome XII. For example, on one homologue, SSU14 had an insertion of the single-copy gene *his3* in a form II rRNA gene cluster and had the wild type allele of *ura4*. The gene *ura4* is centromere-distal to the rDNA (5). On the other homologue, SSU14 had a *leu2* insertion within a form I rRNA gene cluster and a mutant allele of *ura4*. SSU14 was homozygous for *his3* and *leu2* mutations at the normal chromosomal location and thus was effectively heterozygous at the chromosome XII locations.

Since SSU14 was heterozygous for three different auxotrophic markers, we expected to find three types of sectors; leu^-, his^- and ura^-. We found that leu^- and his^- sectors were frequent (0.5% for each class) relative to ura^-

sectors (about 0.1%). As discussed below, we believe the colonies that sectored only for leucine or histidine requirements represent unequal mitotic sister strand recombination events. We decided, therefore, to concentrate our analysis on the ura$^-$ sectors since the events producing such sectors were likely to be non-sister mitotic interactions.

For all ura$^-$ sectors (total of 51), we also examined the sectoring pattern for leu2 and his3. In addition, the prototrophic part of the ura sector was dissected in order to find whether the sector had formed as the result of a reciprocal or non-reciprocal recombination event. Of 51 sectored colonies examined, 21 (41%) were the result of reciprocal recombination and 30 (59%) were the result of mitotic gene conversion. Since ura4 is centromere-distal to the rRNA gene cluster, a recombination event initiating within the rDNA could lead to ura4 becoming homozygous. We found that of the 21 reciprocal recombination events detected with ura4, at least 9 were the result of reciprocal exchange within the rRNA gene cluster. For SSU14, the diagnostic features of reciprocal recombination within the rDNA were that: 1) in the ura$^-$ half of the sectored colony, his3 and ura4 were segregating $0^+:4^-$ spores and leu2 was segregating $2^+:2^-$ spores and 2) in the Ura$^+$ sector, his3 and ura4 were segregating $4^+:0^-$ spores and leu2 was segregating $2^+:2^-$. This result is consistent with a reciprocal mitotic recombination between non-sister rRNA gene clusters, in which the recombination event occured between the leu2 and his3 insertions. This result also suggests that the leu2 insertion is centromere-proximal to the his3 insertion.

Thus, approximately half of the reciprocal recombination events that make ura4 homozygous occur in the rDNA. We also did a meiotic analysis of several of the ura$^-$ sectors produced by gene conversion in order to determine whether gene conversion events initiating in the rDNA could extend to the ura4 locus. In 15 of 30 ura$^-$ sectors analyzed, we detected gene conversion for one or both insertions in the rDNA. For example, in one such class of convertants, in the Ura$^+$ side of the colony, all heterozygous markers were segregating $2^+:2^-$; in the ura$^-$ side of the colony, his3 and ura4 were segregating $0^+:4^-$ and leu2 was segregating $4^+:0^-$. The frequent co-conversion of ura4 and markers within the rRNA gene cluster indicates that mitotic gene conversion is likely to involve very long heteroduplex regions. Similar conclusions have also been reached in other studies (summarized in ref. 1).

In summary, between non-sister rRNA gene clusters, we have detected both reciprocal and non-reciprocal mitotic recombination events. Approximately half of the reciprocal recombination events that make ura4 homozygous occur within the rDNA. We can determine whether this amount of mitotic recombination is that expected for the rRNA gene cluster by the following calculation. The total length of the rRNA gene cluster is 900 kb. The distance from the centromere of chromosome XII to ura4, excluding the rRNA gene cluster, is at least 160 cM (ref. 5); this number represents a minimum estimate since the ura4 gene is not meiotically linked to the rDNA. It has been estimated that one centimorgan represents about 3 kb (ref. 13). Thus, the amount of non-ribosomal DNA between ura4 and the centromere is about 480 kb. Using these estimates, we calculate that about 65% of the DNA between ura4 and the centromere is rDNA. Since about half of the reciprocal recombination events that make ura4 homozygous occur within the rDNA, mitotic non-sister recombination within the rDNA is neither suppressed nor enhanced relative to other chromosomal sequences. As described below, however, unequal mitotic sister strand recombination is a frequent event.

Unequal Mitotic Sister Strand Recombination

Unequal mitotic sister strand recombination events in yeast were first studied by Szostak and Wu (14). By using a cloned leu2 gene inserted within the rDNA, they demonstrated that mitotic loss of the leu2 gene from the leu⁻ side of a sectored colony was correlated with the duplication of the insertion on the Leu+ side of the colony. Since these experiments were done in a haploid strain, these results are consistent with unequal sister strand recombination. The observed frequency of sectoring in these studies was 5×10^{-4}. In similar studies (12,16), we have observed somewhat higher sectoring frequencies, 5×10^{-3} to 10^{-2}. In addition, in about half of the sectored colonies, we were unable to demonstrate a duplication of the insertion correlated with the loss. These events may represent unequal mitotic sister strand conversion although the possiblity of false sectoring cannot be ruled out.

We have also examined the effects of the rad52 mutation on unequal sister chromatid exchange (12). This mutation, which depresses both meiotic and mitotic recombination (discussed in ref. 1), does not significantly influence unequal sister strand recombination in the rDNA. Jackson and Fink (3), examining unequal mitotic recombination between duplicated his4 genes, have reached similar conclusions.

DISCUSSION

In summary, we have detected both meiotic and mitotic recombination events within the rDNA. In meiosis, non-sister strand recombination is suppressed relative to other cellular DNA sequences by approximately two orders of magnitude. In contrast, unequal sister strand recombination occurs in at least 10% of the meiotic cells. In mitosis, non-sister strand recombination occurs at approximately the expected frequency (10^{-3} events per cell cycle) relative to non-rDNA sequences. Unequal sister strand exchange in mitosis is a frequent event, occuring at a 10-fold higher frequency than non-sister mitotic exchange. We have also detected mitotic gene conversion events between non-sister clusters of rRNA genes in which the conversion event extends from the rDNA to the ura4 locus. These non-reciprocal recombination events were approximately as frequent as the reciprocal mitotic exchanges.

The high frequency of unequal recombination within the repeated rRNA genes is likely to be observed in eucaryotes other than yeast. Both Tartof (15) and Schalet (10) have obtained experimental results consistent with unequal recombination within the rDNA of Drosophila.

LITERATURE CITED

1. Esposito, M.S. and Wagstaff, J.E. 1981. Mechanisms of mitotic recombination, p. 341-370. In: J.N. Strathern, E.W. Jones and J.R. Broach(ed.) The Molecular Biology of the Yeast Saccharomyces. Cold Spring Harbor Laboratories, Cold Spring Harbor, N.Y.
2. Hinnen, A., Hicks, J. and Fink, G.R. 1978. Transformation of yeast. Proc. Natl. Acad. Sci. U.S.A. 75:1929-1933.

3. Jackson, J. and Fink, G.R. 1981. Gene conversion between duplicated genetic elements in yeast. Nature 292:306-311.
4. Nakai, S. and Mortimer, R.K. 1969. Studies of the genetic mechanism of radiation-induced mitotic segregation in yeast. Mol. Gen. Genet. 103: 329-338.
5. Petes, T.D. 1979. Meiotic mapping of yeast ribosomal DNA on chromosome XII. J. Bacteriol. 138:185-192.
6. Petes, T.D. 1980. Unequal meiotic recombination within tandem arrays of yeast ribosomal DNA genes. Cell. 19:765-774.
7. Petes, T.D. 1980. Molecular genetics of yeast, p. 845-876. In: E.E. Snell (ed.), Annual Review of Biochemistry, vol. 49. Annual Reviews Inc, Palo Alto, California.
8. Petes, T.D. and Botstein, D. 1977. Simple Mendelian inheritance of the reiterated ribosomal DNA of yeast. Proc. Natl. Acad. Sci. U.S.A. 74: 5091-5095.
9. Petes, T.D., Hereford, L.M. and Skryabin, K.G. 1978. Characterization of two types of yeast ribosomal DNA genes. J. Bacteriol. 134:295-305.
10. Schalet, A. 1969. Exchanges at the bobbed locus of *Drosophila melanogaster*. Genetics 63:133-153.
11. Schweizer, E., Mackechnie, C. and Halvorson, H.O. 1969. The redundancy of ribosomal and transfer RNA genes in *S. cerevisiae*. J. Mol. Biol. 40: 261-277.
12. Smolik-Utlaut, S. 1982. Spontaneous mitotic recombination in the rRNA genes of the yeast *Saccharomyces cerevisiae*. Ph.D. Dissertation, University of Chicago.
13. Strathern, J.N., Newlon, C.S., Herskowitz, I. and Hicks, J.B. 1979. Isolation of a circular derivative of chromosome III. Cell 18:309-319.
14. Szostak, J. and Wu, R. 1980. Unequal crossing over in the ribosomal DNA of *Saccharomyces cerevisiae*. Nature 284:426-430.
15. Tartof, K.D. 1973. Unequal mitotic sister chromatid exchange and disproportionate replication as mechanisms regulating ribosomal RNA gene redundancy. Cold Spring Harbor Symp. Quant. Biol. 38:491-500.
16. Zamb, T.J. and Petes, T.D. 1981. Unequal sister-strand recombination within yeast ribosomal DNA does not require the *RAD52* gene product. Current Genet. 3:125-132.
17. Zamb, T.J. and Petes, T.D. 1982. Analysis of the junction between ribosomal RNA genes and single-copy chromosomal sequences in the yeast *Saccharomyces cerevisiae*. Cell 28:355-364.

7. EVOLUTIONARY CONSERVATION OF THE STRUCTURE OF EUCARYOTIC TELOMERES*

Jack W. Szostak

Department of Biological Chemistry, Harvard Medical School and
Cancer Genetics, Sidney Farber Cancer Institute, Boston, Massachusetts 02115

Rec. Adv. Yeast Mol. Biol. (1982) 1:76-92

Telomeres are special structures at the ends of eucaryotic chromosomes that are extremely stable and are fully replicatable. Neither the physical structure nor the mechanism of replication of telomeres are well understood. In this paper, a possible reason for the evolution of telomeres is discussed, and several model systems that have been studied in recent years are reviewed. Recently reported experiments concerning the cloning and initial characterization of yeast telomeres are summarized. Data concerning the non-snapback nature of yeast telomeres is presented, and efforts to find cdc mutants defective in telomere function are described. Finally two models for telomere resolution are described, along with possible ways of distinguishing between these models.

INTRODUCTION

Eucaryotic chromosomes are extremely long linear DNA molecules. Yeast chromosomes range in length from about 200 kb to over 1000kb (1); chromosomes of higher eucaryotes may range from 10 to 1000 times as long. Each chromosome has a centromere, the point or region of attachment to the spindle, and two ends, the telomeres. The physical structure of telomeres has been a subject of interest for some time.

Two considerations suggest that a telomere must be more than the end of a DNA duplex. The first is a problem of DNA replication that was first pointed out by Watson (2) over 10 years ago. This problem is easy to understand if one inagines a replication fork moving towards the end of a DNA molecule (see Figure 1). There is no problem, at least theoretically, with the leading strand. The DNA polymerase could simply continue until it reached the last nucleotide, then fall off. However, the lagging strand cannot be completed. No known DNA polymerase is able to initiate primer free DNA synthesis. Instead, synthesis is initiated by a short RNA primer. (3, 4, 5) This primer is later degraded, and the resulting gap is then filled in by DNA synthesis which is initiated from the upstream DNA fragment. However, there is no upstream DNA fragment at the end of a DNA molecule; hence a short unreplicated stretch of DNA would remain. Repeated rounds of DNA synthesis would result in progressive loss of DNA from the ends of a chromosome. Therefore, a telomere must be a special structure which is fully replicatable. The second major problem is one of stability. DNA ends are extremely reactive structures that tend to engage in ligation (fusion), degradation, and recombination. In the 1930's and 1940's Muller (6) and McClintock (7,8) showed that new DNA ends produced by breaking chromosomes either with X-rays, or mechanically by using dicentric chromosomes, readily fused with each other. X-rays and gamma rays, which cause double strand breaks in DNA

* Support was obtained from National Institutes of Health grant GM27862 to J.W.S.

are known to cause deletions in many organisms, and to stimulate mitotic recombination. Presumably deletions result from exonucleolytic degradation at the site of a double strand break, followed by fusion of the two ends. Resnick (9,10) has described the effects of ionizing radiation on yeast, and has proposed a model to explain the initiation of mitotic recombination by double strand breaks.

Recent work from this laboratory (11) has demonstrated all three of the above reactions of broken DNA ends in a well defined system, that of yeast transformation with specific plasmids. Replicating plasmids transform yeast at high frequency (12). Restriction enzyme digestion of such plasmids within vector sequences not homologous to chromosomal sequences results in a 10 fold reduction in the recovery of transformants; the plasmids in the transformants that are recovered are circular molecules that must have formed by ligation of the cut DNA ends. (13) These plasmids frequently have deletions at the site of restriction enzyme digestion. When the cut is made in plasmid sequences homologous to genomic sequences, the replicating plasmids are stimulated to recombine with the chromosome, and integrated transformants are recovered at high frequency. The intergration frequency of non-replicating plasmids is also greatly stimulated by restriction digestion with homologous sequences (11).

Since telomeres are extremely stable structures that do not engage in any of these reactions, they must be protected in some way. Evidence discussed below suggests that telomeres are complex structures that have evolved to solve the problems of complete DNA replication, and of stability, of a DNA end.

The Origin of Linear Chromosomes

Given the problems which DNA ends present, it is interesting to speculate about why eucaryotes have linear chromosomes as opposed to the circular chromosomes found in procaryotic cells. Meiosis is an almost universal aspect of the life cycle of eucaryotic organisms; the essence of the meiotic process is the random reassortment of genetic information, so that new combinations of genes are produced in the succeeding generation. The primary mechanism of recombination in meiosis relies on the division of the large eucaryotic genome into a number of separate segments; these segments (chromosomes) assort independently (see reference 14 for discussion). A cell with many chromosomes must have a mechanism to assure the segregation of one copy of each chromosome into each daughter cell after mitosis, as well as a mechanism for correct meiotic segregation. This is, of course, accomplished by having a centromere on every chromosome. The centromere is the part of the chromosome that attaches to the spindle apparatus; the spindle physically puts one daughter chromosome into each daughter cell. However, the existence of a centromere on a chromosome is incompatible with a circular structure for that chromosome, at least in an evolutionary sense. After DNA replication, two sister chromatids are formed. Any odd number of reciprocal exhanges (cross-overs) between two circular sister chromatids will result in a single dicentric circle. Recombination between circular homologues in a diploid could also result in the fusion of two chromosomes to produce a dicentric. These problems are amplified during meiosis, when recombination levels are greatly elevated. Ring chromosomes have been known to be unstable for these very reasons, since the work of McClintock on maize (15) and Muller on Drosophila (1). The same phenomena have recently been observed for ring chromosomes and mini-chromosomes in yeast (16,17). In contrast, any number of cross-overs can take place between linear chromosomes, and no change in the structure of the genome will result. The telomere, therefore, must have evolved either before, or in concert with, the evolution of the fragmented genome and the centromere.

Other Functions of Telomeres

Telomeres appear to have other functions in addition to providing an end that is stable and that is fully replicatable. Cytological evidence from a number of laboratories suggests that telomeres may be sites of chromosomal attachment to the nuclear membrane (see reference 18 for discussion). Electron microscopy of yeast nuclei has shown telomeres apparently attached to the nuclear membrane. (19) The chromosomes of Drosophila salivary cells are arranged in a specific and reproducible three dimensional pattern (D. Agard and J. Sedat, personal communication) in a shell adjacent to the nuclear membrane. The centromeres are clustered together at one pole, and the telomeres are clustered together at the opposite pole. Both genetic studies and cytological evidence support the idea that telomeres may act as sites for the inititiation of chromosome pairing.

The Structure of Telomeres

Genetic observations from Neurospora provide evidence that telomeres are more than simply point structures at the ends of chromosomes, and that the ends of different chromosomes have sequences in common with each other. A frequent class of chromosomal aberration in Neurospora is the quasi-terminal translocation, in which a fragment of one chromosome is attached to the tip of another chromosome. (20) No essential genetic information is lost from the tip region. Quasi-terminal translocations revert precisely at a very high frequency; it is thought that the revertants arise by homologous recombination between sequences at the new tip and sequences at the old tip, suggesting that different chromosomes share certain sequences at or near their telomeres. This has been confirmed for several organisms by in situ hybridization with cloned satellite DNA probes. For example, the 5S RNA sequences of Xenopus are located at the tips of chromosomes (21), and many satellite sequences are present at or near the ends of chromosomes. (22-25) A cloned DNA fragment that hybridizes to the telomeres of all Drosophila chromosomes has been reported.(26) This probe also hybridizes to fine fibers (ectopic pairing fibers) that connect the ends of different chromosomes. The nature and significance of these fibers is not understood. Southern blot restriction mapping revealed that the probe sequences was present in long tandem arrays in the genome; this made it impossible to tell how close to the actual ends of the chromosome these sequences were.

Model Systems

In view of the many interesting properties of telomeres, there has been considerable interest in looking more closely at the molecular nature of telomere structure. This has been quite difficult until recently, because it is not possible to clone a protected piece of DNA such as a telomere in cloning vectors which can only accept fragments of DNA bounded by two simple ends, usually restriction cuts. In an effort to circumvent this problem, various small linear DNA molecules have been examined as model systems.

Adenovirus. The replication of Adenovirus is now well understood. The virus encodes a protein which acts as a primer for DNA synthesis, and which remains attached to the 5' end of each viral strand. (27) The replication of Adenovirus appears to be very different from the replication of normal chromosomal DNA, since full length single strands are produced, and these single strands are converted to double stranded DNA in a separate reaction. There is no evidence for covalently attached proteins at the termini of normal chromosomes, and this is probably not the normal structure of a telomere.

Parvoviruses. The Adeno-associated viruses and parvoviruses are a family of linear viruses that are largely single stranded. (28) They have short double stranded segments at each end (see Figure 2a) that are formed by a Y shaped terminal hairpin loop; i.e. the single strand folds back upon itself for a short region. Near the hairpin loop is a stretch of DNA that is not fully base paired. This sequence exists in two complementary forms (flip and flop), and the detection of these two forms has been taken as strong evidence that these viruses replicate their ends according to a model to telomere replication first proposed by Bateman (29), as a simplification of the Cavalier-Smith model (30) (see below). The terminal hairpin loop is a structural motif that is found in may different eucaryotic linear molecules, and is probably the most fundamental aspect of the structure of a telomere. As discussed below, there are various ways in which one can imagine the complete replication of a molecule the ends in a hairpin loop. Such a loop has the added advantage of explaining the stability of telomeres: there are no DNA ends available for ligation, exonucleolytic degradation, or recombination.

Vaccinia. Vaccinia virus is a long (about 200 kb) linear double stranded DNA virus. At its ends are two 10 kb sequences that are in inverted orientation relative to each other. (31) The ends of the molecule are crosslinked by hairpin loops (32), in which there are 104 bases of unpaired DNA. (33) The hairpin loops have recently been sequenced (33), and have been found to exist in two forms (flip and flop) that are complementary. Near the ends of the molecule are two clusters of 70 bp repeat units. (34) These repeat units contain smaller internal repeats. (33) The number of repeat units found varies between different subcloned viral isolates. (35) The significance of the repeats and of the variation is unknown. The DNA that is isolated from virions is closed, i.e. the two strands of the duplex are crosslinked through the hairpin loops at each end into a continuous circular strand. This causes the viral DNA to behave as snapback DNA - it rapidly renatures after denaturation because of the terminal crosslinks. Apparently, nicks are introduced into the viral DNA at or near the ends shortly after the initiation of infection, however the nature of these nicks is not known.

Paramecium mitochondrial DNA. It has recently been found that a number of mitochondrial DNA molecules are linear. Of these, the mtDNA of Paramecium has been studied in great detail by Cummings et al. (36-38) Replication of this molecule begins at one end, where the two strands of the duplex are crosslinked by a hairpin loop. Replication results in the generation of a head to head linear dimer, which is subsequently processed into two linear monomers. In this case, the hairpin loop consists of about 350 nucleotides of unpaired (single-stranded) DNA; the sequence of this DNA consists of an 11-Fold reiterated 34 nucleotide repeat unit. The adjacent DNA does not contain repeated sequences. The nature of the replication events at the other end of the molecule is not known.

Tetrahymena. Vaccinia virus replicates in the cytoplasm, and carries all of its own replication functions with it. Mitochondrial DNA also exists in an environment very different from that of the nucleus. Neither system, therefore, is ideal for understanding the replication of true chromosomes. Perhaps the best model system is that of Tetrahymena. This organism has the useful property of having two distinct kinds of nucleus, the micronucleus and the macronucleus. During the derivation of the macronucleus from the micronucleus, the micronuclear chromosomes are fragmented into a large number of smaller pieces, some of which are then selectively amplified. One of the smallest fragments is a 21 kb linear molecule that contains the ribosomal DNA (rDNA). The whole molecule is an inverted repeat, and is amplified to a level of about 200 copies per cell.(39,40) Sufficient quantities of this molecule can be purified so that a detailed biochemical analysis of its structure is possible. The structural details of the terminus of the Tetrahymena rDNA (see Figure 2d) were elucidated by E.H. Blackburn and J.G. Gall. (41) The very terminus of the molecule is probably a hairpin loop, although this has never been rigorously proven. Adjacent to the terminus is a long cluster of repeat units of the hexanucleotide sequence 5'-C-C-C-C-A-A-3', with the 3' hydroxyl towards the center of the molecule. The number of repeats varies from 20 to 70 on different molecules; thus terminal restriction fragments show a size heterogeneity of 2-300 bp. The rDNA molecule does not exist as a single closed strand; rather there are a series of single strand interruptions within the cluster of repeats, on both the C_4A_2 strand and the G_4T_2 strand. The interruptions in the C_4A_2 strand are well characterized (41); they are due to the absence of the first C of a C_4A_2 repeat unit. Several of these interruptions are spaced two to three repeat units apart, near the terminal (distal) side of the cluster. The interruptions on the G_4T_2 strand are not as well characterized chemically. Both of these classes of interruptions are substrates for the initiation of nick translation by DNA polymerase I. Since the interruptions occur near both ends of the cluster, nick translation with just two nucleotide triphosphates can proceed through the entire cluster.

The ends of other micronuclear DNA fragments from Tetrahymena (42) and the related holotrichous ciliate Glaucoma (43) appear to have a similar structure to the ends of Tetrahymena rDNA. The macronuclear fragments of hypotrichous ciliate such as Oxytricha end in C_4A_4 repeats. (44) The extrachromosomal rDNAs of Physarum (45) and Dictyostelium (46) have single strand interruptions within repeated CCCTA and $C_{1-8}T$ sequences respectively.

Tetrahymena Ends Function in Yeast

We have recently described experiments that demonstrate that the ends of the Tetrahymena ribosomal DNA plasmid function in yeast. (47) These experiments are summarized in this section.

Since we did not know in advance what would happen to the Tetrahymena rDNA ends in yeast, we attempted to devise an experimental protocol which would clearly tell us whether or not the ends would replicate in yeast, and if so, if they would maintain their original structure. This was done as follows. A circular yeast/E. coli shuttle vector (pSZ213) was constructed. This plasmid contained pBR322 sequences (48) for growth and selection in E. coli, the ars1 segment (49) and the LEU2 gene (50) for replication and selection in yeast, and the HIS3 gene. (51) The HIS3 gene was used to provide a convenient BglII restriction site, and for subsequent genetic purposes. All other BglII and

BamHI sites were removed by in vitro mutagenesis. DNA of pSZ213 was then linearized by digestion with BglII, cutting in the middle of the HIS3 gene. Meanwhile, Tetrahymena rDNA was purified, and a terminal 1.5 kb BamHI restriction fragment was purified by agarose gel electrophoresis.

The restriction cut vector DNA and the Tetrahymena rDNA end fragment were ligated to each other in the presence of both restriction enzymes (BamHI and BglII). This improves the efficiency of the reaction, since any end fragments that ligated to each other would be re-cut with BamHI, and any vector that circularized or dimerized would be re-cut with BglII. Only the desired reaction products, in which end fragments were ligated to the vector, could accumulate because the sequence formed by joining a BamHI site to a BglII site is not a substrate for either enzyme. After the ligation reaction, all of the DNA was re-cut to completion with both restriction enzymes, and then subjected to agarose gel electrophoresis. Three large fragments were visible : unreacted vector DNA, vector plus one end fragment, and vector plus two end fragments. The latter species was eluted from the gel, and the DNA (an estimated 2-5 ng) was used to transform a $leu2^-$ strain of yeast to $LEU2^+$.

The DNA species we had constructed was a linear molecule, with both ends derived from the ends of the rDNA plasmid of Tetrahymena. If these ends functioned in yeast, we expected to recover transformants carrying linear plasmids with the same structure as the transforming DNA. On the other hand, if the ends did not work in yeast, we expected to recover some type of circular plasmid. Circles could arise by the fusion of the two ends of one molecule, by the inability of a circular replication intermediate to resolve to a linear form, or by deletion of the ends of the molecule followed by circularization. We constructed and purified the linear molecules we used prior to transformation in order to simplify the array of plasmids that would be recovered after transformation.

Of the 15 transformants that were recovered, 14 were mitotically unstable, which suggested that they contained autonomously replicating plasmids, as opposed to chromosomally integrated information. DNA was prepared from 5 unstable transformants, and the structure of the plasmids they contained was determined by agarose gel electrophoresis and Southern blot restriction mapping. In every case, the plasmids had a linear restriction map, and the map was identical to that of the DNA used for transformation. Moreover, no molecules that were not linear could be detected in any of these strains. It was therefore clear that the ends of the Tetrahymena rDNA were in fact able to function as ends in yeast, and that they functioned very efficiently. This was quite a surprising result, since Tetrahymena and yeast are not closely related phylogenetically, but are in different kingdoms. (52) Apparently, the structural features required for function as a stable end have been highly conserved through evolution. This conclusion is futher supported by our initial results from the characterization of yeast telomeres (see below).

The linear yeast plasmid with two Tetrahymena ends (pSZ216) has a useful and unexpected property: its average copy number in a selectively grown culture is about 20 copies/cell. Plasmid DNA therefore represents about 0.5-1.0% of the mass of genomic DNA. The plasmid is visible on ethidium stained agrose gels of undigested genomic yeast DNA; this fact simplifies the initial analysis of new linear plasmids, since transformants can be rapidly screened for those containing plasmids of the correct size. A simplified procedure for the construction of new linear plasmids with Tetrahymena ends has recently been described. (53) Rather than constructing the desired plasmid entirely in vitro, the yeast vector is ligated with unpurified Tetrahymena ends, the ligated DNA is re-cut with restriction enzymes to destroy incorrect molecules, and molecules with two functional ends are selected by yeast transformation. Subsequent characterization is by agarose gel electrophoresis and Southern blot restriction mapping.

Once we knew that the Tetrahymena ends worked in yeast, we wished to know which of the unusual structural features of these ends were maintained during replication of the DNA in yeast. We examined the single strand interruptions within the C_4A_2 cluster, and the size heterogeneity of that cluster. (47) If the single strand interruptions had been sealed during replication of the Tetrahymena ends in yeast, the linear plasmid would have behaved as snapback DNA. This was not observed; indeed no snapback molecules were detectable. Subsequent experiments involving nick translation with various combinations of deoxynucleotide triphosphates revealed the presence of nicks or gaps within the C_4A_2 cluster. We do not yet know whether these interruptions have the same structure as the interruptions found on this DNA when it replicates in Tetrahymena. Heterogeneity in the size of the C_4A_2 cluster was tested by examination of small terminal restriction fragments in Southern blot experiments. The terminal 1.3 kb PstI fragment did not run as a sharp band on a gel, but exhibited a heterogeneity of about 200 bp; the average size of the terminal fragment varied from transformant to transformant, and most were about 200 bp larger than the mean size of the same fragment isolated directly from Tetrahymena. While the significance of neither the single strand interruptions nor the size heterogeneity of the C_4A_2 cluster is understood, it is clear that these features are maintained in yeast. It is therefore likely that these structural features in some way reflect the ability of this piece of DNA to function as a stable end.

To test the hypothesis that structural features in addition to the hairpin loop at the terminus of the molecule are required for function as an end, we constructed an artificial hairpin fragment out of a piece of lacZ DNA. A circular plasmid (pSZ211 (54)) was linearized by restriction enzyme digestion, and ligated to itself at high DNA concentration, so as to produce long linear concatemers in which half of the adjacent repeat units were in inverted orientation. Digestion of this DNA with a second restriction enzyme released a DNA fragment that was a perfect inverted repeat; this fragment was purified on a gel. Upon denaturation, these molecules formed two identical half-length molecules terminated at one end by a hairpin loop. These hairpin loop fragments with no sequence information in common with the Tetrahymena end fragment, were used to construct linear molecules consisting of pSZ213 with two hairpin ends by following the same procedure described above for constructing linear molecules with two Tetrahymena ends. When these linear molecules were used to transform yeast, only circular plasmids were recovered. These apparently arose by the initiation of bidirectional replication at some internal site, with the generation of a head to head dimer circle upon the completion of replication. The dimer circles then decayed by recombination between sequences in direct orientation to give monomer circles. One plasmid had the structure expected from precise replication of the entire hairpin loop; others had deletions in that area. Clearly, replication of a hairpin loop is possible in yeast, but additional information is required if the replicated DNA is to resolve into new hairpin loops.

Cloning Yeast Telomeres

The yeast linear plasmid pSZ216 transforms yeast quite efficiently; however, after digestion with a restriction enzyme that cuts the plasmid once, the transformation efficiency becomes extremely poor. Presumably molecules that have only one functional end are degraded in the cell. This provides a powerful selection for the replacement of a missing end, and we used this selection to clone yeast telomeres. Genomic DNA from a strain carrying the linear plasmid pSZ216 was digested with the restriction enzyme PvuI, and the fragments were ligated to each other. In some cases we expected the cut linear plasmid to ligate to a telomere fragment of yeast DNA, and thus regenerate a linear plasmid with two functional ends. A total of 56 mitotically unstable transformants were screened by examining the DNA for the presence of linear plasmids by agarose gel electrophoresis. Three plasmids were found that were smaller than the original vector and could not, therefore, have arisen by religation of the two halves of the vector. All three turned out to have a single PvuI fragment of yeast DNA at one end; in each case these fragments appeared to be identical to each other by restriction mapping.

In order to examine the origin of these yeast fragments, and to look more closely at the structure of normal yeast telomeres, it was necessary to subclone a portion of the yeast fragment in E. coli so that a telomere specific hybridization probe could be prepared. We were unable to recover the linear plasmids in E. coli by transforming directly with intact DNA. This problem was solved by the removal of the Tetrahymena end with one restriction enzyme, and the removal of part of the yeast end with a second enzyme. After circularization of the cut molecules with ligase, E. coli transformants were easily recovered. Southern blots were prepared from wild type untransformed yeast, and hybridized with the telomere specific probe. The probe sequences were moderately repeated in the genome (30-40 X), and complex restriction patterns were obtained. A PvuI digest revealed that a fragment of DNA the same size as the cloned end fragment existed in yeast. Analysis of other digests showed that a site could be defined that acted as a site for digestion with every enzyme we examined; this site corresponded with the end of the cloned yeast fragment, and we concluded that it represented the ends of the chromosomes that hybridized with our probe.

Further evidence that our probe was hybridizing to bona fide telomeres was obtained by comparison of two haploid strains: a wild type, and a ring chromosome III (kindly provided by A. Klar). The ring chromosome strain should lack two telomeres from chromosome III. Southern blot restriction analysis showed that this strain was missing at least one fragment when probed with the telomere probe; since not all fragments were resolved, it is possible that a second missing fragment was not detected.

Are Telomeres Snapback DNA?

Fangman et al. (55,56) have reported that a substantial fraction of carefully prepared high molecular weight yeast DNA behaves as snapback DNA in denaturation/renaturation experiments. They calculated that the presence of two crosslinks per chromosome would be necessary to explain their data, and hypothesized that these crosslinks were due to closed telomeric structures, i.e. continuous single strands crosslinked at the end of the chromosome by a hairpin loop. We have used our telomere specific probe to ask whether yeast telomeres do in fact behave as snapback DNA.

Preliminary experiments showed that the hybrid linear plasmid did not behave as a snapback molecule, suggesting the presence of nicks or gaps at or near the yeast end of the molecule as well as at the Tetrahymena end of the plasmid. Experiments with DNA polymerase I and various combinations of deoxynucleotide triphosphates showed that the yeast end contained single strand interruptions that were a substrate for the initiation of nick translation. Substantial levels of nick translation were obtained with dATP + dCTP, and dGTP + dTTP, just as with the Tetrahymena end.

We then used the telomere probe in a hybridization experiment to examine the telomeres of untransformed yeast. Genomic DNA of several different strains was digested with restriction enzymes (PvuI or HindIII) that gave a known pattern of bands after Southern blotting. Half of each sample was run on an agarose gel as native DNA; the other half was denatured by incubation at 100°C for 2', then run in parallel with the native sample. After Southern blot analysis, we always found that denaturation destroyed all of the bands found with native DNA, and the majority of the DNA seemed to run as full length single strands after denaturation. We have examined DNA from log phase as well as stationary phase cultures; no difference was seen.

There results are quite surprising because it is difficult to imagine how DNA replication could proceed past the single strand interruptions that are present in telomeric DNA. Apparently any molecules in which these lesions are sealed are of such a transient nature that we are unable to detect them by this hybridization analysis. The most likely explanation for the earlier results of Forte and Fangman is that yeast DNA contains a low level of spontaneous internal cross-links that are not related to telomeres.

We have screened several cdc mutants for defects in telomere replication and resolution. Strains were grown and DNA was prepared at 23° and at 37°. The DNA samples were analyzed as above with the telomere probe. Neither dimer-length restriction fragments nor snapback fragments were detected with cdc 2, 6, 13, 16, 20, 23 or 40 (Figure 3).

Telomere Resolution

If we assume that the single strand interruptions in telomere DNA are sealed prior to replication, then replication of a hairpin loop must give rise to an inverted repeat structure (see Figure 1). While we have not been able to detect restriction fragments characteristic of such inverted repeats in yeast, they are seen in the rDNA of Tetrahymena. An important question, then, is how these inverted repeat replication intermediates are resolved to yield hairpin loops at the termini of two new linear molecules. Two experimental approaches to these problems are possible. First, it may be possible to construct new linear plasmids containing progressively shorter fragments of either the Tetrahymena or the yeast telomere. Presumably at some point the efficiency with which these ends function will become impaired and it may be possible to detect replicated but unresolved molecules (inverted repeats) or molecules in which the nicks or gaps have been sealed. A second approach, which is underway in this laboratory, is to construct, in vitro, DNA molecules which are sufficiently similar to the hypothetical replication intermediate that they will be recognized and resolved by the cell. If such molecules can be generated, they may provide a suitable substrate for the in vitro study of the resolution reaction.

Models of Telomere Resolution

The Bateman model of telomere replication was the first to postulate the existence of an inverted repeat as a transient replication intermediate. (29) In this model, the inverted repeat is resolved by the introduction of symmetrically displaced nicks on either side of the center of symmetry; the DNA in between is then denatured, after which the overhanging single stranded ends fold back on themselves to form two new hairpin termini (see Figure 4A). An alternative model has been proposed (47) (Figure 4B) in which the inverted repeat is enzymatically transformed into a cruciform structure. The center of the cruciform is a Holliday junction; this could be enzymatically resolved just as during recombination. This model requires that the two halves of the inverted repeat be sufficiently homologous that pairing is possible. If artificial resolution substrates can be prepared in vitro, it may be possible to test this model directly by making an inverted repeat with imperfect symmetry.

Neither of these models fully explains the structural features of the Tetrahymena rDNA ends, or of the cloned yeast telomeres. In both cases it appears that there is a cluster of short repeat units near the end of the molecule, and that within that cluster are a set of single strand interrruptions on both strands of the duplex. The function of the repeats is not clear, although they could be involved in the initial stages of the formation of a cruciform intermediate, or as the recognition site of enzymes involved in the introduction of nicks or gaps required for resolution. The presence of interruptions on both strands, and the presence of more than one interruption on one strand are even less clear. No doubt models of telomere replication and resolution will become more realistic in the near future as it becomes possible to do more sophisticated experiments.

Acknowledgements

The author would like to thank E.H. Blackburn, C.J. Kenyon, T.L. Orr-Weaver and A.W. Murray for helpful discussions, and T. Claus for excellent technical assistance.

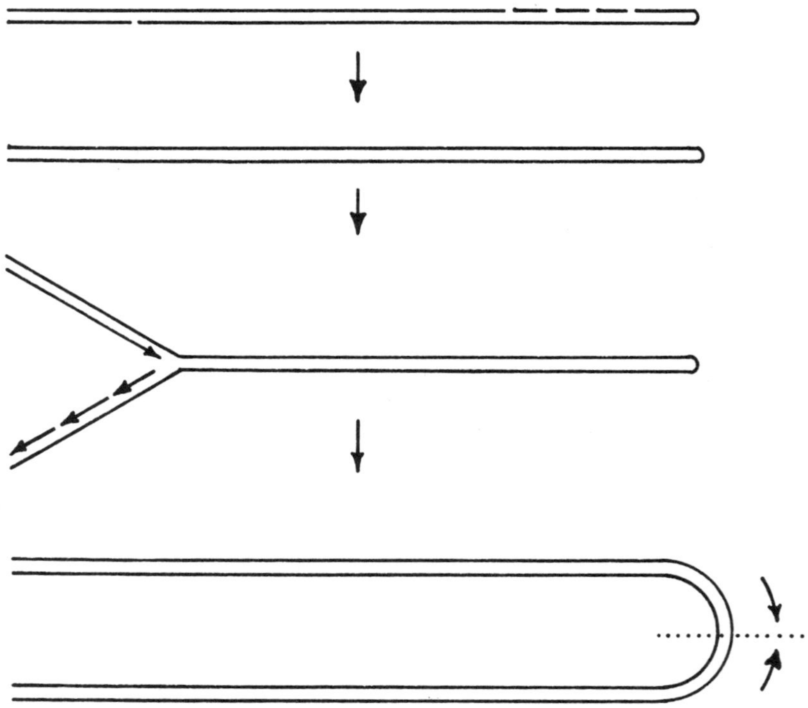

Figure 1. Replication of telomeric DNA. Single stranded interruptions are sealed to yield a closed hairpin loop. DNA replication, initiated from some internal site, proceeds towards the DNA terminus. After replication is complete, the hairpin loop has been transformed into an inverted repeat.

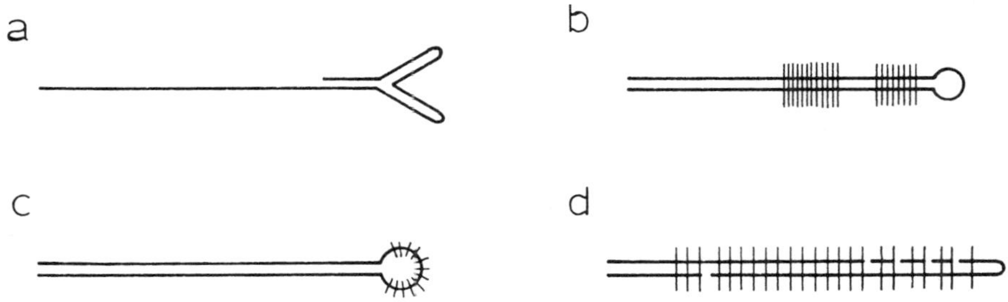

Figure 2. The ends of four linear DNA molecules. a)Adeno-associated viruses and parvoviruses are single stranded with a short duplex region at each end. Internal palindromes result in a Y shaped hairpin loop. b)Vaccinia virus has a 104 base unpaired hairpin loop at each end. Adjacent to the end are two clusters of a 70 base-pair repeat units. c)Paramecium mitochondrial DNA is crosslinked at the replication initiation end by a 350 base unpaired hairpin loop, composed of 11 repeats of 34 bases. d)Tetrahymena ribosomal DNA has a hairpin terminus, an adjacent cluster of C_4A_2 repeats, and a set of single stranded interruptions within the cluster of repeats.

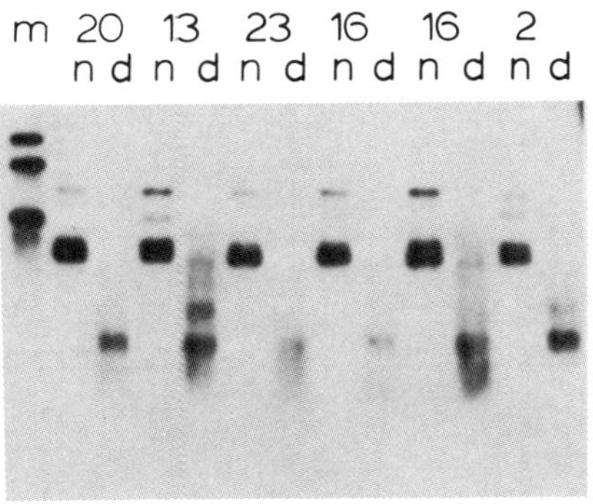

Figure 3. Non-snapback nature of yeast telomeres. Yeast DNA was prepared from cells grown at 23° to a density of 2 x 10^7/ml, then incubated at 37°C for 4 hours. The DNA was digested with the restriction enzyme HindIII, and electrophoresed on an 0.7% gel either native or denatured (2' at 100°C). The gel was blotted to nitrocellulose paper and hybridized with the telomere specific probe described in the text. All bands due to native telomere DNA are absent after denaturation, proving the non-snapback nature of yeast telomeres. Lanes: n, cdc20; 13, cdc13; 23, cdc23; 16, cdc16; 2, cdc2.

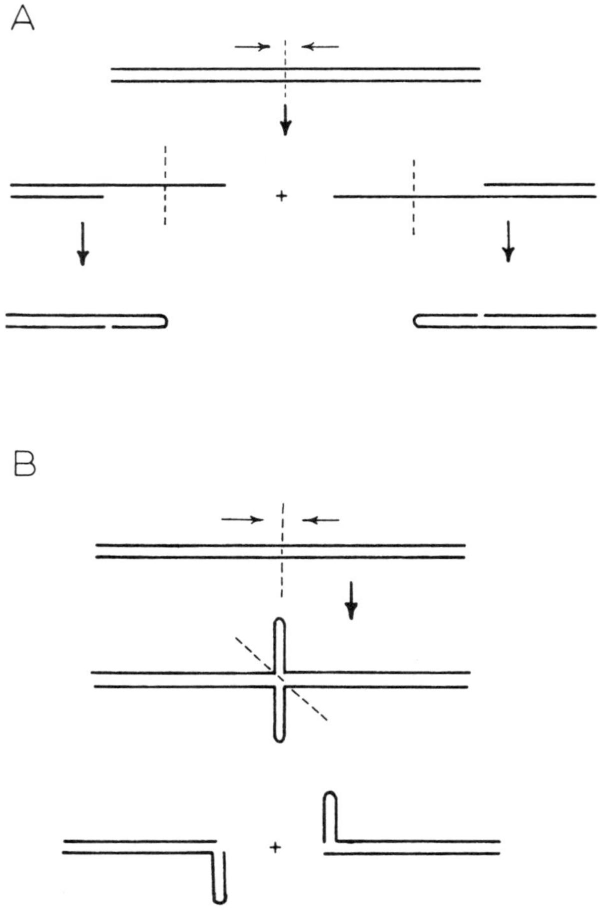

Figure 4. Models of telomere resolution. A. the Bateman model. Symmetrically displaced nicks are introduced on either side of the inverted repeat. After separation of the half-molecules, hairpin loops reform. B. An alternative model in which the inverted repeat is transformed into a cruciform structure, with a Holliday junction in the center. Resolution of the Holliday junction by recombination enzymes yields two half-molecules terminated in hairpin loops.

LITERATURE CITED

1. Petes, T.D. and Fangman, W.L. 1972. Sedimentation properties of yeast chromosomal DNA. Proc. Natl. Acad. Sci. USA 69:1188-1191.

2. Watson, J.D. 1972. Origin of concatameric T7 DNA. Nature New Biol. 239:197-201.

3. Sugino, A., Mirose, S. and Okazaki, R. 1972. RNA-linked nascent DNA fragments in Escherichia coli. Proc. Natl. Acad. Sci. USA 69:1863-1867.

4. Anderson, S., Kaufmann, F. and Depamphilis, M. 1977. RNA primers in SV40 DNA replication: identification of transient RNA-DNA covalent linkages in replicating DNA. Biochemistry 16:4990-4998.

5. Tseng, B.Y. and Goulian, M. 1977. Initiater RNA of discontinuous DNA synthesis in human lymphocytes. Cell 12: 483-489.

6. Muller, H.J. 1938. The remaking of chromosomes. The Collecting Net-WoodsHole 13:181:98.

7. McClintock, B. 1941. The stability of broken ends of chromosomes in Zea mays. Genetics 26: 234-282.

8. McClintock, B. 1942. The fusion of broken ends of chromosomes following nuclear fusion. Proc. Natl. Acad. Sci. USA 28:458-463.

9. Resnick, M.A. and Martin, P. 1976. The repair of double-strand breaks in the nuclear DNA of Saccharomyces cerevisiae and its genetic control. Molec. Gen. Genet. 143:119-129.

10. Resnick, M.A. 1976. The repair of double-strand breaks in DNA: A model involving recombination. J. Theor. Biol. 59:97-106.

11. Orr-Weaver, T.L., Szostak, J.W. and Rothstein, R.J. 1981. Yeast transformation: A model system for the study of recombination. Proc. Natl. Acad. Sci. USA 78:6354-6358.

12. Stinchcomb, D.T., Struhl, K. and Davis, R.W. 1979. Isolation and characterization of a yeast chromosomal replicator. Nature 282:39-43.

13. Orr-Weaver, T.L. and Szostak, J.W. 1982. manuscript in preparation.

14. Smith, J.M. 1978. The Evolution of Sex. Cambridge University Press, Cambridge.

15. McClintock, B. 1938. The production of homozygous deficient tissues with mutant chracteristics by means of the aberrant mitotic behavior of ring-shaped chromosomes. Genetics 23:315-376.

16. Strathern, J.N., Newton, C.S., Herskowitz, J. and Hicks, J.B. 1979. Isolation of a circular derivative of yeast chromosome III: Implications for the mechanism of mating type interconversion. Cell 18:309-319.

17. Clarke, L. and Carbon, J. 1980. Isolation of a yeast centromere and construction of functional small circular chromosomes. Nature 287:504-509.

18. Dancis, B.M. and Holmquist, G.P. 1979. Telomere replication and fusion in eukaryotes. J. Theor. Biol. 78:211-224.

19. Byers, B. 1982. Cytology of the yeast life cycle, p. 59-96. In: J.N. Strathern, E.W. Jones and J.R. Broach (ed.), The Molecular Biology of the Yeast Saccharomyces: Life Cycle and Inheritance. Cold Spring Harbor Laboratory, Cold Spring Harbor, New York.

20. Newmeyer, D. and Galeazze, D.R. 1977. The Instability of Neurospora duplication bp (IL-IR) H4250, and its genetic control. Genetics 85:461-487.

21. Pardue, M.L., Brown, D.D. and Birnsteil, M.L. 1973. Location of the genes for 5S Ribosomal RNA in Xenopus laevis. Chromosoma 42:191-203.

22. Jones, K.W., 1970. Chromosomal and nuclear location of mouse satellite DNA in individual cells. Nature 225:912-915.

23. Pardue, M.L. and Gall, J.G. 1970. Chromosomal localization of mouse satellite DNA. Science 168:1356-1357.

24. Eckhardt, R.A. and Gall, J.G. 1971. Satellite DNA associated with heterochromatin in Rhynchosciara. Chromosoma 32:407-427.

25. Stack, S.M., Clarke, C.R., Cary, W.E. and Muffly, J.T. 1974. Different kinds of heterochromatin in higher plant chromosomes. J. Cell. Sci. 14:499-504.

26. Rubin, G. 1977. Isolation of a telomeric DNA Sequence from Drosophila melanogaster. Cold Spring Harbor Symp. Quant. Biol. Vol. XLII:1041-1046, Cold Spring Harbor, New York.

27. Tamanoi, F. and Stillmann, B.W. 1982. Function of Adenovirus terminal protein in the initiation of DNA replication. Proc. Natl. Acad. Sci. USA 79:2221-2225.

28. Berns, K.I., Hauswirth, W.W., Fife, K.H. and Lusby, F. 1979. Adeno-associated virus DNA replication. Cold Spring Harbor Symp. Quant. Biol. Vol. XLIII p. 781-789. Cold Spring Harbor Press, Cold Spring Harobr, New York.

30. Cavalier-Smith, T., 1974. Palindromic base sequences and replication of eukaryote chromosome ends. Nature 250:467-470.

31. Garon, C.F., Barbosa, F. and Moss, B., 1978. Visualization of an inverted terminal repetition in Vaccinia virus DNA. Proc. Natl. Acad. Sci. USA 75:4863-4867.

32. Geshelin, P. and Berns, K.I., 1974. Characterization and localization of the naturally occurring cross-links in Vaccinia virus DNA. J. Mol. Biol. 88:785-796.

33. Baroudy, B.B., Venkatesan, S. and Moss, B. 1982. Incompletely Base-paired flip-flop terminal locps link the two DNA strands of the Vaccinia virus genome into one uninterrupted polynucleotide chain. Cell 28:315-324.

34. Wittek, R. and Moss, B., 1980. Tandem Repeats within the inverted terminal repetition of Vaccinia virus DNA. Cell 21:277-284.

35. Moss, B., Winters, E., and Cooper, N., 1981. Instability and reiteration of DNA sequences within the Vaccinia virus genome. Proc. Natl. Acad. Sci. USA 78:1614-1618.

36. Goddard, J.M. and Cummings, D.J. 1977. Mitochondrial DNA replication in Paramecium aurelia. Cross-linking of the initiation end. J. Mol. Biol. 109:327-344.

37. Pritchard, A.E., Herron, L.M. and Cummings, D.J., 1980. Cloning and Characterization of Paramecium mitochondrial DNA replication initiation regions. Gene 11:43-52.

38. Pritchard, A.E. and Cummings, D.J., 1980. Replication of linear mito-chondrial DNA from Paramecium: Sequence and structure of the initiation end cross-link. Proc. Natl. Acad. Sci. USA 78:7341-7345.

39. Karrer, K.M. and Gall, J.G., 1976. The macronuclear ribosomal DNA of Tetrahymena pyriformis is a palindrome. J. Mol. Biol. 104:421-453.

40. Engberg, J., Andersson, P., Leick, J. and Collins, J. 1976. Free ribosomal DNA molecules from Tetrahymena pyriformis GL are giant palindromes. J. Mol. Biol. 104:455-470.

41. Blackburn, E.H. and Gall, J.G., 1978. A tandemly repeated sequence at the termini of the extrachromosomal ribosomal RNA genes in Tetrahymena. J. Mol. Biol. 120-33-53.

42. Yao, M.C. and Yao, C.H., 1981. Repeated hexanucleotide C-C-C-C-A-A is present near free ends of macronuclear DNA of Tetrayhymena. Proc. Natl. Acad. Sci. USA 78:7436-7439.

43. Katzen, A.L., Cann, G.M. and Blackburn, E.H., 1981. Sequence specific fragmentation of macronucelar DNA in a holotrichous ciliate. Cell 24: 313-320.

44. Klobutcher, L.A., Swanton, M.T., Donini, P. and Prescott, D.M., 1981. All gene-sized DNA molecules in four species of hypotrichs have the same terminal sequence and an unusual 3' terminus. Proc. Natl. Acad. Sci. USA 78:3015-3019.

45. Johnson, E.M., 1980. A family of inverted repeat sequences and specific single-strand gaps at the termini of the Physarum rDNA palindrome. Cell 22:875-886.

46. Emery, H.S. and Weiner, A.M., 1981. An irregular satellite sequence is found at the termini of the linear extrachromosomal rDNA of Dictyostelium discoideum. Cell 26:411-419.

47. Szostak, J.W. and Blackburn, E.H., 1982. Cloning Yeast telomeres on linear plasmid vectors. Cell 29:245-255.

48. Bolivar, F., Rodriquez, R.L., Greene, P.J., Betlack, M.C., Heyneker, M.L., Boyer, M.W., Crosa, J.M. and Falkow, S., 1977. Construction and characterization of new cloning vehicles II. A multipurpose cloning system. Gene 22:95-113

49. Stinchcomb, D.T., Struhl, K. and Davis, R.W., 1979. Isolation and characterization of a yeast chromosomal replicator. Nature 282:39-43.

50. Ratzkin, B. and Carbon, J., 1977. Functional expression of cloned yeast DNA in Escherichia coli. Proc. Natl. Acad. Sci, USA 74:487-491.

51. Struhl, K., 1980. A physical, genetic and transcriptional map of the cloned his3 gene region of Saccharomyces cerevisiae. J. Mol. Biol. 136:309-332.

52. Whittaker, R.H., 1969. New concepts of kingdoms of organisms. Science 163:150-160.

53. Szostak, J.W., 1982. A rapid procedure for the construction of linear yeast plasmids. Methods in Enzymology, in press.

54. Ruby, S.W., Szostak, J.W. and Murray, A.W., 1982. Cloning regulated yeast genes from a pool of lacZ fusions. Methods in Enzymology, in press.

55. Forte, M.A. and Fangman, W.L., 1978. Naturally occuring cross-links in yeast chromosomal DNA. Cell 8:425-431.

56. Forte, M.A. and Fangman, W.L., 1979. Yeast Chromosomal DNA Molecules have strands which are cross-linked at their termini. Chromosoma 72:131-150.

Part II

Gene Regulation, Transposons, and Secretion

8. THE ROLE OF SITE-SPECIFIC RECOMBINATION IN EXPRESSION OF THE YEAST PLASMID 2 MICRON CIRCLE*

James R. Broach

Department of Microbiology
State University of New York
Stony Brook, New York 11794

Rec. Adv. Yeast Mol. Biol. (1982) 1:93-112

The yeast plasmid 2 micron circle contains a specialized recombination system. There are two regions within the plasmid, each 599 bp in length, which are precise inverted repeats of each other and between which recombination readily occurs in yeast. This recombination, which requires a protein encoded in the plasmid itself at a locus designated FLP, establishes and maintains the coexistence within the cell of two distinct forms of the plasmid, designated A and B. These two forms differ structurally only in the orientation of one unique region with respect to the other. We provide evidence in this paper that FLP mediated recombination between the repeats is site specific. That is, the recombination does not occur throughout the inverted repeat but is limited to a specific sequence of less than sixty base pairs within the center of the repeat. In addition, the ability of a repeat to serve as substrate in this recombination is abolished by a four base pair deletion at the XbaI site in the middle of the repeat. This site lies at the center of an extended dyad symmetry and the possible role of this symmetrical feature in recognition by FLP protein is discussed. We previously proposed that FLP mediated recombination serves as a genetic switch. That is, we suggested that the transcriptional products of the two forms are different. Therefore, converting one form to the other by FLP mediated recombination would alter the composition of 2 micron circle transcripts in the cell. We have investigated this possibility by examining the transcriptional products of various Flp$^-$ plasmids which are frozen in one orientation or the other or which contain large insertions at various locations within the genome. The results of this analysis, presented in this paper, argue that the transcriptional pattern of the plasmid is not altered by interconversion and thus that FLP mediated recombination does not serve to modulate transcription.

INTRODUCTION

The yeast plasmid 2 micron circle is a 6318 bp double stranded, circular DNA species present in most *Saccharomyces* strains at 50 to 100 copies per cell (12, 14). The plasmid has an unusual structure in that it contains two regions, each 599 bp in length, which are precise inverted repeats of each other and which separate the molecule into two unique regions of approximately equal size (12). In yeast recombination readily occurs between these two repeated sequences, which leads to the inversion of one unique region with respect to the other (1). As a consequence, 2 micron circle exists in yeast as a mixed population of two distinct plasmids which differ in the relative orientation of the two unique regions. These two forms have been variously designated A and B, XY' and XY, 14 and 23, or R and L (4). In mitotic cells,

*This work was supported by NIH grant GM27929.

detectable recombination between the repeat sequences requires a protein encoded in the plasmid itself, at a locus which we have designated *FLP* and which corresponds to the A coding region of Hartley and Donelson (5, 11, 12). In cells lacking *FLP* activity, no recombination between the inverted repeats of 2 micron circle is detected. Thus general mitotic recombination makes at most a minimal contribution to the interconverion of 2 micron circle.

No cellular phenotype has been convincing associated with the presence of 2 micron circles in yeast (4). Strains lacking 2 micron circles, designated [cir^{o}] strains, have been obtained either as natural isolates of standard laboratory strains or through competetive exclusion using high copy number hybrid 2 micron circle plasmids (9, 13, 16). Extensive analysis of isogenic [cir^{o}] and [cir^{+}] strains has failed to demonstrate any effect of the presence of 2 micron circles on growth rate or extent, mating ability or mating type interconversion, sporulation, or drug resistance (4). Nonetheless, 2 micron circle contains several genes in addition to *FLP* which are expressed in yeast. Two of these correspond to coding regions B and C and have been designated *REP1* and *REP2*. Mutations in either of these genes prevent high copy number propagation of the plasmid. A third locus involved in high copy number propagation has been localized in the large unique region (M. Jayaram, Y-Y. Li, and J. Broach, manuscript in preparation). This locus, designated *rep3*, does not correspond to a coding region and is active only in cis. The location of the three 2 micron circle genes as well as the origin of replication of the molecule are indicated in Figure 1. Thus, although the plasmid does not provide a detectable cellular phenotype, a large portion of its genome is expressed in yeast, yielding activities apparently involved exclusively with its own maintenance.

On the basis of the transcriptional products of 2 micron circle in yeast, we previously proposed that *FLP* mediated interconversion in 2 micron circle is a requisite component of the expression of the plasmid (6). Specifically, we suggested that the transcriptional products of the two different forms of the plasmid are different. This is diagrammed in Figure 2. The polyA containing RNA species transcribed from 2 micron circles in yeast and the genomic locations from which each of them arise are indicated in Figure 2A. A scheme by which the all of these transcripts can be generated by cleavage of two primary transcripts, one from form A and one from form B, is shown in Figure 2B. In this model, transcription is initiated in the large unique region in both forms of the plasmid and continues through the inverted repeat into and through the smaller unique region. Since the smaller unique region of form A is in the opposite orientation with respect to the larger unique region than that of form B, the form A primary transcript will correspond to a different strand of the smaller unique region than that from form B. Thus transcription of form A and of form B yields different RNA species. Since interconversion would maintain both forms of the plasmid in a cell, each cell would possess both primary transcripts and thus a complete set of 2 micron circle mRNAs. Such a model would allow for the possibility that expression of the plasmid could be modulated through *FLP* mediated recombinational interconversion.

We have examined the extent to which the transcription model described in Figure 2 is correct by determining the *in vivo* transcriptional patterns of various hybrid 2 micron circle plasmids. These have included plasmids frozen in one orientation or the other or plasmids which contain insertions or deletions near the proposed primary promoter in the large unique region. The

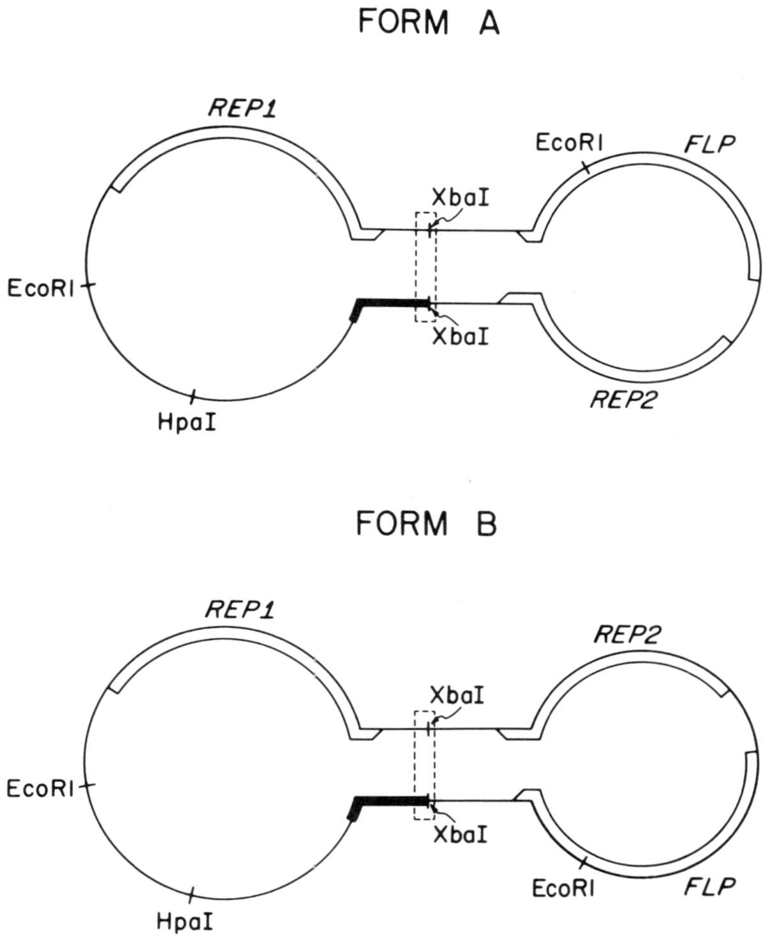

Figure 1. The Yeast Plasmid 2 Micron Circle.

On a schematic diagram of the two forms of the 2 micron circle plasmid are indicated the locations and extents of the three identified plasmid genes (open bars; the taper in each case lies at the 3' end of the gene), the origin of replication (filled bar), and the region within which FLP mediated, site specific recombination occurs (dashed box). The circular portions of the figures represents unique sequences while the linear portions correspond to the inverted repeat sequences.

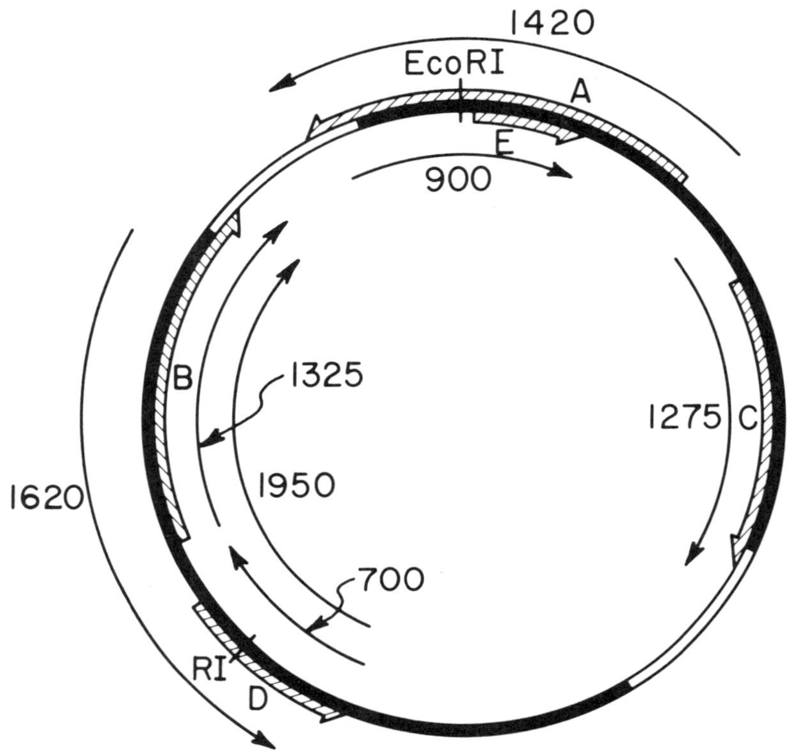

Figure 2. Transcription of 2 Micron Circle.

A. The positions from which a number of cellular polyA containing RNA species are transcribed in vivo from the 2 micron circle are indicated on a diagram of the A form of the plasmid. The transcripts are identified by their lengths in bases. The arrowheads lie at the 3' end of the RNA species and thus denote the direction of transcription of each of the species. The locations and orientations of the five largest open coding regions of the plasmid are indicated by the hatched lines abutting the circle. Regions A, B, and C, correspond to FLP, REP1, and REP2, respectively. No phenotype has been observed for mutations within coding region D. The inverted repeat regions of the plasmid are represented by the unfilled portion of the circle.

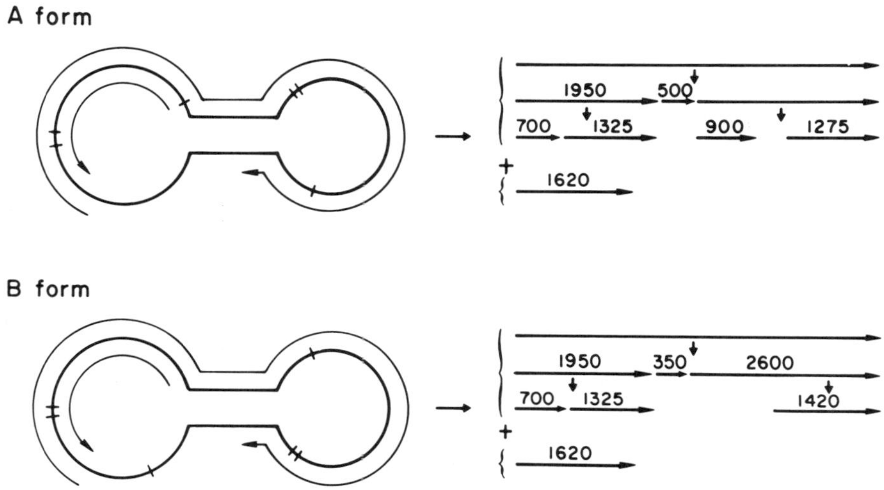

Figure 2. Transcription of 2 Micron Circle.

B. A scheme by which all two micron circle transcripts could be derived by transcription initiated at only two sites on the plasmid is shown. On the left the locations and orientations of postulated primary transcripts from the two forms of the plasmid are superimposed on dumbell shaped representations of the molecules. The EcoRl and HindIII sites are denoted by hatch marks. On the right a pathway by which the hypothetical primary transcripts could be processed to yield all the observed 2 micron circle polyA RNA species is presented.

results from these initial studies are presented in this paper and suggest that the model as proposed is incorrect. The limitations of these experiments are discussed and other potential roles for FLP mediated recombination within 2 micron circle are proposed.

MATERIALS AND METHODS

Isolation of Tn5 Insertions.

Random insertions of Tn5 into plasmid CV20 were obtained by the procedure described by Ruvkun and Ausubel (15). *E. coli* strain JE5507 (*man aroD arg lac spc str gal* su^+) harboring plasmid CV20 was infected with λ::Tn5 (λ*b*221 *rex*::Tn5 *c*I857 *O*am8 *P*am29) at 32° and kanamycin resistant clones were recovered at a frequency of 10^{-5}. Since λ::Tn5 can neither lysogenize nor replicate in JE5507, kanamycin resistant clones can arise only by transposition of Tn5 to either the bacterial chromosome or to the resident plasmid. We reovered members of the latter class by isolating DNA from the pooled kanamycin resistant clones and using it to transform JE5507 to ampicillan and kanamycin resistance. Individual clones were retained, and the location of Tn5 insertion in indivdual plasmids was determined by appropriate restriction enzyme digestion of plasmid DNA isolated from small overnight cultures (3).

Plasmid Constructions.

Plasmid CV20 was derived from plasmid CV19, whose construction has been previously described (5), following propagation of plasmid CV19 in a [cir^+] yeast strain and recovery of individual plasmids in *E. coli*. In one such plasmid, the large unique region of CV19 had been replaced, in the opposite orientation, with that from an endogenous plasmid. Thus, although CV19 has only a single *Eco*R1 site in the 2 micron circle moiety, CV20 has the normal two.

The constructions of plasmids HΔ24, XHinΔ1 and HinΔ14 have been previously described (5). We constructed plasmids XHinΔ2 and and HXΔ by cloning the appropriate *Xba*I restriction fragment of 2 micron circle onto a variant of plasmid CV03, in which the *Bam*H1 site had been converted into an *Xba*I site. Plasmid XΔ2-1 was constructed by nuclease S1 digestion of *Xba*I digested CV7 DNA, followed by religation and recovery in *E. coli*.

Miscellaneous Methods.

E. coli and yeast transformation, preparation of DNA from yeast and *E. coli*, restriction analysis and Southern hybridization were performed as described previously (7). Isolation of polyA containing RNA from yeast, fractionation of RNA on denaturing agarose gels, transfer to diazotized (DBM) paper, and hybridization procedures were as before (6).

RESULTS

2 Micron Circle Recombination Is Site-Specific.

In order to investigate the mechanism of FLP mediated recombination in 2 micron circles, we isolated a set of insertion derivatives of a hybrid plasmid, CV20. Plasmid CV20 consists of pBR322, the LEU2 gene of yeast, and the entire 2 micron circle genome. The pBR322 plus LEU2 sequences are cloned into the EcoR1 site in the small unique region (cf. Figure 1). Since the FLP gene spans this site, CV20 is flp⁻ and thus does not interconvert readily in [cir^{o}] strains. However, in [cir^{+}] strains recombination between the two inverted repeats on the plasmid readily occurs. Isolation of derivatives of CV20, each containing a random insertion of the bacterial transposon Tn5 into the 2 micron circle moiety of the plasmid, was accomplished using a procedure developed by Ruvkun and Ausubel and described in Materials and Methods. The locations of the Tn5 insertion in a number of isolates were determined by restriction enzyme analysis, and those plasmids containing an insertion within an inverted repeat were retained for analysis of FLP mediated recombination as described below.

The rationale of the procedure for determining whether FLP mediated recombination is site-specific and, if so, for localizing that site is diagrammed in Figure 3. In the center plasmid of part A is a diagram of the 2 micron circle moiety of a CV20 plasmid containing an insertion of Tn5 in the middle of an inverted repeat. For clarity, the pBR322 plus LEU2 sequences, which are present in the EcoR1 site in the small unique region, have been omitted. Digestion of this plasmid with EcoR1 plus BamH1 yields three restriction fragments, the electrophoretic fractionation of which are schematically diagrammed in part B. Those fragments containing sequences homologous to Tn5 are indicated on the diagram. As shown in the figure, FLP mediated recombination occurring to the left of the Tn5 insertion in the plasmid I (site a) would generate plasmid II. Thus the cell in which such recombination occurs would contain a mixture of plasmids I and II. Digestion of this mixture of plasmids with EcoR1 plus BamH1 would yield five restriction fragments, as shown in part B, three of which contain Tn5 sequences. If on the other hand, FLP mediated recombination occurred only to the right of the insertion (at site b), then a Flp^{+} cell in which plasmid I was introduced would generate a mixture of plasmids I and III. The digestion of this plasmid mixture with EcoR1 plus BamH1 would also yield five fragments, three of which contain Tn5 sequences. However, as is indicated in the figure, the restriction digestion pattern of the combination of plasmids I and III is clearly distinguishable from that of plasmids I and II. Finally, if FLP mediated recombination can occur anywhere throughout the inverted repeat, then a Flp^{+} cell in which plasmid I was introduced would contain a mixture of all three plasmids. Digestion of this mixture of plasmids would yield four Tn5 containing fragments. Thus by determining the number of Tn5 containing restriction fragments present in a Flp^{+} strain harboring a pCV20 plasmid with a Tn5 insertion in the inverted repeat, we can determine whether FLP mediated recombination is site specific. In addition, if recombination is site specific, then by assessing the relative sizes of these Tn5 containing restriction fragments, we can determine whether the site lies to the right or left of the Tn5 insertion.

The results of this analysis are presented in Figure 4. Individual derivatives of pCV20, each containing an insertion of Tn5 at the location indicated in part A, were used to transform isogenic [cir^{o}] and [cir^{+}] leu2 yeast strains to leucine prototrophy. Total cellular DNA was isolated from the transformants and digested with EcoR1 plus BamH1. The digested DNA was fractionated by electrophoresis on agarose gels, transfered to nitrocellulose, and probed with labeled Tn5 DNA. The autoradiogram of the washed filter for

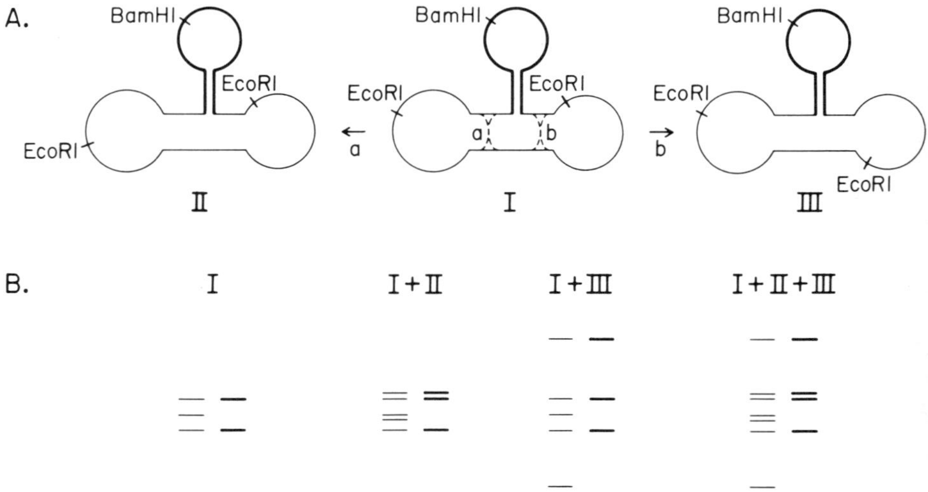

Figure 3. Scheme for Determining the Site of Recombination in 2 Micron Circle.

In the center of the upper (A) portion of the figure is shown the B form of 2 micron circle (thin line) containing a Tn5 insertion (heavy line) within the inverted repeat (plasmid I). The locations of the EcoR1 and BamH1 sites are indicated. In plasmid pCV20 used in this study, pBR322 plus LEU2 sequences are inserted as an EcoR1 fragment in the EcoR1 site in the right hand loop. For clarity, this portion of the molecule has been omitted from the figure. To the left and right are diagrammed the plasmids (II and III) which would arise following recombination between the inverted repeats of 2 micron circle, occurring either to the left or right of the Tn5 insertion, respectively. In the lower (B) portion of the figure are schematic representations of an agarose gel fractionation of restriction fragments which would result from a BamH1 plus EcoR1 digestion of a mixture of the plasmids indicated above each pair of tracks. For each set, the track on the left shows the position of all the fragments which would arise from the digestion. The track on the right shows only those fragments in the digestion which would contain Tn5 sequences.

two of these insertion derivatives is shown in part B. As previously described, no recombination products are observed after propagation of the two plasmids in the [cir^o] strain (track Flp^-). Thus, in the absence of the FLP gene product, recombination does not occur between the inverted repeats of 2 micron circle. However, after propagation of the plasmids in the [cir^+] strain, recombination products are clearly evident (track Flp^+). Significantly, in both cases only one additional Tn5 containing fragment is present. Thus, FLP mediated recombination does not occur throughout the inverted repeat but apparently is restricted to limited region within the inverted repeat. The pattern of Tn5 containing restriction fragments from plasmid pCV20::Tn5 #186, following propagation in the Flp^+ strain, indicates that the site of FLP recombination lies to the left of the Tn5 insertion in this plasmid. This is true for all of the plasmids containing a Tn5 insertion to the right of that in pCV20::Tn5 #186 (data not shown). Similarly, the pattern of Tn5 restriction fragments obtained from the Flp^+ strain containing pCV20::Tn5 #95 indicates that recombination occurs to the right of the Tn5 insertion in this plasmid. Again, this is true for all plasmids containing an insertion to the left of that in pCV20::Tn5 #95 (data not shown). Thus, the site of FLP mediated recombination is delimited to the region bracketed by Tn5 insertion #95 and #186.

We have obtained confirmation of the site specificity of FLP recombination as well as a further definition of the location and extent of the site by analysis of the ability of various hybrid plasmids, each containing different portions of an inverted repeat from 2 micron circle, to participate in FLP mediated recombination. In Figure 5 we present a diagram of various deletion derivatives of plasmid CV7. Plasmid CV7 consists of pBR322, the LEU2 gene, and the small EcoR1 fragment from the B form of 2 micron circle, which spans one of the inverted repeats as well as the 2 micron circle origin of replication. Various deletions extending into the inverted repeat sequence of this plasmid were constructed as described in Materials and Methods and the extent and location of the deletion in each of these derivatives is indicated in Figure 5.

The ability of each of these plasmids to serve as a substrate in FLP mediated recombination was assessed using two different assays. In the first assay we determined whether the deleted plasmid could recombine in yeast with resident, intact 2 micron circles, following transformation of [cir^+] leu2 strains. Those plasmids with a deletion extending into the inverted repeat from the left can only transform [cir^+] strains by recombining with the endogenous circles, since these deletions remove the 2 micron circle replication origin carried on CV7 (5). Thus, the ability of these plasmids to participate in FLP recombination could be recognized merely by determining whether or not they could transform a leu2 [cir^+] strain to leucine prototrophy at high frequency. However, for those plasmids with deletions internal to the inverted repeat or with deletions which enter the inverted repeat from the right, transformation is independent of the recombinational potential of the plasmid. Nonetheless, the extent to which each of the plasmids could recombine with a resident circle could be measured physically. This was accomplished by isolating DNA from a transformant, obtained with the plasmid in question, and determining by Southern analysis whether a species corresponding to a hybrid between the transforming plasmid and a resident circle could be detected.

(a)

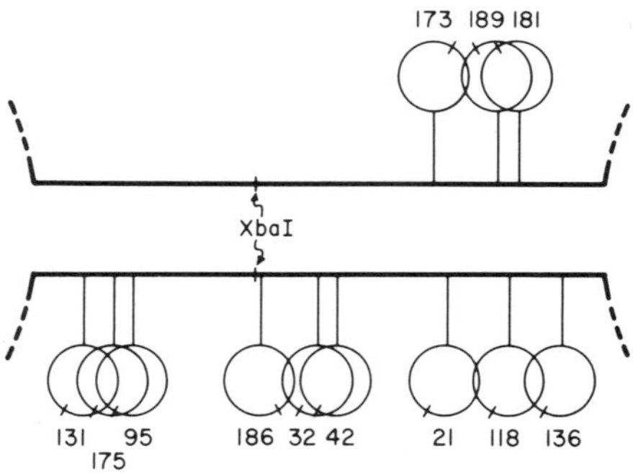

(b)

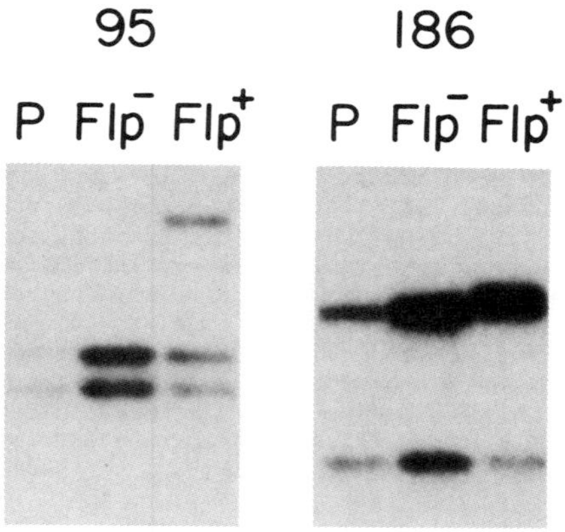

Figure 4. Localization of the Site of 2 Micron Circle Recombination.

A. The positions of individual Tn5 insertions used in this study are indicated as a composite of lollipop structures on a schematic representation of the two inverted repeats of 2 micron circle (heavy line). The large unique region of 2 micron circle lies to the left. The position of the XbaI restriction site in the inverted repeat is indicated. In addition, the position of the single, assymmetric BamH1 site in each of the Tn5 insertions is indicated by the slash in the circular portion of the lollipop.

B. After propagating plasmid CV20, containing the designated Tn5 insertion, either in the yeast strain DC04 [cir^o] (track Flp^-) or in strain DC04 [cir^+] (track Flp^+), total DNA was isolated from the strains, digested with EcoR1 plus BamH1, fractionated on a 0.8% agarose gel, transferred to nitrocellulose, and probed with labelled λ::Tn5 DNA. For reference purposes, purified plasmid DNA was also digested with BamH1 plus EcoR1 and fractionated on the same gel (lane P). The resulting autoradiogram of the filter is shown. The upper band in track Flp^+ for #186 is a doublet, which is clearer on a shorter exposure of the autoradiogram.

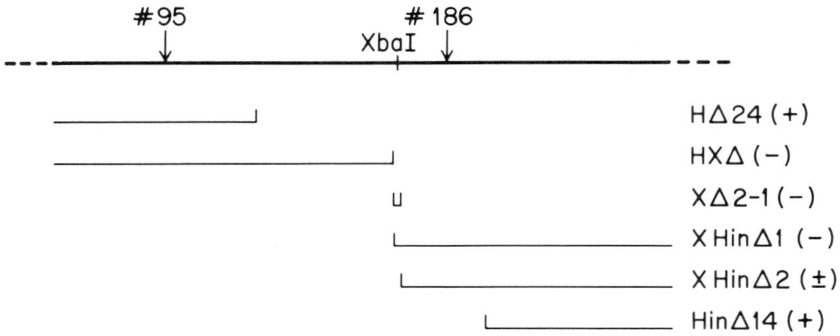

Figure 5. Localization of the Site of 2 Micron Circle Recombination by Deletion Analysis.

The heavy line at the top of the figure represents approximately 200 bp spanning the XbaI site in the 2 micron circle moiety of plasmid CV7. The positions at which Tn5 insertions #186 and #95 map in the equivalent region of plasmid CV20 are indicated. Shown below the line are those sequences deleted in the plasmids designated on the right. In parenthesis next to each plasmid designation is a "+" or "−" to indicate whether that plasmid retains or has lost the site for recombination as determined by the assay described in the text.

The second assay to determine the extent to which each deletion plasmid is a subtrate for FLP recombination was based on the recent observation that chromosomal integration of a segment of 2 micron circle spanning an inverted repeat causes a marked instability of the chromosome in which the insertion occurs (10). This instability is absolutely dependent upon FLP activity and the initial event in chromosome loss is a FLP-dependent recombination between the integrated plasmid and a resident 2 micron circle. Thus, by integrating each of the plasmids diagrammed in Figure 5 at leu2 (located on chromosome III) and then determining the frequency with which chromosome III is subsequently lost, we could could assess the extent to which each plasmid could participate in FLP recombination.

Both assays for the recombination potential of the various plasmids yielded equivalent, unequivical results, which are summarized to the right in Figure 5. Plasmids HΔ24 and HinΔ14 can participate in FLP mediated recombination to the same extent CV7, which contains an intact inverted repeat. All other deletion plasmids tested failed to display any activity as a substrate for FLP recombination. Thus, these results confirm that FLP recombination occurs at a specific site. In addition, we conclude that this site lies to the right of deletion HΔ24 and to the left of deletion HinΔ14, a site consistent with that defined by the Tn5 insertions. Finally, it is clear that the integrity of the sequences immediately spanning the XbaI site in the inverted repeat is crucial for FLP recombination, since a small deletion around this site completely abolishes recombination activity.

Recombination Within 2 Micron Circle Does Not Affect Expression of the Plasmid.

The Patterns of In Vivo Transcription From the A and B Forms of 2 Micron Circle Are Identical. We have previously proposed that transcription of the two forms of 2 micron circle yields different RNA species and that FLP mediated recombination promoted the interconversion of the two forms of 2 micron circle as a means of alternating between these transcriptionally distinct states (6). To test this hypothesis, we examined the in vivo transcriptional products of plasmids CV19 and CV20 in a [cir^o] strain. Plasmids CV19 and CV20 are essentially identical except that the 2 micron circle genome carried on CV19 is in the A form while that of CV20 is in the B form. In both plasmids, however, pBR322 sequences interrupt the FLP gene. Thus both plasmids are flp and, as a consequence, each persists in its original orientation during propagation in [cir^o] yeast strains. We isolated polyA containing RNA from [cir^o] strains in which these plasmids were separately resident. These RNA samples were fractionated on denaturing agarose gels and plasmid specific transcripts were identified by hybridization with labeled 2 micron circle DNA following transfer of the RNA to DBM paper. These results are presented in Figure 6. As can be seen, the same RNA species complementary to 2 micron circle are present in both strains. Clearly evident in the exposure shown are the two relatively abundant transcripts from 2 micron circle -- namely, the 1325 and 1275 base RNAs, which are derived from REP1 and REP2, respectively -- and a larger trancript, which originates from within the pBR322 sequences carried on the two plasmids (unpublished observations). Longer exposures of the filter or hybridization using probes to specific regions of the 2 micron circle genome has failed to reveal any differences in the transcriptional products from the two plasmids. Thus the transcriptional products of CV19 and

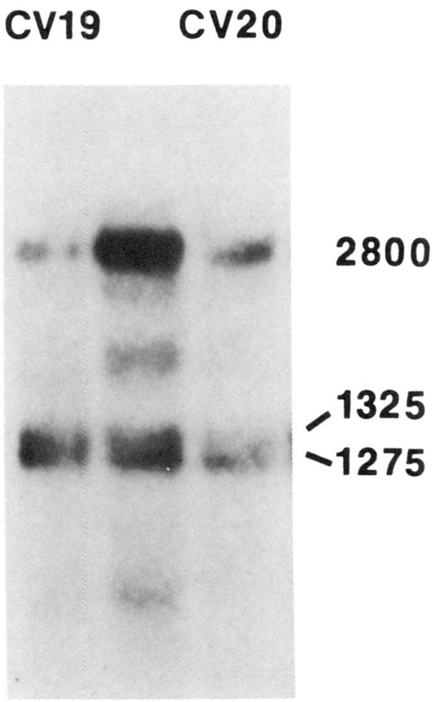

Figure 6. Transcription Pattern from Individual Forms of 2 Micron Circle.

Plasmid CV19 consists of the entire A form 2 micron circle genome with pBR322 plus LEU2 sequences inserted into the EcoRl site which lies in the middle of the FLP gene. Plasmid CV20 is identical to CV19 except that the 2 micron circle sequences are in the B form confiruration. Each of these plasmids were used to transform a [cir^o] strain to leucine prototrophy and polyA containing RNA was obtained from representative transformants. Samples (10 μg) of the RNA from each transformant were fractionated separately or together (middle lane) on methyl mercury agarose gels, transfered to diazotized paper, and probed with labeled 2 micron circle DNA. The sizes of the predominant RNA species evident on the autoradiogram of the hybridized filter are indicated by length designations. The 2800 bp transcript apparently is initiated at an adventitious promoter site within the pBR322 moiety of the plasmid, extends into the 2 micron circle portion, and terminates within the inverted repeat at the normal site of termination of the FLP gene transcript.

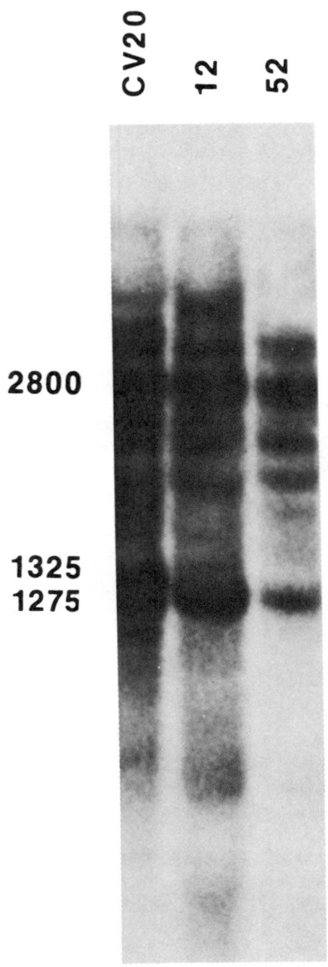

Figure 7. Transcription Pattern of 2 Micron Circle Plasmids with Insertions in REP1.

PolyA RNA was isolated from a yeast strain in which was resident either plasmid CV20 or the CV20 derivative plasmid 12 or 52. Each of these latter plasmids contain a Tn5 insertion in the REP1 coding region. The RNA was fractionated, immobilized and probed with labelled 2 micron circle DNA (ref 6). The predominant species are indicated by size (in bases) designations.

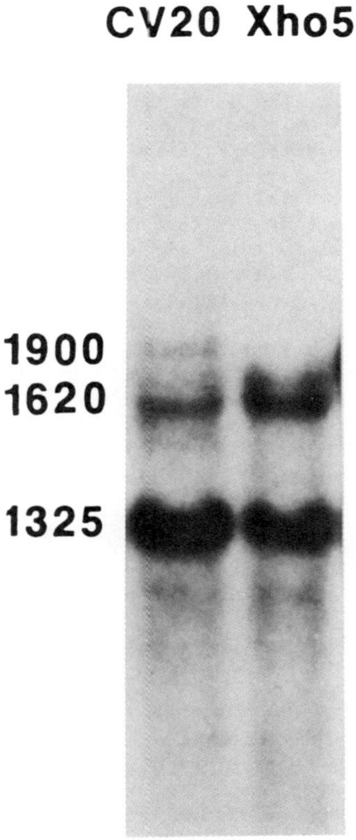

Figure 8. Transcription Pattern of a 2 Micron Circle Plasmid with a Deletion Spanning the Promoter of the 190C Base Transcript.

PolyA containing RNA was isolated from a strain containing CV20 or a derivative of CV20, designated Xho5, in which sequences between the PstI site and the HpaI in the larger unique region of the 2 micron circle genome had been deleted. The RNA was fractionated, transfered, and probed with the nick translated 1314 bp HindIII fragment of 2 micron circle, which spans the 5' portion of REP1. The positions of migration of the three transcripts from this portion of the 2 micron circle genome are indicated.

CV20 are identical, suggesting that the transcriptional products of the two forms of 2 micron circle are also identical.

Insertions and Deletions in the Large Unique Region of 2 Micron Circle Have Little Effect on Plasmid Transcription. A second component of our hypothesis concerning 2 micron circle expression was that almost all transcription of the plasmid is initiated at a single site within the large unique region. To test the validity of this assumption, we have examined the effect on in vivo transcription of insertions within the large unique region, downstream from the proposed initiation site, and of deletions which remove this proposed initiation site. Plasmids CV20::Tn5 #12 and CV20::Tn5 #52 contain insertions of Tn5 within the REP1 coding region in the large unique region. These two plasmids as well as plasmid CV20 were introduced into a [cir^o] yeast strain by transformation and polyA containing RNA was obtained from representative transformants. After fractionation on agarose gels and transfer to DBM paper, the plasmid specific transcripts were identified by hybridization with labeled 2 micron circle DNA. These results are shown in Figure 7. As is evident from the autoradiogram, the 1325 REP1 mRNA is absent from strains containing CV20::Tn5 #12 or #52. Less evident in the Figure, but clear in tracks probed with specific plasmid restriction fragments (data not shown), is the absence of the 1620 base transcript, which is transcribed from the same region but in the opposite direction as the REP1 RNA. However, except for a few minor, high molecular weight transcripts no other difference in the trancriptional products of the insertion containing plasmids versus the parent are apparent. Clearly, the 1275 base REP2 transcript is produced normally from both mutant plasmids. Thus, insertions within the REP1 gene in the large unique region fail to display any defect in the transcription of the small unique region.

By the procedure described in Materials and Methods we constructed a plasmid, designated Xho5, which is essentially identical to CV20 except that it contains a deletion in the large unique region which removes 400bp spanning the site corresponding to the 5' end of the 1950 base transcript (cf. Figure 2). Thus in the model we proposed for transcription of the plasmid, this deletion should have removed the principal 2 micron circle promoter. This plasmid was transformed into a [cir^o] strain and the 2 micron circle specific transcripts examined as before. In the experiment shown in Figure 8 the RNA was probed with the 1314 bp HindIII restriction fragment from 2 micron circle, which covers the 5' portion of the REP1 coding region. As can be seen, plasmid Xho5 does not synthesize the 1950 base transcript. Nonetheless, the REP1 transcript is synthesized in near normal quantities, as is the 1620 base transcript. In addition, the normal contingent of transcripts from the small unique region are also synthsized from Xho5 (data not shown). Thus these results indicate that the 1950 base transcript is not an obligate precursor of the REP1 mRNA and that initiation in the large unique region is not a prerequisite for trancription of the small unique region.

DISCUSSION

By analysis of both insertion and deletion mutations within the inverted repeat region of the 2 micron circle, we have demonstrated that FLP mediated recombination occurs only at a specific site. Our analysis limits this site to a 65 bp region, although the actual site could be much smaller. Indeed, the 4 bp deletion present in plasmid XΔ2-1 prevents this plasmid from participating in FLP mediated recombination.

The sequence of the DNA spanning the site of recombination is shown in Figure 9, within which are indicated the position of several relevant insertions and deletions. As is evident, this region of the plasmid is replete with symmetrical features. The XbaI site lies in the middle of a 22 bp region which is bracketed by a dyad symmetry that extends for 51 bp in either direction. Within each half of this dyad symmetry is a second dyad symmetry, so that this stretch of DNA could assume a cloverleaf structrue if it were single stranded. Viewed in a slightly different light, in the 100 bp region spanning the XbaI site is a sequence of 16 bp in length which is repeated, almost precisely, four times in alternating orientation. That this repeated sequence may play a role in the recombination event is suggested by the fact that this sequence also appears in the inverted repeat of Tn5, which under certain circumstances functions as a substrate for FLP mediated recombination (M. Jayaram and J.R. Broach, maunscript in preparation).

Whatever the actual recognition site is for the FLP recombination system, we can conclude that this site does not appear anywhere else in the yeast genome. We have recently shown that integration of the recognition site for FLP mediated recombination within a chromosome causes a marked instability of that chromosome (10). Therefore, if a site for FLP recombination existed in a chromosome, that chromosome would display significant instability. Since such an instability is not seen in normal yeast strains, we can assume that such a site does not exist.

Although the data presented in this paper demonstrate that FLP mediated recombination is initiated at a specific site, they do not address the mechanism of the recombination event. It is possible that recombination is initiated at or near the XbaI site by a single stranded nick. This could induce strand invasion, leading to the formation of a crossed strand intermediate which would be free to migrate throughout the inverted repeat before being resolve by appropriate strand scission. Evidence for the existence of such crossed strand intermediates, at least in meiotic cells, has been presented by Bell and Byers (2). As an alternative mechanism, FLP gene product could coordinately initiate and resolve the recombination event by, for example, a site-specific, double strand exchange. Additional experimental data will be needed to resolve this issue.

The function of FLP mediated interconversion of 2 micron circles is unknown. It is formally analogous to specialized recombination systems in procaryotic cells -- such as G-loop inversion in phage mu (8, 17) and phase variation in Salmonella (18) -- which function as genetic switches to permit alternate expression of different sets of genes. This analogy lent credibility to our previous hypothesis, based on the transcriptional pattern of 2 micron circle, that recombination functions as a genetic switch to interconvert transcriptionally distinct states of the plasmid. This hypothesis makes two testable predictions. First, the transcriptional products of the A form of the plasmid are distinct from those of the B form. Second, termination or inhibition of initiation of transcription in the large unique region should abolish transcription of the smaller unique region.

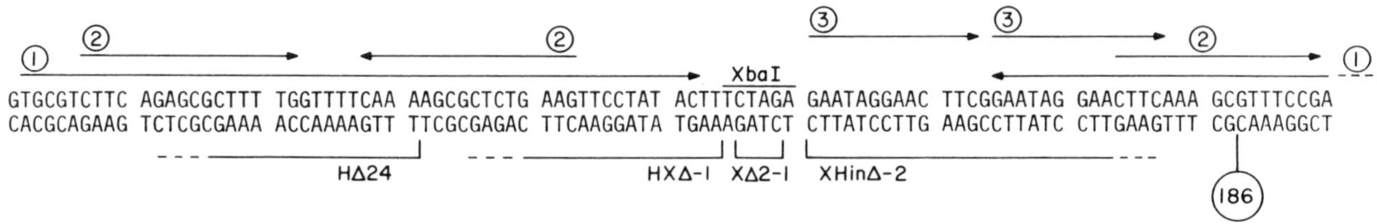

Figure 9. Sequence of the Site for Recombination in 2 Micron Circle.

The DNA sequence from part of the second inverted repeat of the A form of 2 micron circle (position 3891 to position 3990 in the numbering system of Hartley and Donelson, ref 12) is shown. Above the sequence are indicated regions of symmetry (numbered arrows) and the location of the *Xba*I restriction site. Below the sequence are indicated the location of Tn*5* insertion #186 and the regions deleted in several of the plasmids used in this study. The site required for recombination lies between the endpoint of deletion HΔ24 and position of Tn*5* #186.

In this paper we have presented results from transcription experiments which address these two predictions. First, we demonstrated that the in vivo transcriptional products of two hybrid plasmids, each containing the entire 2 micron circle genomes but frozen in opposite orientations, are identical. Additionally, we demonstrated that neither insertion of Tn5 sequences in the large unique region nor removal of the putative primary promoter of the plasmid abolished transcription of the small unique region. Thus, it would appear that our previous hypothesis is incorrect, and that an alternative explanation for FLP promoted recombination within 2 micron circle must be sought. It should be noted, though, that there is a caveat to the transcription experiments described in this paper. Namely, both pBR322 and Tn5 contain sequences which fortuitously function as promoters in yeast. As a consequence, it is possible that the 2 micron circle transcripts which should have been eliminated, due to the absence of recombination or to the insertion of Tn5, are being synthesized as a result of transcription initiated within one of these bacterial elements. Thus in spite of the data presented in this paper, the possibility that FLP mediated interconversion alters the transcriptional state of the plasmid cannot be definitively eliminated. Nonetheless, the transcriptional data presented in this paper is sufficiently suggestive that we are compelled to entertain other explanations for the function of FLP mediated recombination. One possible role for this recombination could be to ensure uniform segregation of the plasmid. That is, the recombination system could provide a mechanism by which catenated molecules arising following replication of the circular plasmid could be untangled. This possibility is being explored.

REFERENCES

1. Beggs, J.D. (1978) Transformation of yeast by a replicating hybrid plasmid. Nature 275 104-109.
2. Bell, L. and B. Byers (1979) Occurrence of crossed strand-exchange forms in yeast DNA during meiosis. Proc. Natl. Acad. Sci. USA 76 3445-3449.
3. Birnboim, H.C., and J. Doly (1979) A rapid alkaline extraction procedure for screening recombinant plasmid DNA. Nuc. Acids Res. 7 1513-1523.
4. Broach, J.R. (1981) "The Yeast Plasmid 2 Micron Circle" in Molecular Biology of the Yeast Saccharomyces: Life Cycle and Inheritance (J. Strathern, E. Jones, and J. Broach, eds) Cold Spring Harbor Laboratory Press, Cold Spring Harbor, New York, pp. 445-470.
5. Broach, J.R. and J.B. Hicks (1980) Replication and recombination functions associated with the yeast plasmid, 2 circle. Cell 21 501-508.
6. Broach, J.R., J.F. Atkins, C. McGill, and L. Chow (1979) Identification and mapping of the transcriptional and translational products of the yeast plasmid 2 micron circle. Cell 16 827-839.
7. Broach, J.R., J.N. Strathern, and J.B. Hicks (1979) Transformation in yeast: development of a hybrid cloning vector and isolation of the CAN1 gene. Gene 8 121-133.
8. Bukhari, A. and L. Ambrosio (1978) The invertible segment of bacteriophage Mu DNA determines the absorption properties of Mu particles. Nature 271 575-577.
9. Dobson, M.J., Futcher, A.B., and B.S. Cox. (1980) Control of Recombination Within and Between DNA Plasmids of Saccharomyces cerevisiae. Curr. Genet. 2 193-200.

10. Falco, C., Li, Y.-Y., Broach, J.R., and D. Botstein. Genetic properties of chromosomally integrated 2 plasmid DNA in yeast. Cell 29 in press
11. Gerbaud, C., P. Fournier, H. Blanc, M. Aigle, H. Heslot, and M. Guerineau (1979) High frequency of yeast transformation by plasmids carrying part or entire 2 micron yeast plasmid. Gene 5 233-253.
12. Hartley, J.L., and J.E. Donelson (1980) Nucleotide sequence of the yeast plasmid. Nature 286 860-864.
13. Livingston, D.M. (1977) Inheritance of 2 micron DNA plasmid from Saccharomyces. Genetics 86 73-85.
14. Clark-Walker, G.D., and G.L.G. Miklos (1974) Localization and Quantification of Circular DNA in Yeast. Eur. J. Biochem. 41 359-364.
15. Ruvkun, G.B., and F.M. Ausubel (1981) A general method for site-directed mutagenesis in prokaryotes. Nature 289 85-88.
16. Toh-e, A., and R.I. Wickner (1981) J. Bacteriol. 145 1421-1424.
17. van de Putte, P., S. Cramer, and M. Giphart-Gussler (1980) Invertible DNA determines host specificity of bacteriophage Mu. Nature 286 218-222.
18. Zeig, J., and M.I. Simon (1980) Analysis of the nucleotide sequence of an invertable controlling element. Proc. Natl. Acad. Sci. USA 77 4196-4200.

9. PERIODIC SYNTHESIS OF HISTONE PROTEINS THROUGH THE CELL CYCLE OF *Saccharomyces cerevisiae* AS DETERMINED BY CENTRIFUGAL ELUTRIATION

James R. Ludwig II† and Calvin S. McLaughlin*

Department of Biological Chemistry
California College of Medicine
University of California, Irvine
Irvine, California 92717

Rec. Adv. Yeast Mol. Biol. (1982) 1:113-121

The methods of centrifugal elutriation, dual isotopic labelling, and Triton Acid Urea/SDS acrylamide gel electrophoresis have been used to study the synthesis of histone proteins through the cell cycle of the budding yeast, *Saccharomyces cerevisiae*. All three histone proteins studied, H2A, H2B, and H3, showed periodic patterns of synthesis. In each case the peak of histone protein synthesis substantially preceded that for the synthesis of DNA. The significance of the timing of histone gene expression, and its relationship to replication is discussed.

INTRODUCTION

The budding yeast, *Saccharomyces cerevisiae*, possesses a cell cycle similar to those of higher eukaryotes. It contains the same four major divisions of G1, S, G2, and M. Those differences which exist between *Saccharomyces cerevisiae* and higher forms are, for the most part, confined to mitosis. In the budding yeast, the nuclear membrane does not break down during mitosis (1), nor does the chromatin condense (2). However, the other stages of the cycle, including S, are very comparable to those of higher cells. Also, yeast chromatin is similar to that of higher cells (3,4). If yeast chromatin is digested with appropriate nucleases, the DNA fragments produced show a subunit structure similar to that of chromatin from higher eukaryotes (5,6,7). Like other fungi, yeast has a shorter DNA subunit repeat (6), which may reflect a shorter linker region (8).

Histone proteins H2A, H2B, H3, and H4 have all been characterized in yeast (9,7,10). They have been shown to be associated in normal nucleosomal structures with nuclear DNA and to have amino acid compositions comparable to those established from higher eukaryotes (11,12). However, the existence of a yeast H-1 protein has not been demonstrated clearly. Although an H-1-like protein has been observed (11), its properties are also similar to those of HMG (high mobility group) proteins (13,14,15,16,17,18).

* We wish to thank Lynna Hereford for introducing us to AUT/SDS gel eletrophoresis and for ongoing discussions on the problem of histone gene expression. We would also like to thank Court Saunders and Jim Davie for the gift of purified histone proteins. This work was supported by grant GM28772 from the National Institute of General Medical Sciences and grant AI16252 from the National Institute for Allergy and Infectious Diseases and represents publication No. 32 from the Mycology Center.

† Present Address: Department of Biochemistry, University of Manchester Institute of Science and Technology, Manchester, England M60 1QD, United Kingdom.

In higher eukaryotes, the stage of the cell cycle defined by the period of DNA synthesis involves more than DNA replication. It includes the synthesis of histone and non-histone chromosomal proteins and their association with DNA to produce faithful and efficient replication of the chromatin unit (19). There is much evidence in higher eukaryotes for a coordination of DNA and histone synthesis. Many studies demonstrate that inhibiting the formation of one component hinders the synthesis of the other (20,21,22,23,24). In yeast, inhibition of protein synthesis with cycloheximide disrupts DNA synthesis during a short period at the beginning of S phase (25,26,27). This suggested a decreased temporal correspondence between histone synthesis inhibition and inhibition of replication and left open the possibility of continuous histone synthesis through the cycle. The data obtained by Moll and Wintersberger (19) argue for predominant, if not exclusive, synthesis of histones in S phase. However, the pulses used to measure synthesis of histone proteins were twenty minutes in duration, a full sixth of a cell cycle for many strains and equivalent to the entire period of S phase (1). More importantly, the methods used to obtain cell cycle specific fractions involved induction synchrony by alternating rounds of nutritional starvation and recovery (27,28,29). Such synchrony techniques may generate patterns of gene expression which are not observed when the cell cycle is studied by a non-stressful, cell-selection technique, such as centrifugal elutriation (30).

We report here a study of histone synthesis using centrifugal elutriation, a specific double isotopic labelling technique, and high resolution, Triton Acid Urea/SDS two-dimensional gel electrophoresis.

MATERIALS AND METHODS

Strains

The diploid strain of <u>Saccharomyces cerevisiae</u> SKQ2n was used in all experiments described. It was obtained from Brian Cox (Oxford University) and has the following genotype:

a/α; ade 1/+; +/ade 2; +/his 1

Chemicals

Purified yeast histone proteins were a gift from Court Saunders and Jim Davie at Oregon State University. DAPI (4′,6-Diamidino-2-phenylindol dihydrochloride) was obtained from Boehringer-Mannheim. All other chemicals were obtained from standard chemical sources. All radioactive labels were obtained from Schwarz-Mann.

Growth and Labelling of Cells

Cells were grown to midlog phase at 23°C on a rotary shaker. In all cases the medium used contained per liter: 6.7 g Yeast Nitrogen Base minus amino acids (Difco), 12.5 mg of each of the twenty amino acids naturally occurring in proteins except methionine, cysteine, and lysine, 20 g glucose, and 10 mg adenine and uracil. The doubling time of SKQ2n in this medium is 2 hours. In all cases, cultures used were 125 ml in volume, containing cells in mid-log phase (80 Klett units). The relative rates of synthesis for histones and other basic proteins were determined by a double-label procedure. Cells were labelled for 4 hours with [^{14}C] yeast protein hydrolysate (7.6 μCi/ml), then pulse-labelled for 10 minutes with [^{3}H] lysine (25 μCi/ml). Very heavily labelled cells for the identification of histone spots on gels were obtained by labelling a 10 ml culture for 20 minutes with [^{14}C] yeast protein hydrolysate (12 μCi/ml). In all cases incorporation was halted by the addition of ice to the medium. The cells were immediately centrifuged at 5,000 rpm for 10 minutes at 0°C, washed once, and resuspended in ice-cold, distilled, deionized water.

Cell Cycle Fractionation and Fluorescent Staining

The procedure used for fractionation was described previously (30,31). Briefly, cells grown as described above were harvested on ice, resuspended in ice-cold, distilled, deionized water, sonicated to disrupt clumps and loaded at 2°C into a Beckman JE-6 rotor, spinning at 3000 rpm. After loading at 9 ml/min, 10 fractions (150 ml each) were collected and represent the entire cell cycle. Sodium chloride was added to each fraction to facilitate pellet formation, and the cells were collected by centrifugation.

The quality of separation obtained was determined by staining cells with the DNA specific stain DAPI, using the post-vital staining method of Williamson and Fennell (32).

Two-Dimensional Gel Electrophoresis

Triton Acid Urea/SDS two-dimensional gel electrophoresis was used to resolve basic proteins. 20 μl volumes of an RNase-DNase solution were added to cell pellets at 0°C - 4°C in 1.5 ml microfuge tubes (33). An additional 180 μl sterile, deionized water was added and the cell suspensions transferred to 3 ml sterile glass tubes, where oven baked glass beads (Glasperlen, 0.45 mm) were added to the meniscus. After lysis the extracts were transferred back into microfuge tubes and the beads washed twice with 200 μl of sterile, deionized water. Washes were pooled with the extracts and to each of these 0.5-0.6 ml volumes were added 1 ml of 50% TCA. The samples were allowed to precipitate on ice for ten minutes, then centrifuged for 2 minutes in a Beckman microfuge at 0°C. The resultant pellets were washed with acetone-HCl, then acetone precipitated (1 ml volume) for ten minutes on ice. After a 2 minute centrifugation, the resultant pellets were allowed to dry at 37°C for 1 hour. To these samples were added 25 μl AUT Sample Buffer and 25 μl acetic acid 6.7% in glycerol. Electrophoresis was performed as described previously (34,35).

Identification of Histone Spots on Gels

Co-electrophoresis of 10-20 μg of purified histone protein with total yeast protein extracted from 0.1 ml of an 80 Klett (midlog) culture of cells labelled for 20 minutes with [^{14}C] yeast protein hydrolysate (12.5 μCi/ml) was performed. This gave protein patterns on staining with Coomassie Blue in which yeast cell proteins were not discernable, but histone proteins were intensely stained. Autoradiograms of such gels showed the total yeast pattern after a two-week exposure. These autoradiograms were then used to locate spots to be cut out of gels. Spots were cut from the gels, solubilized for 48 hours using a toluene-based scintillant containing protosol (3.5%) and counted using a Beckman LS 8000 Scintillation Counter.

RESULTS

Resolution of the Rotor

We have used several methods to ascertain the accuracy of the separation obtained by centrifugal elutriation (31,33). Biochemically, the peak of DNA synthesis (Fig.3) has been localized to fraction number four, a fraction which is composed almost exclusively of small budded cells (33). The morphologic composition of each fraction has been determined by post-vital staining of nuclei with the DNA-specific fluorescent dye DAPI (32). Data obtained from this method are shown in Table 1.

Table I

Cell Separation by Centrifugal Elutriation

Percentage of Cell Types

Fraction No.	Dead	Unbudded	Small Bud	Nuclear Migration	Doublet
1	26	66	1		
2	3	97			
3	2	29	69		
4		7	88	5	
5		13	62	18	7
6		16	24	45	15
7		10	39	28	23
8		8	24	26	42
9		8	17	24	51
10		9	18	26	47

The purity of fraction four discussed above is evident as is that for fraction two, which contains exclusively unbudded cells. Fraction three, however, represents a breakpoint between two morphological classes, containing 70% small budded and 30% unbudded cells. Subsequent fractions show varied distributions with fractions six and nine showing extreme enrichment for nuclear migrants and doublets, respectively. Dead cells are limited primarily to fraction one, which was therefore omitted from study.

Identification of Histone Proteins on Gels

The histones are low molecular weight, basic proteins. They are, therefore, best resolved by electrophoresis under acid conditions and high acrylamide concentrations. The system of choice was one described by Alfageme (35), which involves an acid extraction followed by a Triton Acid Urea first dimension. The addition of triton-X100, a non-ionic detergent, enhances the resolution of the histones, presumably by exploiting differences in hydrophobicity between otherwise similar molecules (35,36). Histones (H2A, H2B, H3, and H4), as well as a number of ribosomal proteins are well resolved (38). Histone H-1 was not identified in this system. Locations of the individual histone proteins were determined by co-electrophoresis of 10-20 μg of purified histone proteins with a low concentration of protein extract from cells which had been previously labelled with [^{14}C] yeast protein hydrolysate to a high specific activity. This allowed identification of spots by comparison between stained gels, where only the histone standards were visible.

Histone Synthesis Through the Cell Cycle

Through the use of a specific double-labelling protocol and the Triton Acid Urea/SDS two-dimensional gel electrophoresis, it has been possible to determine the nature of histone synthesis through the yeast cell cycle. The basis for such labelling and the analysis of plots obtained was described by Elliott and McLaughlin (33). Briefly, the plots represent the ratio of the synthetic rate (dA/dt) to accumulation (A) versus progression through the cell cycle (t). For exponential synthesis, this ratio is a constant and the plot is therefore a horizontal line. For periodic synthesis, there is a peak in the ratio corresponding to the period of synthesis, surrounded by regions of near zero ratio.

A pattern of periodic synthesis is observed for histones: H2A, H2B, and H3 (Fig.1b,c,d). Histone protein H4 ran very near to the front of migration in this system. Accurate data could not be obtained. H2A and H2B exhibit periodic synthesis, with sharp peaks at fraction three. H3, similarly, shows periodic synthesis, with a peak in early fractions. For every experiment, ribosomal proteins run in the same gels were cut out, solubilized, and counted in exactly the same manner as the histone proteins. None of the ribosomal proteins showed periodicity (Fig.1a), rather they displayed exponential pattern of synthesis observed for these proteins in earlier studies (37). A most interesting aspect of the histone synthetic data is the clear distinction in time between the synthesis of histones and the synthesis of DNA (Fig.2). DNA synthesis peaks in fraction four in this system while all three histone proteins showed clear peaks of synthesis prior to fraction four.

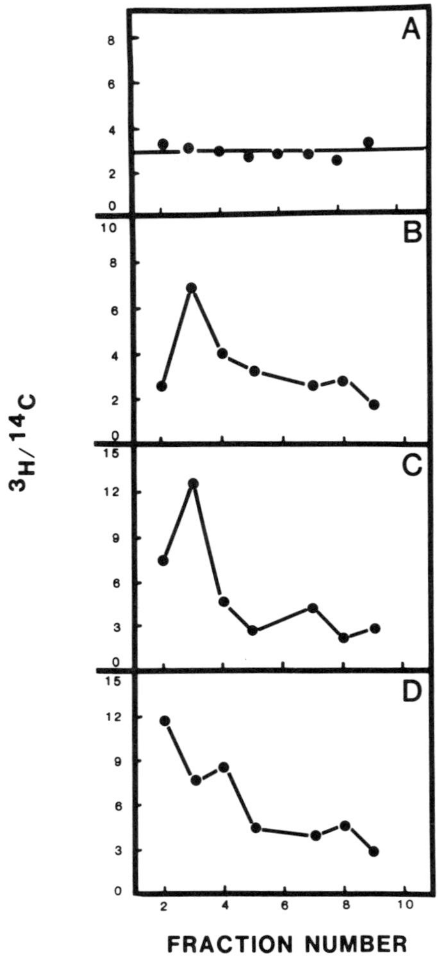

Fig. 1. Ratio of Pulse to Long-Term Radioactivity (dA/dt/A) through the cell cycle. Cells were labelled for 10 minutes. The cells were separated by elutriation and the synthesis of individual proteins determined as described. A corresponds to a representative ribosomal protein; B corresponds to histone H2A; C corresponds to histone H2B; and D corresponds to histone H3.

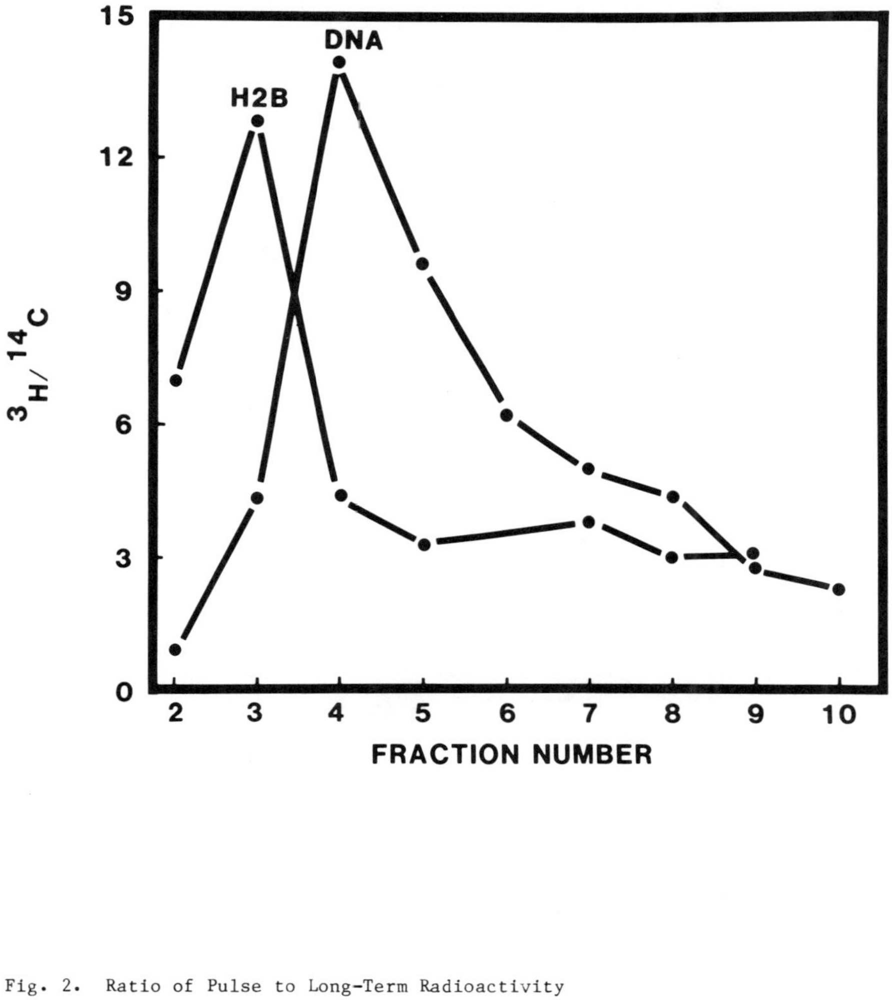

Fig. 2. Ratio of Pulse to Long-Term Radioactivity (dA/dt/A) through the cell cycle for histone H2B and DNA. Cells were labelled and analyzed for histone synthesis as described in Fig.2 and for DNA synthesis as described by Elliott and McLaughlin (33).

DISCUSSION

Using elutriation, we have shown that the synthesis of histone proteins varies periodically through the cell cycle. This demonstrates the sensitivity of our methods to periodicities which are truly cell cycle specific. It is significant that these data are similar to those obtained using classic induction synchrony techniques (19). This suggests that although synchrony methods may generate periodicities which are independent of the cell cycle, they do not obliterate a set of periodicities which are truly of a cell cycle origin and required for cell division.

The high resolution of our cell selection method allows a precise localization of histone synthesis. We find that this synthesis does not persist through the whole of S phase, but is restricted to a period at the beginning of S phase. This point is significant in terms of a possible dependence of DNA synthesis on histone synthesis. The experiments of Hereford and Hartwell (26), as described in the INTRODUCTION showed that blocking protein synthesis early in S blocked DNA synthesis, while a later block in protein synthesis did not affect replication. If histone synthesis occurred throughout S, this finding would suggest the absence of a dependence relation. The restriction of histone synthesis to the beginning of S suggests a re-evaluation of this conclusion. Further, the coincidence of the period of required protein synthesis with that of maximal histone synthesis suggests the possibility that the required protein synthesis is mainly a requirement for histone synthesis.

Our observation that the peak of histone synthesis precedes that for DNA by approximately one-tenth of a cell cycle is consistent with those obtained by Hereford et al. (38). That study employed Northern analysis and *in vitro* translation of mRNA extracted from cells separated by elutriation. Taken together these studies indicate that transcriptional rather than translational control is important during the initiation period of histone synthesis. However, the interrelationship between DNA synthesis and histone synthesis is complex. Moll and Wintersbrger (19) concluded from experiments involving the use of hydroxyurea to block DNA synthesis that DNA synthesis was required for histone synthesis. We have shown that the inhibition of DNA synthesis leads to a rapid degradation of histone mRNA (38). On the other hand we find the highest level of histone synthesis occurs well before the peak of DNA synthesis. By the time the peak of DNA synthesis occurs histone synthesis has returned to near basal levels for H2A and H2B. This suggests that the mechanisms that control the amount of histone produced have to be fairly complex. They involve an initiation event for histone mRNA synthesis prior to the bulk of DNA synthesis, regulation of the amount of histone produced and a termination event which involves some post transcriptional control prior to completion of DNA synthesis (38). These mechanisms are under study.

We have described a method for biochemically sequencing events in a normal cell cycle. This method avoids abnormal growth conditions as well as physical and nutritional stresses. The sensitivity of the system has allowed precise characterization of the pattern of synthesis for a group of periodic proteins, the histones, and has allowed accurate positioning of these synthetic events within the cell cycle. These periodic molecular events in the cell cycle provide interesting insights into the nature of the mechanisms that control cell division and suggest that further study of the periodic molecular events of the cell cycle will be most profitable.

LITERATURE CITED

1. Hartwell, L.H. 1974. Bacteriological Reviews 38: 164-198.
2. Gordon, C.N. 1977. J. Cell Sci. 24: 81-93.
3. Elliott, S.G. and McLaughlin, C.S. 1978. Proc. Natl Acad. Sci., USA 77: 4384-4388.
4. Pinon, R. and Salts, Y. 1977. Proc. Natl Acad. Sci., USA 74: 2850-2854.
5. Lohr, D. and Holde, K.E. van 1975. Science 188: 165-166.
6. Lohr, D., Koviac, R.T., and Holde, K.E. van 1977. Biochemistry 16: 463-471.
7. Nelson, D.A., Beltz, W.R., and Rill, R.L. 1977. PNAS (USA) 74: 1343-1347.
8. Noll, M. 1976. Cell 8: 349-355.
9. Thomas, J.O. and Furber, V. 1976. FEBS Lett. 66: 274-280.
10. Wintersberger, U., Smith, P., and Letnansky, K. 1973. Eur. J. Biochem. 33: 123-130.
11. Sommer, A. 1978. Molec. Gen. Genetics 161: 323-331.
12. Goff, C.G. 1976. J. Biol. Chem. 251: 4131-4138.
13. Goodwin, G.H. and Johns, E.W. 1973. Eur. J. Biochem. 40: 215-219.
14. Goodwin, G.H., Nicolas, R.H., and Johns, E.W. 1975. Biochem. Biophys. Acta 405: 280-291.
15. Mohlberg, J. and Rusch, H.P. 1969. Arch. Biochem. Biophys. 134: 577-589.
16. Kuehl, L., Lyness, T., Dixon, G.H., and Levy-Wilson, B. 1980. J. Biol. Chem. 255 (3): 1090-1095.
17. Weber, S. and Isenberg, I. 1980. Biochemistry 19: 2236-2240.
18. Weisbrod, S., Groudine, M., and Weintraub, H. 1980. Cell 19: 289-301.
19. Moll, R. and Wintersberger, E 1976. Proc. Natl Acad. Sci. USA 73 (6): 1863-1867.
20. Borun, T.W., Scharff, M.D., and Robbins, E. 1967. Proc. Natl Acad. Sci. USA 58: 1977-1983.
21. Sadgopal, A. and Bonner, J. 1969. Biochem. Biophys. Acta 18: 349-357.
22. Weintraub, H. 1972. Nature 240: 449-453.
23. Weintraub, H. and Holtzer, H. 1972. J. Molec. Biol. 66: 13-35.
24. Seale, R.L. and Simpson, R.T. 1975. J. Molec. Biol. 94: 479-501.
25. Williamson, D.H. 1973. Biochem. Biophys. Res. Commun. 52: 731-740.
26. Hereford, L.M. and Hartwell, L.H. 1973. Nature New Biol. 244: 129-131.
27. Golombek, J. and Wintersberger, E. 1974. Exp. Cell Res. 86: 199-202.
28. Williamson, D.H. and Scopes, A.W. 1960. Exp. Cell Res. 20: 338-349.
29. Golombek, J., Wolf, W., and Wintersberger, E. 1974. olec. Gen. Genetics 32: 137-145.
30. Ludwig, J.R., Foy, J.J., Elliott, S.G., and McLaughlin, C.S. 1982. Molec. Cell. Biol. 2: 117-126.
31. Gordon, C.N. and Elliott, S.G. 1977. J. Bacteriol. 129 (1): 97-100.
32. Williamson, D.H. and Fennell, D.J. 1976. Meth. Cell Biol. 12: 335-351.
33. Elliott, S.G. and McLaughlin, C.S. 1978. Proc. Natl Acad. Sci. USA 75: 4384-4388.
34. Woolford, J.L., Hereford, L.M., and Roshbash, M. 1979. Cell 18: 1247-1259.
35. Alfageme, C.R., Zweidler, A., Mahowald, A., and Cohen, L.H. 1974. J. Biol. Chem. 249: 3729-3737.
36. Zweidler, A. and Cohen, L.H. 1972. Fed. Proc. 31: 926.
37. Elliott, S.G., Warner, J.R., and McLaughlin, C.S. 1979. J. Bacteriol. 137 (2): 1048-1050.
38. Hereford, L.M., Osley, M.A., Ludwig, J.R., and McLaughlin, C.S. 1981. Cell 24: 367-375.

10. CONTROL OF AMINO ACID BIOSYNTHESIS IN YEAST*

H. Greer, M. Penn, B. Hauge, and B. Galgoci

Cellular and Developmental Biology Department
Harvard University, Cambridge, Massachusetts 02138

Rec. Adv. Yeast Mol. Biol. (1982) 1:122-142

In yeast, there is coordinate regulation of different amino acid biosynthetic pathways. In this paper we describe the isolation and characterization of 43 amino acid analog sensitive mutations which define four unlinked complementation groups, AAS101, AAS102, AAS103, and AAS104, two of which identify new genes involved in general control. We have shown that the AAS^+ genes effect regulation at the mRNA level. Analysis of an aas101-complementing clone indicates that it (i) contains sequences that are repeated in the yeast genome, and (ii) codes for an RNA transcript that is derepressed under histidine starvation conditions. We have also demonstrated the existence of a transcript homologous to an open reading frame in the 5' flanking region of the HIS4 gene. A possible regulatory function for this transcript and/or polypeptide is discussed.

INTRODUCTION

In prokaryotes, functionally related genes are organized into common transcriptional units called operons which are controlled by a single regulatory region. In eukaryotes, however, no operons or polycistronic mRNAs have been identified (5, 10,14). Moreover, in several fungi it has been demonstrated, both genetically and biochemically, that, in addition to coordinate regulation of the different unlinked genes within a given pathway, there is also co-regulation of genes in completely different amino acid biosynthetic pathways (2, 3, 5, 11). This phenomenon is called "general control of amino acid biosynthesis" (4).

In *Saccharomyces cerevisiae*, it has been shown that starvation for a single amino acid causes derepression of the biosynthetic enzymes for all branched amino acids (aromatic and aspartate family), for the basic amino acids histidine, lysine and arginine, and also for serine and valine (9, 13, 19). (However, the extent of derepression of different enzymes within a given pathway may differ). The isolation of several regulatory mutations in this general control system has been reported. The mutations ndr1 (RH428), ndr2 (RH487) (for non-derepressing) and aas1, aas2 (for amino acid analog sensitive) are unable to derepress enzymes in different amino acid biosynthetic pathways (9, 13, 19). These mutations define three genes. It is postulated that these genes encode positive regulatory elements of the general control

*Support was obtained from National Institutes of Health grant 2R01 GM25207 to H.G.

system. Tra3 (for triazole alanine resistance) mutations have constitutive levels of these same biosynthetic enzymes and it is presumed to be a negative controlling element (19). The mechanism of action of these genes has thus far remained elusive.

In this paper we present information which may bear on the molecular nature of this general control system. We discuss: (i) the identification and characterization of two new genes which are necessary for derepression of many of these amino acid biosynthetic enzymes, and which may themselves be regulated by conditions of amino acid imbalance, and (ii) the existence of a short transcript (which has the capacity to code for a small polypeptide) in the 5' flanking sequence of one of the genes under general control, the HIS4 gene.

We have attempted to saturate the yeast genome with amino acid analog sensitive mutations to determine the number and mode of action of positive regulatory genes involved in this general control of amino acid biosynthesis. We have isolated 43 independent aas mutations which define four unlinked complementation groups (AAS101, AAS102, AAS103, and AAS104), two of which identify new genes involved in general control. These AAS genes effect amino acid derepression by regulating mRNA levels. By screening a yeast genomic library, genes which can complement different aas mutations have been cloned. A preliminary characterization of several of these clones is discussed.

In addition, analysis of the 5' coding regions of several cloned yeast biosynthetic genes subject to general control, HIS4, HIS3 and TRP5, indicates that there are open reading frames preceding each of these structural genes. It is striking that the position of the open reading frame preceding TRP5 is almost identical to that of HIS4. We have demonstrated by both S1 protection and primer extension experiments that a discrete transcript corresponding to this 5' HIS4 region exists.

RESULTS

I. AAS Mutants

AAS Mutant Isolation

Forty-three independent aas mutants were isolated in the wild type strain S288C by mutagenizing and screening for enhanced sensitivity to several amino acid analogs. The rationale for this selection is that cells which become sensitive to amino acid analogs may result from an inability to derepress the amino acid biosynthetic enzymes which are necessary to overcome the deleterious effects of the analogs.

YEPD cultures of S288C (MATα) grown at 30° C were plated on YEPD medium and irradiated with ultraviolet light to produce 50% killing. After two days growth at 30° C, approximately 2×10^5 colonies were replicated to both minimal and minimal + 0.75 mM 5-methyl tryptophan (5MT) medium. Forty-seven 5MT sensitive colonies were identified. Upon further screening, 43 of these mutants (giving a final mutation frequency of 2×10^{-4}), also exhibited sensitivity to analogs of the end products of additional amino acid biosynthetic pathways: 3-amino-1,2,4-triazole (AT), canavanine (can), and ethionine (eth), analogs of histidine, arginine, and methionine, respectively. These putative general control mutations are designated aas for amino acid analog sensitivity, in accordance with the nomenclature of Wolfner et al. (19). About one-half of the mutants are also petites.

Genetic Analysis of the aas Mutants

Dominance. All of the aas mutations are recessive. Dominance was tested by crossing each of the aas mutants to an AAS^+ strain. The resulting heterozygous diploids show the phenotype of the AAS^+ parent when tested for amino acid analog sensitivity.

Complementation groups. The 43 aas mutations define four complementation groups called AAS101, AAS102, AAS103, and AAS104. The distribution of the mutations is: four alleles in aas101 (-1 to -4), 20 alleles in aas102 (-1 to -12, -14 to -21), 18 alleles in aas103 (-1 to -18) and one allele in aas104 (-1). These groups were determined by crossing the 43 aas mutants to previously isolated mutants which also show sensitivity to various amino acid analogs: ndr1-2 (RH428), ndr2-1 (RH487) (9, 13) and aas1 and aas2 (19). Representatives of the newly isolated aas mutations in the opposite mating type (obtained by crossing original mutants to wild type), were also tested in pairwise combinations with the original 43 aas mutants. Complementation was determined by scoring analog sensitivity on minimal + AT (15 mM) plates (see Table 1). These results indicate that the currently existing mutations define five amino acid analog sensitive complementation groups: I (AAS101), II (AAS102, NDR2, AAS1), III (AAS103, NDR1), IV (AAS104), and V (AAS2). Thus, two of the groups isolated in this study, AAS101 and AAS104, define two new genes involved in general control of amino acid biosynthesis.

Segregation. The aas mutations segregate as single mutations. Crosses of aas101-1, aas102-1, aas103-1, and aas104-1 to AAS^+ strains show 2:2 segregation for the aas mutation.

Linkage. Meiotic linkage data is consistent with the complementation results described above. Pairwise crosses between many of the aas mutants were subjected to tetrad analysis. Two classes emerged: (i) 101 x 102, 101 x 103, 101 x 104, 102 x 103, 103 x 104, 102 x 2, and 103 x 2; and (ii) 102 x ndr487, 102 x 1, and 1 x ndr487. The mutations paired from class (i) are unlinked to each other: PD (parental ditype) = NPD (nonparental ditype) for each cross. However, the mutations paired from class (ii) do not give any recombinants between the respective two mutations: all tetrads are PD (four AT sensitive spores/tetrad). This result indicates that the mutation pairs from (ii) are tightly linked, supporting the complementation test results which indicated that these mutations are in the same gene.

In addition, the AAS101 gene shows tight linkage to the URA3 gene on chromosome V.

Double mutants. Combinations of the aas mutations were constructed in haploid strains from the crosses discussed in the linkage section above: 101-102, 101-103, 101-104, 102-103, 103-104, 102-2, and 103-2. All of these double mutation haploid strains are viable, and are sensitive to the amino

Table 1. Complementation of _aas_ Mutations.

	aas101-1	aas102-1	aas103-1	aas104-1	aas1	aas2	ndr1	ndr2
aas101 (-1 to -4)	−	+	+	+	+	+	+	+
aas102 (-1 to -12, -14 to -21)	+	−	+	+	−	+	+	−
aas103 (-1 to -18)	+	+	−	+	+	+	−	+
aas104 (-1)	+	+	+	−	+	+	+	+
aas1	+	−	+	+	−	+	+	−
aas2	+	+	+	+	+	−	+	+

acid analogs.

Homozygous diploids. The phenotype of the aas mutations in homozygous diploids was also analyzed. The diploids aas101-1/aas101-1, aas102-1/aas102-1 and aas103-1/aas103-1 were constructed; they are sensitive to the same analogs as the haploids. The 101, 102, and 103 diploids sporulate normally.

Analog Sensitivity of the aas Mutants

The aas mutants exhibit enhanced sensitivity to various amino acid analogs relative to their parent S288C. All 43 aas mutants were screened for analog sensitivity by replica plating to minimal media supplemented with a range of concentrations of AT, 5MT, can, or eth, at 23°, 30°, and 37° C (see Table 2). The aas101 mutants are the most sensitive, and aas104 is the least sensitive.

Different mutations within a complementation group confer roughly the same degree of sensitivity. However, in general, the petites are more sensitive than the grande mutants within a given group. The largest variation in sensitivity is seen in different aas102 and aas103 alleles. The most sensitive grande mutations in these genes are aas102-14, -15, and -16, and aas103-12, -15, -16, and -17.

All the aas101, aas102, aas103, and aas104 mutants are temperature sensitive in their analog response: increasing the temperature lowers the concentration needed to inhibit growth. (For some of the analogs, S288C shows a somewhat weaker ts phenotype.)

In addition, an adenine effect was observed. The 43 aas mutants were replica plated to minimal media supplemented with adenine as well as AT, 5MT, can, or eth. The effect of adenine depends on the analog being tested: (i) adenine increases the sensitivity to can or eth, (ii) adenine decreases the sensitivity to AT, and (iii) adenine has no effect on 5MT sensitivity.

Revertants of the aas Mutants

Almost all of the aas mutations give rise to spontaneous revertants as measured by papillation when colonies are replicated to minimal + AT plates. The most stable alleles are aas101-1, aas102-12, 102-19, 103-7, and 103-11. Spontaneous revertants of several alleles of the aas mutations were sought by growing 10 independent clones from each mutant in YEPD media and plating onto minimal + 10mM AT to select for AT resistant clones. The frequency of revertants was $<1.6 \times 10^{-8}$ for aas101-1, 9.8×10^{-6} for aas102-1, and 5.3×10^{-6} for aas103-1. Further testing on 5MT, can, and eth indicated a wide range of resistance among the revertants to these additional analogs; some of the revertants exhibited normal wild type resistance.

Aas Mutations Prevent Derepression of the HIS4 Gene

Growth of his4C^{ts} strains. Aas101-1, aas102-1, aas103-1, and aas104-1 were crossed into a strain carrying a leaky his4 mutation, his4C-207^{ts}. His4C-207^{ts} strains are unable to grow on miminal medium at 37° C but can grow

Table 2. Amino Acid Analog Sensitivity of aas Mutants.

	5 MT				AT			Can				Eth			
	.5 mM	7.5 mM	1.0 mM	1.25 mM	10 mM	15 mM	20 mM	1.25 γ/ml	2.5 γ/ml	5.0 γ/ml	7.5 γ/ml	2.5 γ/ml	5.0 γ/ml	7.5 γ/ml	10.0 γ/ml
23°															
S288C	+	+	+	+	+	+	+	+	+	+	+/−	+	+	+	+
101's	+	+	+	+/−	−	−	−	+	+/−	−	−	+	+	+/−	+/−
102's	+	+	+	+	+/−	+/−	+/−	+	+	+/−	−	+	+	+	+
103's	+	+	+	+	+/−	+/−	+/−	+	+	+/−	−	+	+	+	+
104	+	+	−	−	−	+	+	+	+	−	−	+	+	+	+
30°															
S288C	+	+	+	+	+	+	+	+	+/−	−	−	+	+	+	+
101's	−	−	−	−	−	−	−	+/−	−	−	−	+	+/−	−	−
102's	+/−	−	−	−	−	−	−	+	+/−	−	−	+	+	+/−	+/−
103's	+/−	−	−	−	−	−	−	+	+/−	−	−	+	+	+	+
104	+	+/−	+/−	+/−	−	−	−	+	+	−	−	+	+	+	−
37°															
S288C	+	+	+	+	+/−	+/−	−	−	−	−	−	+	−	−	−
101's	−	−	−	−	−	−	−	−	−	−	−	−	−	−	−
102's	−	−	−	−	−	−	−	−	−	−	−	+	−	−	−
103's	−	−	−	−	−	−	−	−	−	−	−	+	−	−	−
104	−	−	−	−	−	−	−	+	−	−	−	+	+/−	−	−

on minimal media at 30° C because the cells compensate for the partially defective HIS4 enzyme by derepressing the histidine enzymes. However, the presence of any of the aas mutations (101, 102, 103, or 104) in a his4C-207^{ts} strain prevents growth on minimal media at 30° C. It can therefore be inferred that the wild type products of AAS101, AAS102, AAS103, and AAS104 are required for derepression of at least the HIS4C enzyme.

HIS4C enzyme levels. The effect of the various aas mutations on the level of HIS4C enzyme activity, histidinol dehydrogenase, was determined. The aas mutants and wild type cells were grown under repressing or derepressing conditions: (i) minimal + all 20 amino acids (min AA), (ii) minimal + histidine (min his), (iii) minimal (min), and (iv) minimal + 10mM AT (min AT). Wild type cells do not derepress on minimal media alone because the internal histidine pool is very high (5mM), and this pool does not get depleted given the basal wild type level of synthesis of the histidine biosynthetic enzymes (8). However, histidine starvation conditions can be achieved by growth of wild type cells in the presence of the analog AT. AT inhibits the sixth step in the histidine biosynthetic pathway. Wild type cells overcome this inhibition by derepressing the histidine enzymes, as well as other enzymes under general control. As Table 3 shows, the enzyme levels for wild type S288C are essentially the same when grown in min AA, min his, or min. However, on min AT, the levels of HIS4C enzyme levels are derepressed about six fold.

For the aas mutants, the levels on min AA, min his, and min, are the same, or slightly lower, if compared to S288C. However, when starved on min AT media, there is no derepression at all of HIS4C in the aas101, aas102, and aas103 strains; the aas104 strain consistently derepresses to only about one-half the maximal level of S288C. Thus, the aas strains are unable to overcome the AT inhibition. It can be concluded that all four AAS^{+} genes are required for normal derepression of the levels of HIS4 enzyme activity to occur.

Table 3. HIS4C Enzyme Assays of aas Mutants Grown in Different Media

	AA	H	M	AT
S288C	1.0	1.2	1.2	6.3
aas101	.77	.80	.87	.40
aas102	.72	1.2	1.3	.55
aas103	.74	1.0	1.1	.74
aas104	.83	1.1	1.1	3.9

HIS4 mRNA levels. Northern blot analysis indicates that the aas mutations prevent derepression of the HIS4C enzyme by affecting mRNA levels. RNA was isolated from S288C, aas101, aas102, aas103, and aas104 strains, grown at 30° C on min AA, min his, min, and min AT. The RNA's were electrophoresed on a 1.5% agarose formaldehyde gel, transferred to a nitrocellulose filter, and hybridized with ^{32}P-labelled pMG1 (EcoRI-BglII fragment internal to the HIS4 structural gene) and YIp5 (pBR322 - URA3) probes. The URA3 gene was used as an internal standard to normalize for the amount of RNA in each lane. The mRNA levels for S288C and aas101 are shown in Figure 1. In the case of S288C, on minimal media supplemented with all the amino acids, a basal level of HIS4 mRNA is seen. A small derepression effect is observed on minimal or histidine media, whereas maximal derepression occurs when starved for histidine (min AT): the HIS4 mRNA is increased 5-10 fold. In the aas101 strain it is striking that little or no derepression of the HIS4 mRNA above basal level is detectable. Aas102 and aas103 show only intermediate derepression under starvation conditions, whereas aas104 shows the smallest effect (data not shown). These results suggest that the role of the AAS^+ genes is to regulate the mRNA levels of the genes that they control.

Cloning the AAS Genes

Selection. We have constructed a yeast genomic library by ligating Sau3A partially digested DNA into the unique BamHI site of YEp24. The plasmid YEp24 contains both the yeast URA3 gene and part of the yeast 2μ plasmid, which permits episomal maintenance. This bank was used to clone genes which complement either the aas101 or aas103 mutation.

An aas101-1 ura3-52 strain (MP40A-8B) was transformed with the YEp24-S288C library. (Ura3-52 is a non-reverting allele). Ura^+ transformants were selected in the presence of all 20 amino acids, pooled, and tested for aas101-complementing genes by growth on minimal + 10mM AT plates. Two independent AT resistant clones, called 101G and 101M, were identified from about 11,000 Ura^+ transformants. The 101G insert is about 2.7 kb and the 101M insert is about 10 kb. Both of these clones also confer resistance to .75mM 5MT. In a second experiment, the library was used to transform an aas103-1 ura3-52 strain (MP42-2A). Two independent AT resistant clones were isolated from 8,000 Ura^+ transformants. Both of these putative 103 clones were identical, and only one, 103R3, was used for further study. The 103R3 insert is about 5.5 kb.

The analog resistance property of these three clones resides on the plasmid in the transformed stains: spontaneous Ura^- segregants of 101G, 101M, and 103R3 simultaneously become aas^-. In the 103R3 clone, some AT sensitive Ura^+ clones spontaneously arise, but these show altered restriction patterns of the YEp24 insert, suggesting the occurrence of some type of deletion within the plasmid.

Restriction maps. No similarities in restriction maps have been noted between 101G and 101M. This is consistent with our observation that no cross homology exists between the inserted sequences. However, the 101M and 103R3 clones have several contiguous restriction fragments in common. It is possible that this common DNA sequence may complement both aas101 and aas103.

Homologous Genomic DNA. In order to confirm that the 101G cloned sequence represents a contiguous genomic sequence, rather than a cloning artifact, the following experiment was done. Genomic DNA from S288C was digested

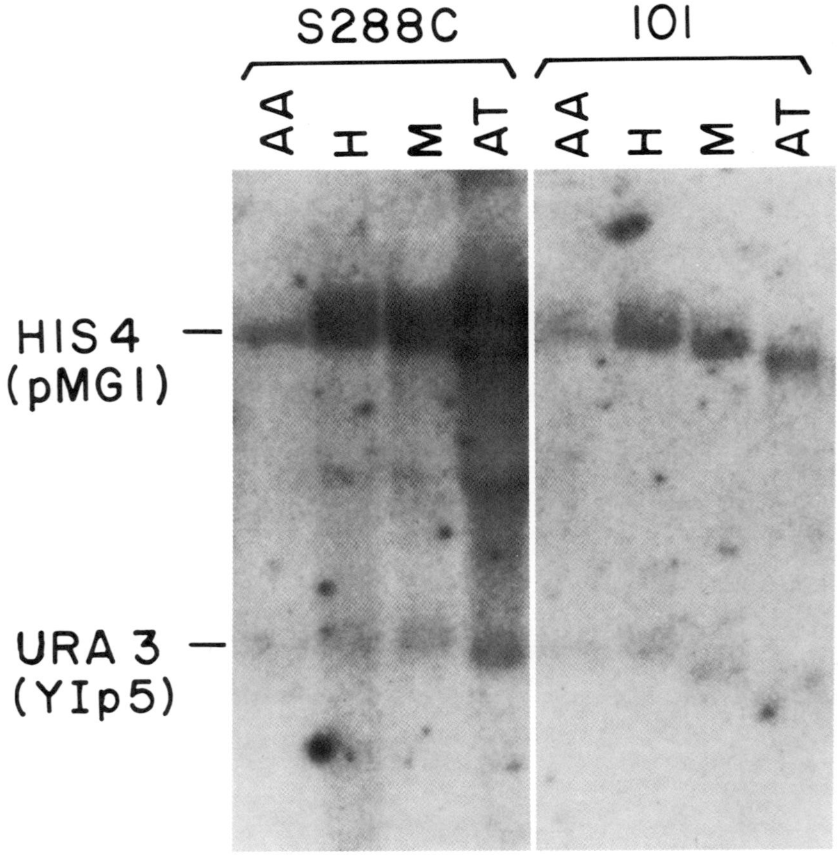

Figure 1. Regulation of HIS4 mRNA levels in S288C and aas101-1.

RNA was prepared from wild type S288C and aas101-1 cultures grown to O.D.$_{550}$ = 0.500 in minimal supplemented with all 20 amino acids (AA), minimal histidine (H), minimal (M) and minimal + 10 mM aminotriazole (AT) media. Extraction of RNA was by a modification of the procedure described in Sripati and Warner (16). Approximately 15 μg of RNA was loaded per lane onto 1.5% agarose formaldehyde gels. RNA was transferred to a nitrocellulose filter (18). Bound RNA was hybridized with ^{32}P nick translated pMG1 (a 1.0 kb fragment containing coding sequences in HIS4) and YIp5 (URA3). Hybridization was in 50% formamide, 5XSSC, 45° C. The extent of hybridization with the URA3 probe was used to normalize for the amount of RNA within each lane.

with BamHI and PstI, electrophoresed on an agarose gel, and blotted onto nitrocellulose paper. The transferred DNA fragments were hybridized with ^{32}P-labelled BamHI-PstI fragment from 101G (see Figure 2). Figure 3 shows that a single band of 1.9 kb hybridizes strongly, as predicted if the cloned BamHI-PstI sequence is contiguous in the genome.

The BamHI-PstI fragment was used to probe a variety of genomic digests with restriction enzymes which fail to cut within the insert. As expected, only one band hybridized strongly in each digest. Surprisingly, however, several other weakly hybridizing bands are apparent. Many of these bands are smaller than the strongly hybridizing bands, indicating that they are not due to partial digestions. Therefore, sequences contained within the BamHI-PstI fragment are cross homologous to other sequences in the genome.

RNA Homologous to 101G. The level of RNA in various strains corresponding to the 101G clone was determined by Northern blot analysis. RNA was isolated from S288C, aas101-1, aas102-15, aas103-15, and aas104-1 strains grown at 30° C on min AA, min his, min, and min AT. The RNAs were electrophoresed on a 1.5% agarose formaldehyde gel, transferred to a nitrocellulose filter, and hybridized with ^{32}P-labelled SphI-SphI fragment from 101G and YIp5 (pBR322-URA3). The URA3 gene was used as an internal standard to normalize the amount of RNA in each lane.

The S288C levels of 101G RNA are roughly equivalent on min AA, min his, and min (see Figure 4). However, under conditions of his starvation, min AT, the 101G mRNA level is very greatly increased. Therefore, the gene homologous to the 101G clone is itself being derepressed under conditions of histidine starvation. This derepression of the 101G transcript is also observed in the aas101-1, aas102-15, and aas103-15 strains, and to a lesser extent in the aas104-1 strain. This result implies that derepression of the 101G mRNA itself is not dependent upon the AAS101$^+$, AAS102$^+$, or AAS103$^+$ genes (and also probably the AAS104$^+$ gene).

II. 5' HIS4 Transcript

5' Open Reading Frames

Examination of the nucleotide sequences of several yeast genes under general amino acid control, HIS4, TRP5, and HIS3, reveals the existence of a sequence which contains an open reading frame upstream from the transcriptional start site of these genes (see Figure 5). In the HIS4 gene, this open reading frame is 198 basepairs (bp); it starts with an ATG at position -383 and ends with an inframe TGA at position -185. If this region is transcribed and translated it has the capacity to encode a polypeptide of 66 amino acids. It is striking that the position of the open reading frame preceding TRP5 is very closely coincident with that of HIS4, -355 to -185. These upstream regions may be candidates for regulatory elements involved in the control of their linked structural genes.

S1 Nuclease Mapping

To determine the possible involvement of this 5' open reading frame in

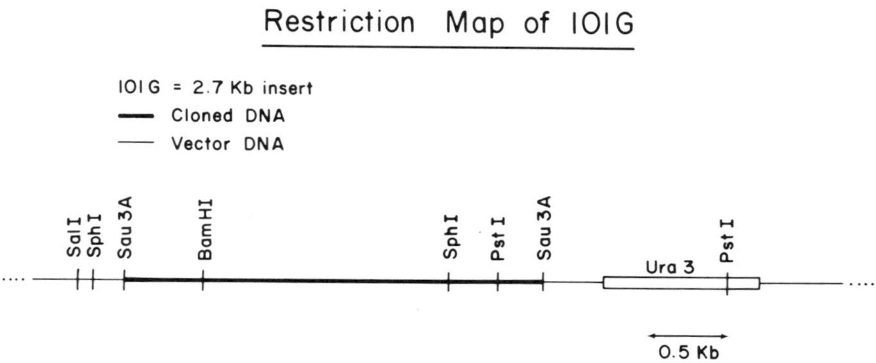

Figure 2. Restriction map of 101G.

The *Sau3A* sites indicated in the figure are not unique, but rather indicate the limits of the inserted genomic DNA fragment. *BglII*, *EcoRI*, *HindIII*, *SalI*, *SmaI* and *XhoI* do not cut within the insert.

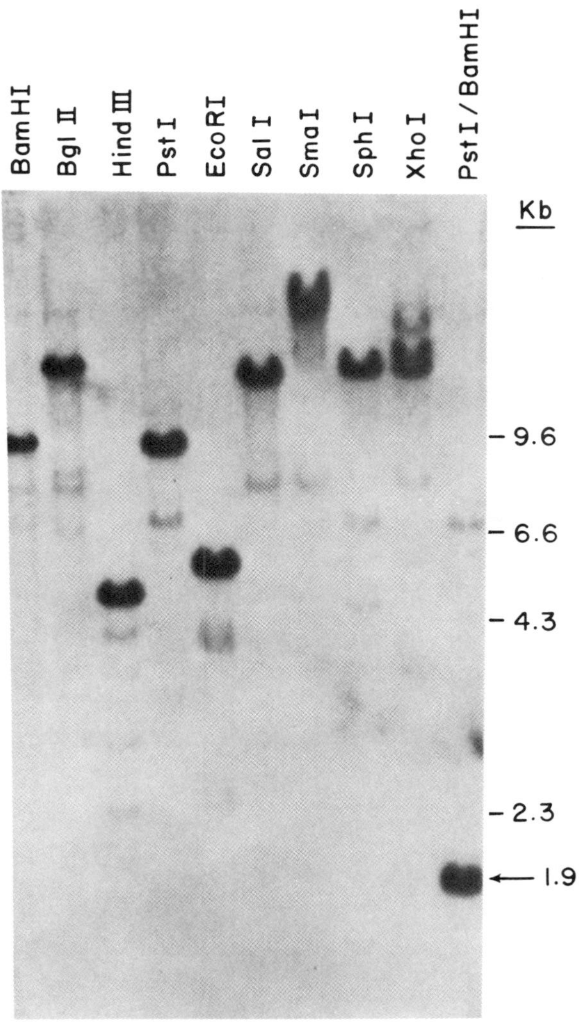

Figure 3. Southern blot of S288C genomic DNA probed with the BamHI-PstI fragment from 101G.

S288C DNA was digested with the restriction enzymes indicated and electrophoresed on a 0.8% agarose gel. DNA was transferred to a nitrocellulose filter according to the method of Southern (15). The filter-bound DNA was hybridized to ^{32}P nick translated (12) 1.9 kb BamHI-PstI fragment from 101G (see Fig. 2). Hybridization was carried out at 65°C in 5XSSC.

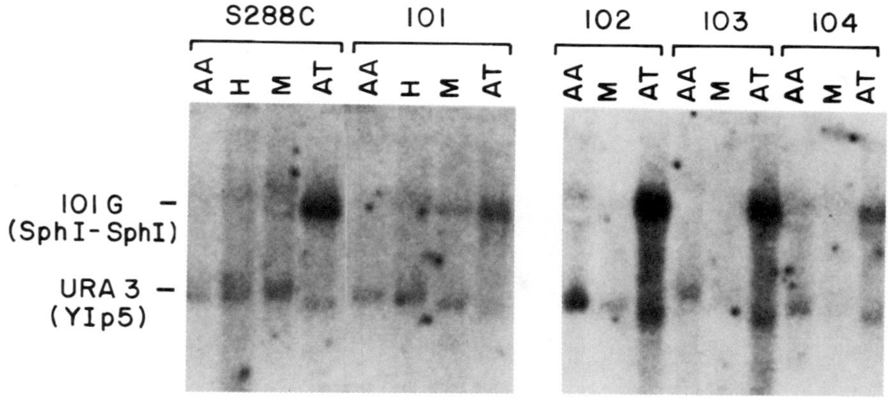

Figure 4. Regulation of 101G encoded RNA levels in wild type and *aas* mutant strains.

RNA was prepared from wild-type S288C, *aas101-1*, *aas102-15*, *aas103-15*, and *aas104-1*, electrophoresed, and blotted as described in Figure 1. Nitrocellulose bound RNA was hybridized with ^{32}P nick-translated DNA (12) from the *SphI*-*SphI* fragment of 101G and with YIp5 (*URA3*). Hybridization conditions were as described in Figure 1. The extent of hybridization with the *URA3* probe was used to normalize for the amount of RNA within each lane.

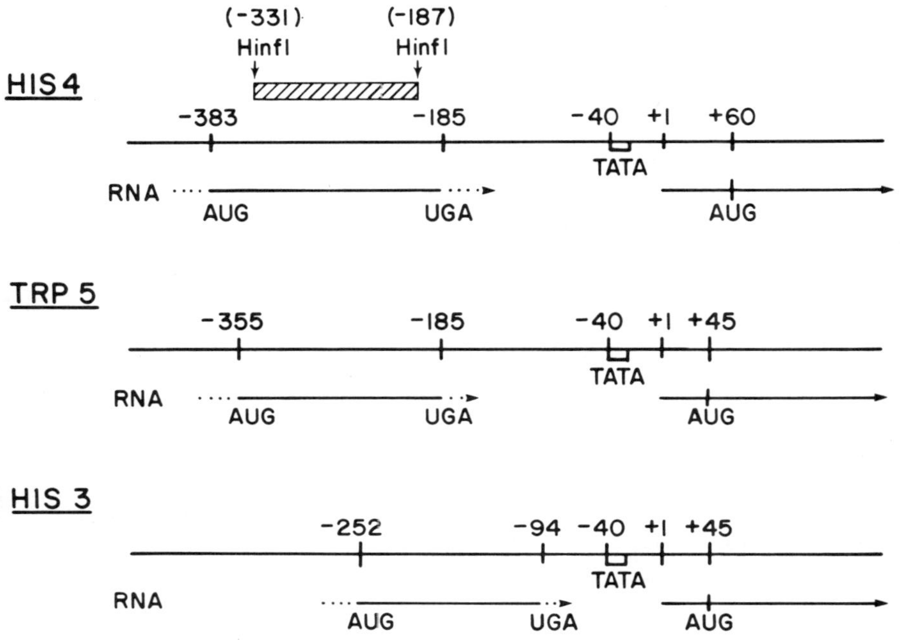

Figure 5. Comparison of the yeast 5' flanking regions of HIS4, TRP5 and HIS3.

The 5' regulatory regions are drawn to scale relative to the transcription initiation sites (+1) for their respective structural genes. The positions of the open reading frames are indicated by the solid black lines; the boundaries are delineated by ATG and TGA codons. Positions of potential TATA boxes are included for comparative purposes. The 148 bp HinfI fragment used in the S1 protection and the primer extension experiments is indicated. The sequence data is from P. Farabough and G. R. Fink, pers. comm. and (17, 20).

HIS4 gene expression, we first ascertained that this region was transcribed. A 148 bp HinfI fragment (see Figure 5), extending from -331 to -187 (internal to the sequences encoding the presumptive polypeptide), was isolated from a polyacrylamide gel and labelled at the 5' end with [γ-^{32}P] ATP. The labelled DNA was denatured and separated into fast and slow migrating strands on a 5% polyacrylamide gel. Total yeast RNA was incubated with each of the strands to allow homologous regions to anneal. This was followed by digestion of single stranded regions with nuclease S1. The protected annealed complexes were resolved on an 8% denaturing polyacrylamide gel and autoradiographed. As seen in Figure 6, there is specific protection from S1 digestion of the fast migrating DNA strand (lanes C, E, and F) whereas the slow migrating strand is unprotected (lane B). This result implies that the fast migrating strand is the coding strand. (This assignment is being confirmed by direct sequencing of the fast migrating strand). The protection of the fast migrating strand is dependent upon the formation of DNA-RNA complexes and is not the result of secondary structure of the probe itself, since complete digestion of the probe does in fact occur in the absence of homologous yeast RNA (lane D).

Cross Hybridizing Regions

To confirm that the fast migrating coding strand is protected from S1 digestion by a transcript originating in the 5' HIS4 region as opposed to elsewhere on the genome, we probed the yeast genome for cross hybridizing sequences. Total yeast DNA was digested with either SalI or EcoRI, electrophoresed on an agarose gel, and the fragments blotted onto nitrocellulose. The ^{32}P labelled 148 bp HinfI fragment was used as the hybridization probe under conditions of low stringency. The autoradiogram in Figure 7 shows that the probe hybridizes only to the predicted 1.5 kb SalI and 3.0 kb EcoRI fragments of the HIS4 region. The absence of cross hybridizing regions from other parts of the genome argues that the fast migrating strand is protected from S1 digestion by a transcript from the 5' HIS4 region.

Primer Extension

The existence of a transcript corresponding to the 5' flanking region of HIS4 was further confirmed by primer extension experiments. The 5' [γ-^{32}P] labelled fast migrating strand was used as a primer for reverse transcriptase in a reaction with total genomic yeast RNA (6). When the reaction was run on an 8% denaturing polyacrylamide gel, two discrete bands of increased molecular weight appeared (data not shown). The appearance of these bands is dependent upon the presence in the reaction mix of the fast migrating strand. No bands appear when the slow migrating strand is used as a primer. A more detailed analysis is in progress to define the *in vivo* transcriptional start site(s) of the upstream HIS4 transcript(s).

DISCUSSION

AAS Genes

We have described the isolation and characterization of 43 *aas* mutants. These mutations define four unlinked genes, AAS101, AAS102, AAS103, and AAS104. Complementation analysis indicates that AAS101 and AAS104 are two new

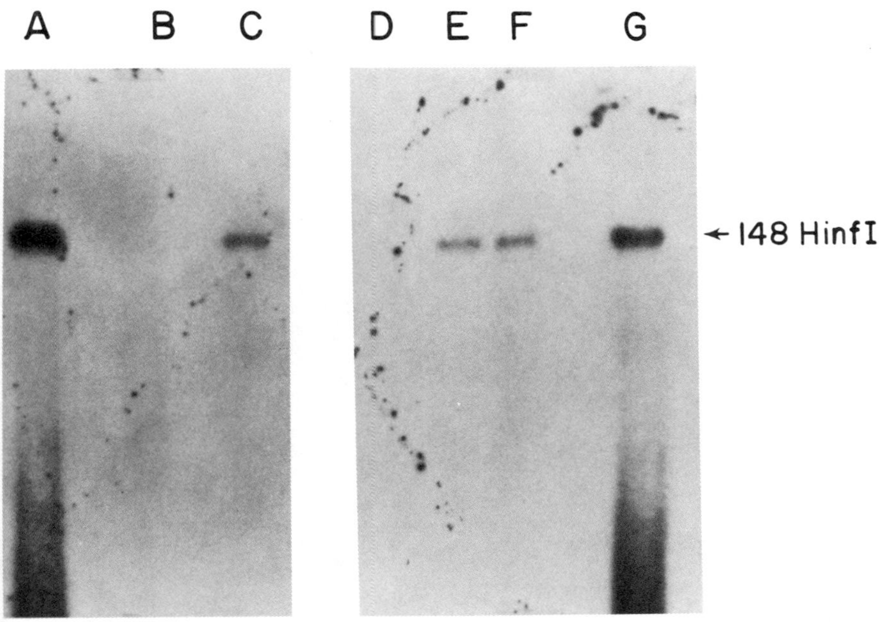

Figure 6. Specific protection of the 148 base HinfI fast migrating strand from S1 nuclease.

The 148 bp HinfI fragment was labelled at its 5' end with [γ-^{32}P], denatured and separated into fast and slow migrating strands on a 5% acrylamide gel (7). Total yeast RNA was hybridized to each of the strands, digested with nuclease S1, and the annealed complexes resolved on an 8% denaturing gel (1). The control lane A contains the 148 bp HinfI fragment, lane B contains the hybridization with the 148 base slow strand, and lane C contains the hybridization with the 148 base fast strand.

In a separate experiment the 148 base fast migrating strand, 5' end labelled with ^{32}P, was hybridized to total yeast RNA from a culture grown in the presence of 0.3 mM histidine (lane F) and a culture starved for histidine with AT (lane E). In lane D, the hybridization was carried out in the absence of homologous RNA. After S1 digestion, the complexes were resolved on an 8% denaturing gel. Lane G is an undigested control. The exposure was at -80°C with an intensifying screen.

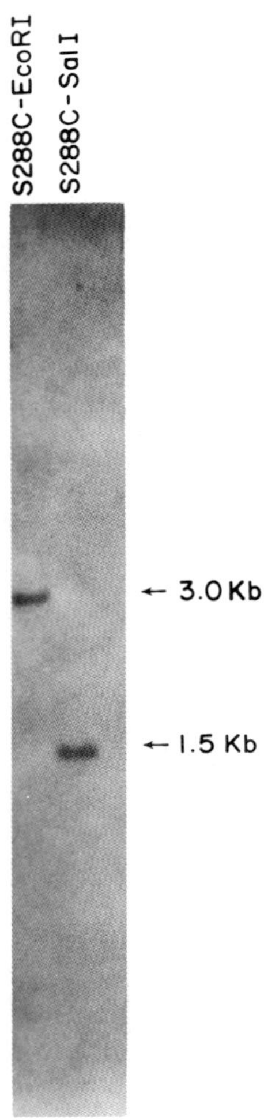

Figure 7. Southern blot of S288C genomic DNA probed with the *HinfI* fragment from the *HIS4* 5' flanking region.

DNA isolated from strain S288C, cut with *SalI* or *EcoRI*, was electrophoresed through a 1% agarose gel, and transferred to nitrocellulose according to the method of Southern (15). A 148 bp *HinfI* fragment, (-335 to -187), was labelled at the 5' end with $[\gamma\text{-}^{32}P]$ ATP by the method of Maxam and Gilbert (7), and was used as the hybridization probe. Hybridization was carried out under conditions of low stringency, 55°C in 5 x SSC. The exposure was for 4 days at -80°C with an intensifying screen.

genes involved in general control, whereas AAS102 and AAS103 are analogous to genes identified in previous studies as (NDR2 and AAS1) and (NDR1), respectively (9, 13, 19). The aas101 and aas104 mutants may not have been isolated in earlier searches because they appear at a very low frequency: only four mutations were obtained in AAS101 and only one in AAS104, out of a total of 43 isolated in this study.

The aas101 mutants have the most severe phenotype, the aas104 has the weakest, and aas102 and aas103 are intermediary. This is apparent from the degree of their sensitivity to various amino acid analogs, their growth rates, and their HIS4C enzyme activity and mRNA levels under starvation conditions. All four AAS genes appear to be involved to varying degrees in regulating derepression of the enzymes of many amino acid biosynthetic pathways. The aas mutations seem to prevent derepression of at least the histidine, tryptophan, arginine, and methionine pathways, as evidenced by their acquired sensitivity to analogs from these four pathways. Specifically, we have assayed the HIS4C enzyme levels and have shown that under histidine starvation conditions, aas101, 102, and 103 mutants do not derepress this enzyme at all; aas104 shows only partial derepression. Northern blot analysis of these mutants for the same growth conditions indicates that the failure to derepress the HIS4C enzyme is a result of lowered HIS4 mRNA levels in the aas mutants. This result suggests that the positive regulatory role of these AAS^+ genes could be to promote transcription of genes coding for amino acid biosynthetic enzymes. Alternatively, the regulation could be exerted at the level of degradation of mRNA.

One aas103-complementing clone, 103R3, and two different aas101-complementing clones, 101G and 101M, were isolated. Clearly only one of the 101 clones can be the AAS1C1 structural gene. The 101G clone has been looked at in greater detail. Northern blot analysis of wild type S288C RNA indicates that under histidine starvation conditions, the 101G mRNA is itself greatly induced. This implies that the 101G gene is regulated in a way that directly relates to its function. If the AAS genes are positive controlling elements needed for derepression of genes under general control, an increase in AAS mRNA levels under amino acid starvation conditions is certainly logical. The Northern blot results also suggest that the regulation of the 101G gene is not mediated by the AAS102 or AAS103 gene products since the derepression of 101G mRNA occurs in strains which are mutated in these genes. (This is probably also true for the AAS104 gene, but the results are somewhat ambiguous.)

Southern blot analysis indicates that at least part of 101G sequences are present in more than one copy in the genome. Whether this reflects multiple gene copies, or the possible presence of some reiterated element, remains to be determined. Another observation is that 101M and 103R3 clones contain roughly 5 kb of inserted sequences in common. It is possible that this sequence is responsible for the observed complementation response in aas101 and aas103 mutants. Further analysis of the cloned sequences is necessary to ensure that the phenotypes which these plasmids confer are not high copy number artifacts.

5' Transcript

In this paper we have determined the existence of a transcript in the 5' regulatory region of one of the genes under general control, HIS4. The S1 nuclease experiments indicate that an *in vivo* transcript exists for only one of the strands of this region. Our preliminary primer extension data indicates that transcripts of this region with two distinct start sites may exist. We are currently defining precisely where these transcriptional start sites map. The upstream HIS4 transcript contains an open reading frame and has the capacity to encode a polypeptide of 66 amino acids. The fact that similarly positioned open reading frames exist upstream from the structural genes of two other amino acid biosynthetic genes, TRP5 and HIS3, strongly suggests that these putative polypeptides may have some regulatory function in the general control of amino acid biosynthesis.

Inspection of the nucleotide sequence of the 5' flanking regions reveals an intriguing property of the open reading frames. If these regions are translated, they exhibit an extreme bias for rare and infrequently used codons. These rare codons tend to be homologous to the anticodons of the minor yeast isoacceptor tRNA species. This suggests a number of regulatory mechanisms which would allow the yeast cell to monitor the charged state of the tRNAs.

Experiments are in progress to determine the involvement of the 5' transcript in HIS4 gene regulation. We are constructing inframe lacZ fusions to this region which will be used in conjunction with Northern blot analysis to correlate mRNA and protein levels with HIS4 gene expression. Another approach for determining the role of the 5' transcript, or its translation product, on HIS4 gene expression, is to construct deletions of the coding region. Typically, deletion analysis of the 5' region of yeast genes has been performed by generating deletions with an exonuclease, such as Bal31, by initiating at an upstream restriction site and deleting bases for varying distances in the direction of the promoter. Deletions generated in this manner remove all genetic homology 5' to the promoter. Moreover, the promoter is now flanked by sequences which are normally well separated. This sets up a situation which is clearly subject to artifacts. To minimize potential artifacts, we are constructing a small deletion which is totally internal to the putative coding sequences, leaving all flanking sequences intact. Analysis of a small deletion which generates a frameshift will allow us to assess the biological role of the putative translation product.

Depending upon the outcome of these experiments, a model for the role of the transcript that we have identified should become apparent. What, if any, the relationship between the AAS^+ genes and this transcript is, remains to be addressed.

ACKNOWLEDGEMENTS

We especially wish to thank George Thireos for many helpful discussions and for comments on the manuscript.

LITERATURE CITED

1. Berk, A.J. and Sharp, P.A. 1978. Spliced early mRNAs of simian virus 40. Proc. Nat. Acad. Sci. USA 75:1274-1278.

2. Carsiotis, M. and Jones, R.F. 1974. Cross-pathway regulation: tryptophan-mediated control of histidine and arginine biosynthetic enzymes in *Neurospora crassa*. J. Bacteriol. 119:889-892.

3. Carsiotis, M., Jones, R.F., and Wesseling, A.C. 1974. Cross-pathway regulation: histidine-mediated control of histidine, tryptophan, and arginine biosynthetic enzymes in *Neurospora crassa*. J. Bacteriol. 119:893-898.

4. Delforge, J., Messenguy, F., and Wiame, J.M. 1975. The regulation of arginine biosynthesis in *Saccharomyces cerevisiae*: The specificity of arg R^- mutations and the general control of amino acid biosynthesis. Eur. J. Biochem. 57:231-239.

5. Greer, H. and Fink, G.R. 1975. Isolation of regulatory mutants in *Saccharomyces cerevisiae*, pp. 247-272. In: D. Prescott (ed.), Methods in Cell Biology, Vol. XI, Yeast Cells. Academic Press, New York.

6. Hagenbüchle, O., Bovey, R., and Young, R. 1980. Tissue-specific expression of mouse α-amylase genes: nucleotide sequence of isoenzyme mRNAs from pancreas and salivary gland. Cell 21:179-187.

7. Maxam, A. and Gilbert, W. 1977. A new method for sequencing DNA. Proc. Nat. Acad. Sci. USA 74:560-564.

8. Messenguy, F. and Fink, G.R. 1973. Feedback-resistant mutants of histidine biosynthesis in yeast, pp. 85-95. In: A. Srb (ed.), Genes, Proteins and Evolution. Plenum Press, New York.

9. Niederberger, P., Miozzari, G., and Hutter, R. 1981. Biological role of the general control of amino acid biosynthesis in *Saccharomyces cerevisiae*. Mol. and Cell Biol. 1:584-593.

10. Petersen, N.S. and McLaughlin, C.S. 1973. Monocistronic messenger RNA in yeast. J. Mol. Biol. 81:33-45.

11. Piotrowska, M. 1980. Cross-pathway of regulation of ornithine carbamoyltransferase synthesis in *Aspergillus nidulans*. J. Gen. Microbiol. 116:335-339.

12. Rigby, P. W., Diekmann, M., Rhodes, C., and Berg, P. 1977. Labeling deoxyribonucleic acid to high specific activity *in vitro* by nick translation with DNA polymerase I. J. Mol. Biol. 113:237-251.

13. Schurch, A., Miozzari, J., and Hutter, R. 1974. Regulation of tryptophan biosynthesis in *Saccharomyces cerevisiae*: mode of action of 5-methyl-tryptophan and 5-methyl-trytophan-sensitive mutants. J. Bacteriol. 117:1131-1140.

14. Sherman, F. and Stewart, J.W. 1982. Initiation of translation. In: J. Strathern, E. Jones and J. Brooch (eds.), Molecular Biology of the Yeast *Saccharomyces*. Cold Spring Harbor Laboratory, Cold Spring Harbor, N.Y.

15. Southern, E. M. 1975. Detection of specific sequences among DNA fragments separated by gel electrophoresis. J. Mol. Biol. 98:503-517.

16. Sripati, E. and Warner, J. 1978. Isolation, characterization, and translation of mRNA from yeast, pp. 61-81. In: D. Prescott (ed.), Methods in Cell Biology, Vol. XX. Academic Press, New York.

17. Struhl, K. 1981. Deletion mapping a eukaryotic promoter. Proc. Nat. Acad. Sci. USA 78:4461-4465.

18. Thomas, P. T. 1980. Hybridization of denatured RNA and small DNA fragments transferred to nitrocellulose. Proc. Nat. Acad. Sci. USA 77:5201-5205.

19. Wolfner, M., Yep, D., Messenguy, F. and Fink, G.R. 1975. Integration of amino acid biosynthesis into the cell cycle of *Saccharomyces cerevisiae*. J. Mol. Biol. 96:273-290.

20. Zalkin, H. and Yanofsky, C. 1982. Yeast gene TRP5: structure, function, regulation. J. Biol. Chem. 257:1491-1500.

11. PROTEIN SECRETION AND ORGANELLE ASSEMBLY IN YEAST*

Randy Schekman, Peter Novick, Susan Ferro-Novick, Brent Esmon, William Hansen, Tina Etcheverry, and Tom Stevens

Department of Biochemistry
University of California, Berkeley
Berkeley, California 94720

Rec. Adv. Yeast Mol. Biol. (1982) 1:143-155

Protein secretion is a major aspect of cell metabolism and provides a mechanism for assembly of internal organelles and the cell surface. The cellular functions which execute the secretory process in yeast have been identified genetically by the isolation of temperature-sensitive lethal mutants that block the secretory pathway at one of four stages. Three of these stages are defined by class A <u>sec</u> mutants which accumulate secretory glycoproteins inside one of three distinct organelles: endoplasmic reticulum (ER), Golgi bodies, or secretory vesicles. Glycoproteins and secretory organelles accumulate at a nonpermissive temperature (37°C) and proceed to a succeeding stage in the pathway when cells are returned to a permissive temperature (25°C) even in the absence of new protein synthesis. Another type of <u>sec</u> mutant (class B) fails to produce active secretory enzymes even though secretory polypeptides are synthesized. Some of the mutants in this class are blocked in the translocation of secretory polypeptides across the ER membrane.

The secretory pathway is responsible for localization of major yeast plasma membrane surface proteins. The export of at least four permease activities and six externally labeled plasma membrane proteins is blocked thermoreversibly in the <u>sec</u> mutants. The transport organelles, secretory vesicles in particular, may carry secreted enzymes with only a subset of plasma membrane proteins. Purified secretory vesicles do not contain chitin synthetase and vanadate-sensitive ATPase, two bonafide integral plasma membrane proteins. Another vesicle may be responsible for transport of these and other membrane proteins.

Part of the secretory pathway is also responsible for localization of vacuolar glycoproteins. Carboxypeptidase Y (CPY) is synthesized as an inactive proenzyme which is matured in the vacuole by cleavage of an 8 Kd amino-terminal propeptide. Proenzyme forms of CPY accumulate thermoreversibly in <u>sec</u> mutants that are blocked in movement from the ER or from the Golgi body, but not in mutants that block transport of secretory vesicles. Vacuoles isolated from <u>sec</u> mutant cells do not contain the proCPY produced at 37°C. These results suggest that vacuolar and secretory glycoproteins require the same cellular functions for transport from the ER and from the Golgi body. The Golgi body represents a branch point in the pathway: from this organelle vacuolar proenzymes are transported to the vacuole for proteolytic processing and secretory proteins are packaged into vesicles.

*Support was obtained from National Science Foundation grant PCM-8019415 and from National Institutes of Health grants GM-26755 and ES-01896.

Introduction

Eukaryotic cells contain a number of membrane-bounded organelles which do not, on simple visual examination, seem to be related. Studies in the 1950's and 1960's by Palade, Porter, de Duve, and Claude led to our current view that compartmentalization provides the eukaryotic cell with a wide range of possibilities for regulation of metabolism. Recent progress on the study of the assembly of cellular organelles, in particular on the mechanism of localization of specific protein constituents, has raised the possibility of intimate connections among the organelles. The secretory process has emerged as a common theme for transport of proteins and lipids to all parts of the cell.

Protein secretion occurs in almost all cell types. Despite the wide range of activity that this implies, the stages in the secretory process are quite similar in all organisms. In eukaryotes, the sequence ER → Golgi → vesicle → cell surface is the generally accepted mode of transport for soluble and membrane proteins. Although prokaryotes clearly do not have specialized secretory organelles, polypeptide penetration across the bacterial cytoplasmic membrane occurs by processes quite analogous to those employed in protein import into the endoplasmic reticulum and into the mitochondrion.

In addition to a role in cell surface assembly, the secretory process may contribute to the assembly of the lysosome, the mitochondrion, and the nucleus. Lysosomal glycoprotein precursors are translocated into ER membranes by the same system used for secretory and plasma membrane proteins (2). Furthermore, both in histochemical and in organelle fractionation studies, lysosomal enzymes are detected in ER and Golgi cisternae (27). Mitochondrial membranes are assembled with lipids synthesized in the ER. Although many mitochondrial enzymes derive from soluble cytoplasmic precursors, the ER and mitochondrial outer membranes appear to share a number of integral proteins (29). Finally, the nuclear envelope is continuous with the ER. This is most apparent during interphase, when the nuclear envelope is reconstructed by outgrowth from the ER. Certain soluble nuclear components may also be derived from secretory organelles. The observation that chromatin-associated high mobility group proteins (HMGs) contain N-glycosidically-linked, complex oligosaccharides suggests that these proteins may gain access to the nucleus via the Golgi body (25). Clearly, secretory organelles play a major role in cell architecture and metabolism.

While a considerable amount is known about the gross features of secretory organelles, and about the structure and covalent modifications of molecules that are transported through the organelles, it has been much more difficult to define cellular functions involved in protein transport. New approaches involving genetic and biochemical techniques will be essential for appreciating the mechanism of transport. In this regard, recent advances in identifying proteins involved in the penetration of secretory and membrane polypeptides across a membrane (32,17,35), and in the discrimination of lysosomal and secretory proteins (13,14,26) provide important examples of these cellular functions.

Identifying the full range of functions required for protein transport can be achieved with a genetic approach. For this and other reasons, my laboratory has undertaken a study of the secretory process in the yeast *Saccharomyces cerevisiae*. Although much less is known about the secretory process in yeast than in mammalian cells, and in some ways there are special technical difficulties in the use of yeast as an experimental system, the potential for a combined genetic and biochemical approach may prove a crucial advantage.

Organization of the Yeast Cell Surface

The yeast cell surface consists of at least three layers: The cell wall which contains mannoproteins and structural polysaccharides (β-1,3 and β-1,6-linked glucan), a periplasm that contains mannoproteins, and a plasma membrane. Most of the soluble secreted enzymes, such as invertase and acid phosphatase, are located in the periplasm or in the cell wall where they are accessible to low molecular weight substrates. Certain smaller non-glycosylated proteins, such as α-factor and killer toxin, are secreted through the cell wall into the culture medium.

Secretion is correlated topologically to the region of cell surface growth. Invertase and acid phosphatase are secreted into the bud portion of a growing cell which corresponds to the point of cell surface addition during most of the division cycle (31, 8). The correlation between secretion and budding is best accounted for by an exocytic mechanism of surface growth. Secretory vesicles may fuse with the inner surface of the bud and deliver mannoproteins to the periplasm and membrane precursors to the plasma membrane. The available cytologic evidence strongly supports this notion. Electron microscopic thin section and freeze fracture views show 50-100 nm vesicles which fuse with the bud plasma membrane (18). Histochemical staining of cells secreting acid phosphatase has shown enzyme-specific stain of the bud-localized vesicles, the ER, and a Golgi-like organelle (16).

Although the yeast secretory process appears to resemble the mechanism used by plant and animal cells, one striking difference is the low level of secretory organelles revealed by standard EM thin section analysis. This low level is consistent with a rapid transit time for export of invertase (20), and a low level of invertase export precursors (22). The small internal pool of secretory precursors provides a sensitive experimental system for the evaluation of mutants that block the secretory pathway and cause an accumulation of secretory enzymes and organelles.

Isolation and Characterization of Secretory Mutants

Given the possibility that the secretory process contributes generally to yeast cell surface growth, Peter Novick assumed that secretory mutants would be lethal. To get around this problem, Novick screened a collection of temperature-sensitive growth mutants for ones that failed to export active invertase and acid phosphatase at the nonpermissive temperature (37°C), but which performed normally at the permissive temperature (25°C). Mutants representing two complementation groups (*sec1*, *sec2*) were found which accumulated secretory enzymes in an intracellular pool (22). A large number of additional *sec* mutants have been isolated based on the observation that *sec1* cells become dense at 37°C. Susan Henry showed that during inositol starvation of an auxotrophic strain, net cell surface growth stopped while cell mass increased (12). Starved cells could be resolved from normal cells on a Ludox density gradient. Similarly, *sec* mutant cells can be enriched from a mutagenized culture by incubation at 37°C followed by Ludox density gradient sedimentation.

The *sec* mutants are of two types. Class A *sec* mutants (192 total) are like *sec1* and *sec2*, in that active secretory enzymes accumulate in an intracellular pool (21). Class B *sec* mutants (23 total) do not secrete or accumulate active secretory enzymes, yet protein synthesis continues at a near normal rate for several hours at 37°C. Complementation analysis has revealed 23 *sec* loci in the

A class and 4 sec loci in the B class. The distribution of mutant alleles suggests that more of both classes could be found.

Many of the class A sec mutants secrete a large fraction of the invertase that accumulates at 37°C when cells are returned to the permissive temperature. In most cases the secretion of accumulated invertase is insensitive to cycloheximide. This implies that the affected gene product is reversibly inactivated by the temperature shift. The result demonstrates that ongoing protein synthesis is not essential for post-translational transit of secretory enzymes, and thus excludes the possibility that newly-synthesized secretory protein forces the flow of the export process. Mutations in one gene (sec7) allow thermoreversible secretion only in growth medium containing a low concentration of glucose.

Perhaps the most dramatic feature of the class A sec mutants is that they accumulate or exaggerate specific secretory organelles. Mutations in ten groups produce 80-100 nm vesicles at 37°C that are distributed throughout the cytoplasm, unlike the bud-localized vesicles seen in wild-type cells. Mutations in another nine genes produce exaggerated endoplasmic reticulum. In these mutants the ER lines the inner surface of the plasma membrane and winds through the cytoplasm where multiple connections with the nuclear envelope are seen. The lumen of both the ER and the nuclear envelope is wider than the corresponding wild-type structure. Due to the high density of ribosomes in the background, it has not been possible to determine if the exaggerated ER is in the rough or smooth form. A third class of mutant, represented by two genes, produces a different organelle depending on the growth medium in which the cells are incubated at 37°C. Mutations in the sec7 gene cause the accumulation of typical Golgi-like structures when mutant cells are incubated at 37°C in medium containing 0.1% glucose. The same mutant, when incubated at 37°C in medium with 2% glucose, accumulates cup- and toroid-shaped organelles that we have called Berkeley bodies. This change in organelle morphology correlates with the effect of glucose on thermoreversible secretion mentioned earlier, and suggests that the Golgi-body structure is a more natural intermediate. In each case where reversible secretion is observed, a return to the permissive temperature allows the accumulated organelle to diminish in abundance. Histochemical staining of mutants representing each of the distinct cytologic types has shown that the secreted enzyme, acid phosphatase, is contained in the lumen of the accumulated organelle (3).

Susan Ferro-Novick has found that the class B sec mutants produce enzymatically inactive forms of invertase (5). In two of the mutants (sec53 and sec59), immunoreactive forms of invertase are produced at 37°C which appear to remain embedded in the ER membrane (6). Perhaps as a result of this aberrant accumulation, the ER in these mutants appears fragmented in contrast to the smooth, thin ER envelope seen in wild-type cells. These mutants also show greatly reduced incorporation of [^{3}H]-mannose into a total glycoprotein fraction at 37°C, although oligosaccharide synthesis is not directly affected. Reduced mannose incorporation appears to be due to a defect in translocation of nascent polypeptide chains to the lumenal surface of the ER membrane where oligosaccharid are transferred to protein. Surprisingly, protease protection experiments have indicated that a significant portion of the invertase polypeptide (11 Kd of a 60 Kd mature polypeptide length) is inserved into and perhaps across the ER membrane in mutant cells at 37°C (7). Furthermore, as with the class A sec mutants sec53 and sec59 are thermoreversible. Upon return to 25°C, in the presence of cycloheximide, the membrane-bound form of invertase is transferred into the lumen of the ER, glycosylated, and transported to the cell surface through the normal pathway. Thus, the sec53 and sec59 gene products define functions required for completion, but not initiation of protein penetration across the ER membrane.

Order of Events in the Pathway

A simple technique exists for the ordering of events along a linear irreversible pathway in which distinct intermediates accumulate in different mutants. In this circumstance a double mutant will accumulate the intermediate prior to the first block that is encountered. This analysis has been performed with mutants representing each of the four stages that are identified by the *sec* mutations. Double mutants containing *sec53* or *sec59* together with any of the class A *sec* mutants fail to accumulate active invertase (6). Thus, these class B *sec* mutants are epistatic to the other mutants. Among the class A *sec* mutants, the ER-accumulating phenotype is epistatic to the Golgi body- and vesicle-accumulating phenotypes, and a Golgi body-accumulating mutant is epistatic to all of the vesicle-blocked mutants (20). The order of events determined by this analysis is shown in Figure 1.

An independent line of evidence supports this order of events. Brent Esmon and Susan Ferro-Novick have analyzed the extent of glycosylation of invertase accumulated at 37°C in the *sec* strains and identified at least two stages in oligosaccharide assembly (31,6). Very little carbohydrate is present on the membrane-bound form of invertase accumulated in *sec53* and *sec59*, consistent with a defect in the translocation of secretory polypeptide into the ER lumen. During normal transport into the ER, invertase acquires 9-10 N-glycosidically-linked oligosaccharides which have a composition identical to the mammalian "high-mannose" oligosaccharide ($Glc_3Man_9GlcNAc_2$) (4). The three glucose residues are removed at some post-translational step in the ER because *sec* mutants blocked in movement from the ER have high mannose oligosaccharides with a composition of $Man_9GlcNAc_2$ (4). In addition to N-linked carbohydrate, yeast mannoproteins have O-linked mannotriose and -tetraose. The O-linked sugars appear to be formed in the ER; O-linked mannose and mannobiose accumulate in the ER-blocked mutants. Oligosaccharide maturation is completed either in transit to or within the yeast Golgi body. Mutants that block movement from the Golgi body, or the later mutants that block discharge of secretory vesicles, accumulate N-linked oligosaccharides that have a complete outer chain structure and O-linked mannotriose and -tetraose. Furthermore, incompletely glycosylated molecules that accumulate at an ER block are processed when cells are returned to the permissive temperature. The sequence of carbohydrate maturation events is summarized in Figure 2. Taken together the cytologic and biochemical results demonstrate that the yeast secretory pathway is essentially identical to the mammalian process.

Plasma Membrane Assembly

The initial characterization of the *sec* mutants suggested a general block in secretion and cell-surface growth at the nonpermissive temperature. In addition to an immediate halt in bud growth, export of a number of secreted proteins (invertase, acid phosphatase, L-asparaginase, α-galactosidase, α-factor, killer toxin) and plasma membrane permease activities ($SO_4^=$ permease, arginine permease, galactose permease, proline-specific permease) is blocked. A more general probe of surface assembly that was used to examine the export and turnover of macrophage plasma membrane proteins (15) has now been adapted by Novick to examine surface assembly in yeast (23). Modification of cell surface amino groups with trinitrobenzenesulfonate (TNBS) followed by precipitation with TNP antibody allows analysis of newly-exported proteins. In this procedure, wild-type and mutant

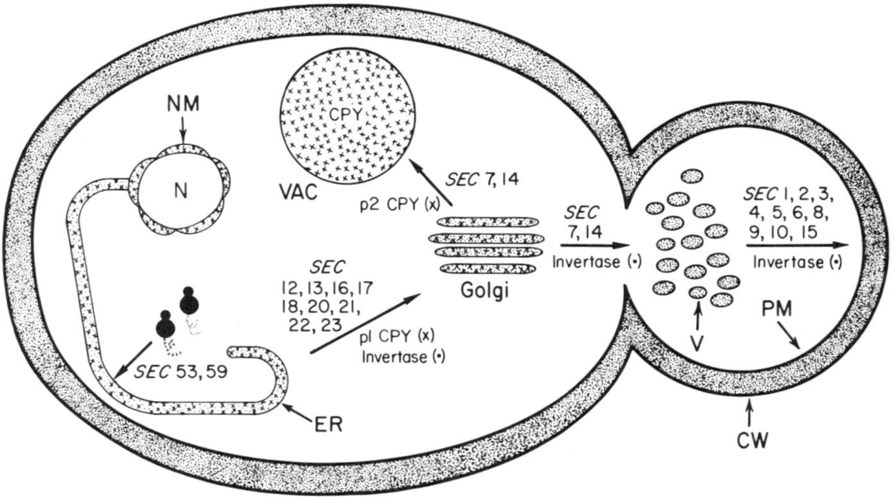

FIGURE 1. Secretory and vacuolar protein transport pathways in yeast. N: nucleus; NM: nuclear membrane; ER: endoplasmic reticulum; SEC: wild-type gene; VAC: vacuole; V: vesicle; PM: plasma membrane; CW: cell wall; CPY: 61 Kd mature carboxypeptidase Y; pl CPY: a 67 Kd proenzyme form of CPY; p2 CPY: 2 69 Kd proenzyme form of CPY.

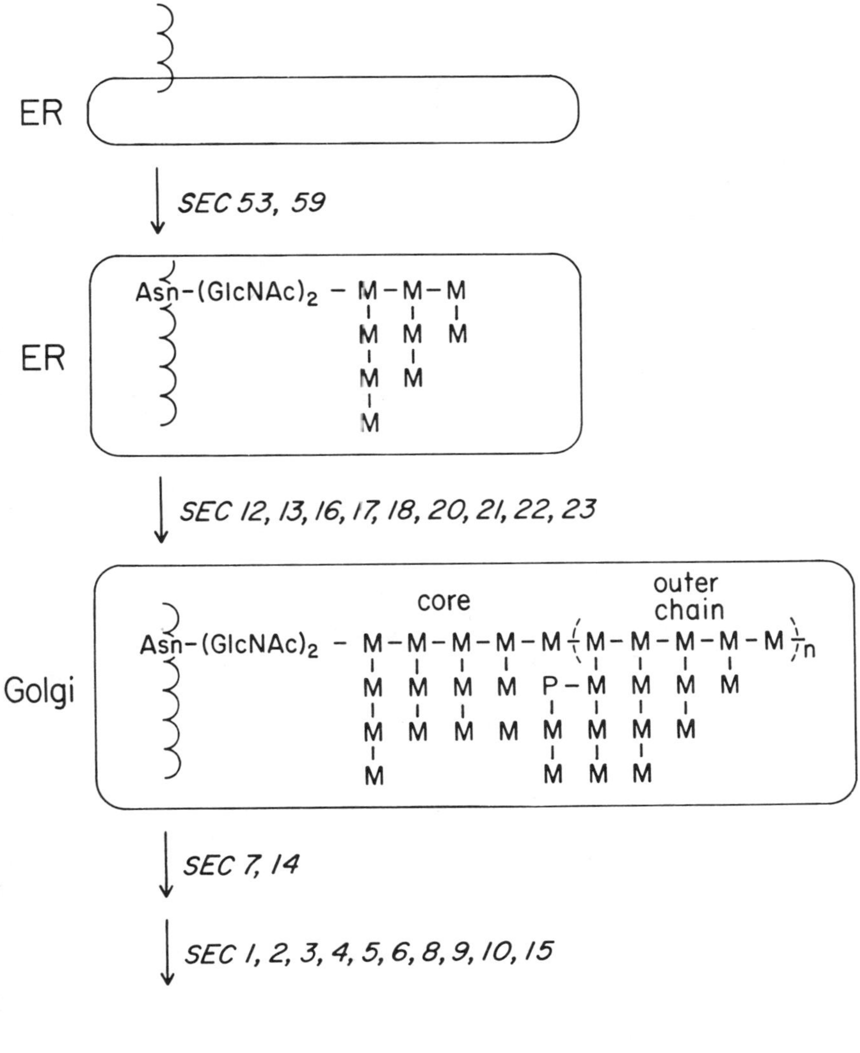

FIGURE 2. Compartmentalized assembly of mannoprotein oligosaccharide. Asn: asparagine; GlcNAc, N-acetylglucosamine; M: mannose; ⌇: polypeptide.

cells are labeled with protein synthesis precursors at 37°C and then tagged with TNBS at 0°C. Under these conditions, TNBS does not penetrate into the cell. Both secreted and plasma membrane surface proteins are tagged in this procedure and can be examined separately: secreted proteins are released when cells are converted to spheroplasts, and tagged membrane proteins are recovered in a sedimented fraction from lysed spheroplasts. Wild-type cells export distinct sets of membrane and secreted proteins as revealed by SDS-gel electrophoresis of TNP-antibody precipitates. The major proteins in both fractions are not exported in _sec_ mutant cells at 37°C, but are at 24°C. Furthermore, accumulated surface proteins are exported when radiolabeled _sec_ mutant cells are returned to 24°C. These results suggest that the secretory process is responsible for the localization of most cell surface proteins in yeast.

In order to test the possibility that secretory organelles that accumulate in the _sec_ mutants contain secretory and plasma membrane proteins in the same structures, mutants representing each stage in the pathway have been used as a source of material in cell fractionation experiments. _sec_ Mutant cells are induced for invertase synthesis at 37°C so that all new activity is contained within the accumulated organelle. Bill Hansen has devised a procedure for lysis of spheroplasts that retains invertase within secretory organelles. Intact organelle are monitored during membrane fractionation using latent invertase as an enzyme marker. Tina Etcheverry has developed a method for isolation of invertase-containing vesicles from the _sec1_ mutant. The procedure involves velocity and density sedimentation followed by electrophoresis on an agarose gel. The final fraction is free of contamination by other known cytoplasmic and membrane markers, while at least one other secretory enzyme, acid phosphatase, fractionates along with invertase. Of particular note is that two integral plasma membrane proteins, chitin synthetase and vanadate-sensitive ATPase (1,34), are synthesized in a sedimentable form in _sec1_ cells at 37°C but fractionate away from secretory vesicles. These two membrane enzymes, and probably others, may be transported to the cell surface in a distinct vesicle which does not behave like the invertase containing particle. The two separable organelles may nevertheless share a requirement for _sec_ gene products involved in transport to the cell surface.

Vacuole Assembly

The yeast vacuole contains a number of hydrolytic glycoprotein enzymes and is thus analogous to the mammalian lysosome (33). Figure 3 lists several soluble and one membrane enzyme known to be localized in the vacuole. The serine-protease carboxypeptidase Y (CPY), is a vacuolar enzyme that has been studied extensively. CPY has a single subunit of 61 Kd, 10 Kd of which is carbohydrate in the form of four N-glycosidically-linked oligosaccharides (10). Hasilik and Tanner (9) found that CPY is synthesized as a proenzyme that is converted _in vivo_ to the mature form with a halftime of 6 min. A 69 Kd proenzyme form of CPY was found by Hemmings _et al_. (11) to accumulate in a mutant, _pep4_, which is pleiotropically defective in the maturation of a variety of vacuolar enzymes. Hemmings _et al_. (11) also found that a chain-terminating mutation in the CPY structural gene which results in a shortened polypeptide in a _PEP4_ strain, produces a fragment that is 8 Kd larger in a _pep4_ strain. These data suggest that proenzyme maturation requires the _PEP4_ gene product and is achieved by cleavage of an amino-terminal 8 Kd propeptide. Proenzyme maturation and transport does not require glycosylation. Yeast cells treated with tunicamycin, a drug which blocks the synthesis of high-mannose oligosaccharides, synthesize, transport, and proteolytically convert a nonglycosylated form of CPY (10,19).

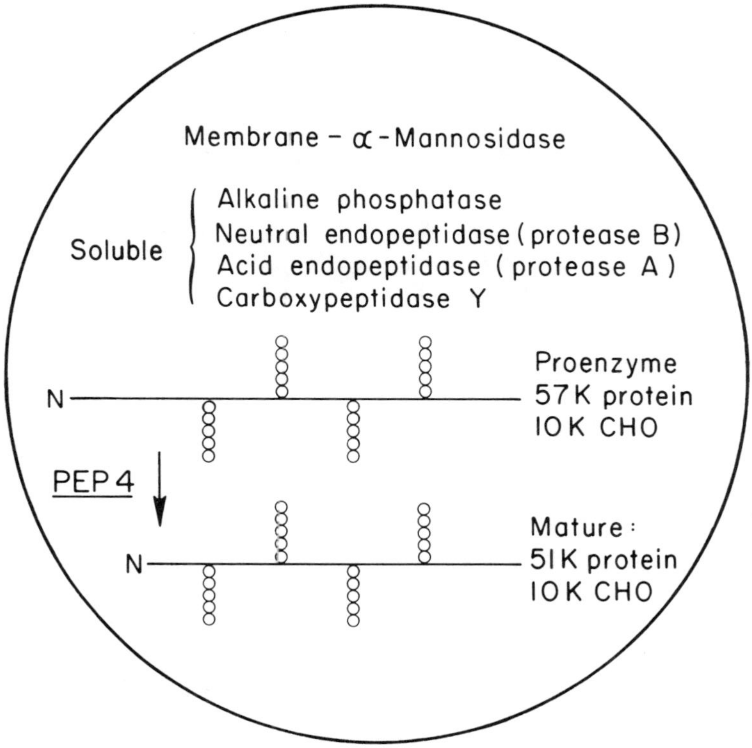

FIGURE 3. Yeast vacuole and major enzyme constituents; maturation of carboxypeptidase Y requires the PEP4 gene product.

Proenzyme proteolytic cleavage is not required for transport. Tom Stevens has shown the proCPY produced in pep4 cells is localized properly in the vacuole (30). Thus, the PEP4 gene product is required for maturation, and not for transport. If proenzyme maturation occurs in the vacuole, and if vacuolar proteins require the secretory pathway for transport, proenzyme forms will not be localized and processed at 37°C in the sec mutants.

Ferro-Novick and Hansen have shown that unglycosylated proCPY remains associated with the ER membrane in sec53 and sec59 when cells are labeled at 37°C (6). Stevens has shown that class A sec mutants that block movement from the ER or from the Golgi body accumulate proCPY in some place other than the vacuole, presumably in the accumulated organelle (30). Upon return to the permissive temperature the accumulated proenzyme forms become processed normally. Mutants that block after the Golgi step have no effect on CPY localization. These results suggest that vacuolar and secretory proteins travel together from the ER to the Golgi body where sorting may occur (Figure 1). These results rule out a secretion-recapture mechanism of localization such as has been suggested in studies on mammalian lysosomal enzyme transport (19).

A mannose-6-P determinant on N-linked oligosaccharides has been implicated in the targeting of lysosomal enzymes in human fibroblasts. Although sorting of lysosomal and secretory proteins in the Golgi body may rely on a carbohydrate structure, the ultimate source of discrimination lies in an amino acid sequence or structural feature of the targeted protein. Carbohydrate does not serve this role in yeast because at least two vacuolar proteins, CPY and an alkaline phosphatase, are synthesized and activated normally in the absence of oligosaccharide synthesis (10,24). Mutant alleles of CPY that result in misdirection of an otherwise normal proenzyme may reveal the signal responsible for normal localization.

Conclusions

The transport of cell surface macromolecules requires a large number of cellular functions. Lesions in these essential functions lead to an interruption of plasma membrane and secretory protein export at one of four stages in a linear pathway (Figure 1). Mutants blocked early in the pathway have shown that glycoprotein carbohydrate synthesis is compartmentalized: core oligosaccharides are added in the ER, and the outer chain structure is extended in the Golgi body (Figure 2). These same mutants have revealed that part of the pathway is responsible for localization of at least one vacuolar glycoprotein.

The sec mutants have provided a new method for tracing the pathway of protein localization in a eukaryote, and have in a limited sense defined the cellular functions required for transport. In vitro reactions that require the sec gene products and thus reflect portions of the secretory pathway will be needed to understand the mechanism of transport.

I gratefully acknowledge R. Roon, R. Buckholz, D. Julius, H. Bussey, J. Tschopp, P. Call, and W. Courchesne for their communication of preliminary results concerning the behavior of sec mutants.

References

1. Durán, A., B. Bowers, and E. Cabib. 1975. Chitin synthetase zymogen is attached to yeast plasma membrane. Proc. Natl. Acad. Sci. USA 72:3952-3955.

2. Erikson, A. H., and G. Blobel. 1979. Early events in the biosynthesis of the lysosomal enzyme, cathepsin D. J. Biol. Chem. 254:11771-11774.

3. Esmon, B., P. Novick, and R. Schekman. 1981. Compartmentalized assembly of oligosaccharide chains on exported glycoproteins in yeast. Cell 25:451-460.

4. Esmon, B., and R. Schekman. In preparation.

5. Ferro-Novick, S., C. Field, and R. Schekman. In preparation.

6. Ferro-Novick, S., P. Novick, W. Hansen, and R. Schekman. In preparation.

7. Ferro-Novick, S., and R. Schekman. In preparation.

8. Field, C., and R. Schekman. 1980. Localized secretion of acid phosphatase reflects the pattern of cell-surface growth in Saccharomyces cerevisiae. J. Cell Biol. 86:123-128.

9. Hasilik, A., and W. Tanner. 1978a. Biosynthesis of the vacuolar yeast glycoprotein carboxypeptidase Y: Conversion of the precursor into the enzyme. Eur. J. Biochem. 85:599-608.

10. Hasilik, A., and W. Tanner. 1978b. Carbohydrate moiety of carboxypeptidase Y and perturbation of its biosynthesis. Eur. J. Biochem. 91:567-575.

11. Hemmings, B. A., G. S. Zubenko, A. Hasilik, and E. W. Jones. 1981. Mutant defective in processing of an enzyme located in the lysosome-like vacuole of Saccharomyces cerevisiae. Proc. Natl. Acad. Sci. USA 78:435-439.

12. Henry, S. A., K. D. Atkinson, A. I. Kolat, and M. R. Culbertson. 1977. Growth and metabolism of inositol-starved Saccharomyces cerevisiae. J. Bacteriol. 130:472-484.

13. Hickman, S., and E. F. Neufeld. 1972. A hypothesis for I-cell disease: Defective hydrolases that do not enter lysosomes. Biochem. Biophys. Res. Commun. 49:992-999.

14. Kaplan, A., D. T. Achord, and W. S. Sly. 1977. Phosphohexosyl components of a lysosomal enzyme are recognized by pinocytosis receptors on human fibroblasts. Proc. Natl. Acad. Sci. USA 74:2026-2030.

15. Kaplan, G., J. C. Unkeless, and Z. A. Cohn. 1979. Insertion and turnover of macrophage plasma membrane proteins. Proc. Natl. Acad. Sci. USA 76:3824-3828.

16. Linnemans, W. A. M., P. Boer, and P. F. Elbers. 1977. Localization of acid phosphatase in Saccharomyces cerevisiae: A clue to cell wall formation. J. Bacteriol. 131:638-644.

17. Meyer, D. I., E. Krause, and B. Dobberstein. 1982. Secretory protein translocation across membranes: The role of the "docking protein". Nature: in press.

References (continued)

18. Moor, H. 1967. Endoplasmic reticulum as the initiator of bud formation in yeast. *Arch. Mikrobiol.* *57*:135-146.

19. Neufeld, E. F., G. N. Sando, A. J. Garvin, and L. H. Rome. 1977. The transport of lysosomal enzymes. *J. Supramolec. Struct.* *6*:95-101.

20. Novick, P., S. Ferro, and R. Schekman. 1981. Order of events in the yeast secretory pathway. *Cell* *25*:461-469.

21. Novick, P., C. Field, and R. Schekman. 1980. Identification of 23 complementation groups required for post-translational events in the yeast secretory pathway. *Cell* *21*:205-215.

22. Novick, P., and R. Schekman. 1979. Secretion and cell-surface growth are blocked in a temperature-sensitive mutant of *Saccharomyces cerevisiae*. *Proc. Natl. Acad. Sci. USA* *76*:1858-1862.

23. Novick, P., and R. Schekman. In preparation.

24. Onishi, H. R., J. S. Tkacz, and J. O. Lampen. 1979. Glycoprotein nature of yeast alkaline phosphatase: Formation of active enzyme in the presence of tunicamycin. *J. Biol. Chem.* *254*:11943-11952.

25. Reeves, R., D. Chang, and S.-C. Chung. 1981. Carbohydrate modification of the high mobility group proteins. *Proc. Natl. Acad. Sci. USA* *78*:6704-6708.

26. Reitman, M. L., and S. Kornfeld. 1981. Lysosomal enzyme targeting. N-Acetylglucosaminylphosphotransferase selectively phosphorylates native lysosomal enzymes. *J. Biol. Chem.* *256*:11977-11980.

27. Rome, L. H., A. J. Garvin, M. M. Allietta, and E. F. Neufeld. 1979. Two species of lysosomal organelles in cultured human fibroblasts. *Cell* *17*:143-153.

28. Schwaiger, H., A. Hasilik, K. von Figura, A. Wiemken, and W. Tanner. 1982. Carbohydrate-free carboxypeptidase Y is transferred into the lysosome-like yeast vacuole. *Biochem. Biophys. Res. Commun.* *104*:950-956.

29. Shore, G. C. 1979. Synthesis of intracellular membrane proteins *in vitro*. Relation between rough endoplasmic reticulum and mitochondrial outer membrane. *J. Cell Sci.* *38*:137-153.

30. Stevens, T., B. Esmon, and R. Schekman. 1982. Early stages in the yeast secretory pathway are required for transport of carboxypeptidase Y to the vacuole. Submitted.

31. Tkacz, J. S., and J. O. Lampen. 1973. Surface distribution of invertase on growing *Saccharomyces* cells. *J. Bacteriol.* *113*:1073-1075.

References (continued)

32. Walter, P., I. Ibrahimi, and G. Blobel. 1981. Translocation of proteins across the endoplasmic reticulum. I. Signal recognition protein (SRP) binds to *in vitro*-assembled polysomes synthesizing secretory protein. II. Signal recognition protein (SRP) mediates the selective binding to microsomal membranes of *in vitro*-assembled polysomes synthesizing secretory protein. III. Signal recognition protein (SRP) causes signal sequence-dependent and site-specific arrest of chain elongation that is released by microsomal membranes. *J. Cell Biol.* *91*:545-561.

33. Wiemken, A., M. Schellenberg, and K. Urech. 1979. Vacuoles: The sole compartment of digestive enzymes in yeast (*Saccharomyces cerevisiae*). *Arch. Microbiol.* *123*:23-25.

34. Willsky, G. R. 1979. Characterization of the plasma membrane Mg^{2+}-ATPase from the yeast, *Saccharomyces cerevisiae*. *J. Biol. Chem.* *254*:3326-3332.

35. Zwizinski, C., and W. Wickner. 1981. Purification and characterization of leader (signal) peptidase from *Escherichia coli*. *J. Biol. Chem.* *255*: 7973-7977.

12. MUTANTS OF YEAST OVERPRODUCING ISO-2-CYTOCHROME c*

Bradley E. Kosiba, Beverly Errede†, Thomas S. Cardillo†† and Fred Sherman

Department of Radiation Biology and Biophysics
University of Rochester School of Medicine and Dentistry
Rochester, New York 14642

Rec. Adv. Yeast Mol. Biol. (1982) 1:156-172

The following mutations cause overproduction of iso-2-cytochrome c, the minor form of cytochrome c in the yeast Saccharomyces cervisiae: CYC7-H mutations that are cis-dominant and that have extended alterations in the 5' region adjacent to the structural gene CYC7; and recessive mutations at any of the unlinked loci cyc8, cyc9, cyc10 and cyc11. Three CYC7-H mutants, CYC7-H1, CYC7-H2, and CYC7-H3 all overproduce 20-30 times the normal amount of iso-2-cytochrome c, but each have different genetic alterations that result in different abnormal sequences at different sites along the 5' region. CYC7-H1 is a reciprocal translocation with a breakpoint between the ATG initiaton codon and a XhoI site at nucleotide position -140. CYC7-H2 contains 5.5 kbp Ty element inserted at position -185. CYC7-H3 contains a deletion that extends from position -223 to approximately 5 kbp. The recessive mutants contain 3 to 7 times the normal amount of iso-2-cytochrome c although certain strains containing two of the recessive mutations overproduce approximately 15 times the normal level. Some of the recessive mutations also cause pleiotropic phenotypes not obviously related to the overproduction of iso-2-cytochrome c. The CYC7-H mutants and the four recessive mutants have higher amounts of the CYC7 transcript which approximate the overproduction of iso-2-cytochrome c. We suggest that the overproduction of iso-2-cytochrome c is due to enhancement of normal transcription either by the presence of abnormal sequences at the 5' regulatory region or by mutation of components involved in transcription of the CYC7 and possibly other genes.

---

*This investigation was supported in part by Public Health Service Research Grant GM12702 from the National Institutes of Health and in part by the U.S. Department of Energy Contract No. DE-AC02-76EV3490 at the University of Rochester, Department of Radiation Biology and Biophysics. This paper has been designated Report No. UR-3490-2197.

†Present address: Dept. of Chemistry
University of North Carolina
Chapel Hill, NC 27514

††Present address: Bio Sciences Division
Research Laboratory
Eastman Kodak Company
Rochester, NY 14650

INTRODUCTION

The iso-cytochrome *c* system in the yeast *Saccharomyces cerevisiae* has been used in numerous diverse investigations, including various aspects of gene expression and regulation. As part of the studies on gene expression, we have been examining alterations that lead to increased levels of gene products. Because of a convenient selective system, mutations causing up to 30 times the normal amount of iso-2-cytochrome *c* have been obtained. Genetic analysis of these mutants and DNA analysis of regions encompassing the mutations revealed that overproduction is caused either by gross alterations in front of the translated region of the structural gene or by mutations at any of four loci unlinked to the structural gene. In this paper, we briefly summarize essential features of the iso-cytochrome *c* system and we describe these types of mutations that cause overproduction of iso-2-cytochrome *c*.

THE ISO-CYTOCHROME *c* SYSTEM

Iso-1-cytochrome *c* and iso-2-cytochrome *c* normally constitute, respectively, 95 percent and 5 percent of the total cytochrome *c* complement in the yeast *Saccharomyces cerevisiae* grown aerobically under derepressed conditions. Both iso-cytochromes *c* apparently carry out equivalent functions in mitochondrial oxidative phosphorylation (13) and both are more or less coordinately regulated during derepression by nonfermentable carbon sources or during induction by oxygen; however, variations in their proportions as well as in their absolute amounts occur during induction by oxygen and during the transition from a repressed state to a derepressed state (18,21). Genetic analyses of strains containing mutationally altered forms of cytochrome *c* have established that the primary structure of iso-1-cytochrome *c* is determined by the *CYC1* gene (24), which is located on the right arm of chromosome X (12), and that the primary structure of iso-2-cytochrome *c* is determined by the *CYC7* gene (4,5), which is located on the left arm of chromosome V (22). DNA sequences have been determined for the translated regions as well as the adjacent regions of both the *CYC1* (26) and *CYC7* (15) loci. The detailed physical information and the procedures for selecting and detecting forward mutations, reverse mutations (23) and mutations causing overproduction (4,27); make iso-cytochrome *c* ideally suited for investigating gene expression.

ISO-2-CYTOCHROME *c* OVERPRODUCERS

Mutants completely deficient in any of the mitochondrial components required for functional respiration are unable to utilize nonfermentable substrates as carbon and energy sources. Thus the double mutant, *cyc1 cyc7*, deficient in both iso-1-cytochrome *c* and iso-2-cytochrome *c*, is unable to grow on media containing, for example, ethanol, glycerol or lactate as the sole carbon source (4). Revertants containing as low as 1 percent of the normal amount of total cytochrome *c* can grow, although poorly, on glycerol or ethanol media but are still unable to grow on a special semi-synthetic lactate medium. In fact, *cyc1 CYC7+* mutants lacking iso-1-cytochrome *c* but containing the normal 5 percent level of iso-2-cytochrome *c* do not grow or barely grow on lactate medium (23,25). Thus, a variety of revertants can be obtained by plating high densities of *cyc1 CYC7+* cells on lactate medium. If the *cyc1* allele is revertible or suppressible, revertants containing iso-1-cytochrome *c* can arise because of intragenic mutations or suppressor mutations. However, if the cyc1 allele is not revertible or suppressible, such as in the case with cyc1 deletions, then the only revertants arising on lactate medium are those containing increased levels of iso-2-cytochrome c. The overproduction of iso-2-cytochrome c in the various revertants ranges from barely detectable increases to approximately 30 times the normal level. Genetic analysis revealed that iso-2-cytochrome c overproduction can be due either to mutations within or adjacent to the CYC7 structural gene or to mutation at any of at least four unlinked loci. The mutations at the CYC7 locus, denoted CYC7-H mutations, occur at very low frequencies and are associated with gross alterations that produce dominant effects. In contrast, mutations at the unlinked loci arise at frequencies typically observed for forward mutations that inactivate genes. These unlinked mutations are recessive (Table 1).

Table 1. Mutants Overproducing Iso-2-cytochrome c

Wild-type alleles	Chromosomal location	Mutant allele	Lesion	Fold increase of Iso-2
CYC7+	V left	CYC7-H1	Translocation	20-40
		CYC7-H2	Ty insertion	15-30
		CYC7-H3	Deletion	15-30
CYC8	II right	cyc8	Point	3-6
CYC9	III right	cyc9	Point	3-6
CYC10	Unknown	cyc10	Point	5-10
CYC11	Unknown	cyc11	Point	3-6

THE CYC7-H OVERPRODUCERS

A genetic and biochemical analysis of the three mutants, CYC7-H1, CYC7-H2 and CYC7-H3 indicate that all three of these mutants overproduce 20 to 30 times the normal amount of iso-2-cytochrome c, all are dominant, and all have gross alterations adjacent to the CYC7 structural gene. The CYC7-H1 mutation involves a reciprocal translocation, the CYC7-H2 involves the insertion of a Ty element and the CYC7-H3 involves a deletion. The alterations adjacent to the CYC7 locus were initially revealed from the analysis of genomic yeast DNA probed with a CYC7+ fragment as shown in Fig. 1. EcoRI or HindIII digests of each of the CYC7-H mutants contain one or two restriction fragments that hybridize to the CYC7+ probe and that differ in size from each other and from the CYC7+ fragment. The different sizes of the restruction fragments suggest that an extensive region near the CYC7 locus is altered in each of the CYC7-H mutants. Genetic analysis of the three CYC7-H mutants and restriction mapping of the cloned CYC7-H2 and CYC7-H3 genes resulted in the characterization of their corresponding aberrations.

The CYC7-H1 Mutation

An analysis of various crosses demonstrated that the CYC7-H1 mutation is a reciprocal translocation with a breakpoint at the CYC7 locus on the left arm of chromosome V and a breakpoint on the right arm of chromosome XVI (20). The reciprocal translocation was identified from the pattern of spore inviability

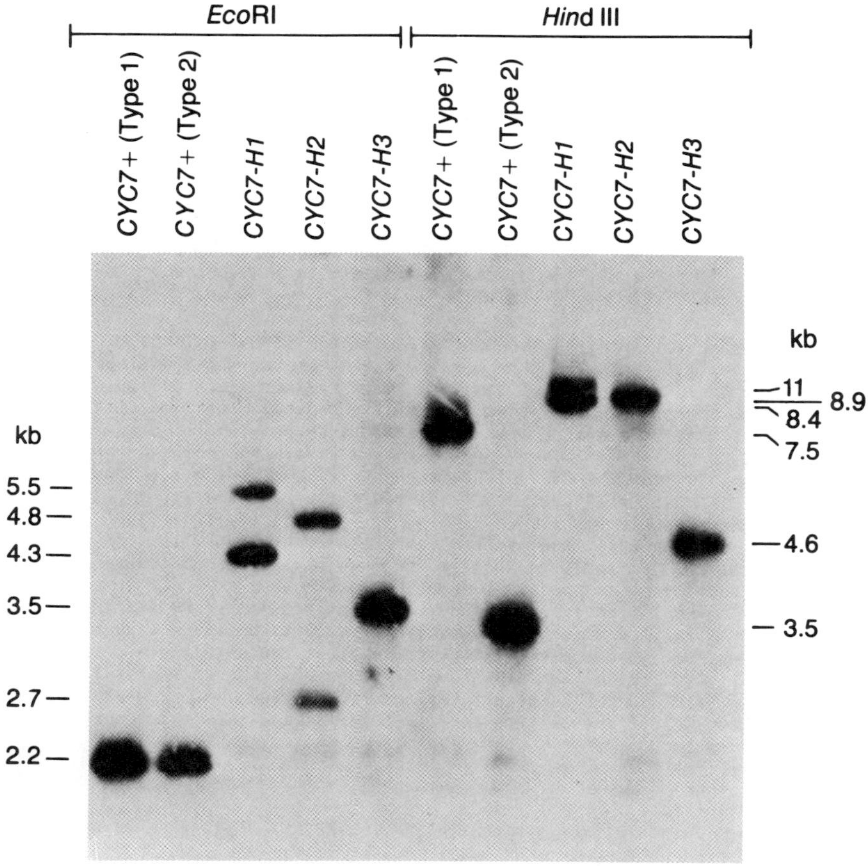

Figure 1. Southern Blot of CYC7-H Mutants. Genomic DNA samples from CYC7+, CYC7-H1, CYC7-H2 and CYC7-H3 strains were digested with either EcoR1 or HindIII, electrophoresed in 1 percent agarose, transfered to a nitrocellulose sheet and hybridized to a nick-translated probe consisting of an EcoR1 DNA fragment containing the CYC7+ region. Type I strains lack a HindIII site 3' to the CYC7 gene, which is present in Type II strains. All of the CYC7-H mutants were derived from CYC7+ Type II strains. The autoradiogram of these digests demonstrates that fragments complementary to the CYC7+ region differ greatly in size, indicating that each of the CYC7-H1, CYC7-H2 and CYC7-H3 mutants contain different alterations within the CYC7 region. (Adapted from 19).

and abnormal linkage in heterozygous CYC7-H1/CYC7+ crosses. In addition, the chromosomal constitution was further characterized by the analysis of crosses that were homozygous for the translocation but heterozygous for alterations in the CYC7 structural gene. The results of the genetic analysis resulted in the genetic maps of the translocated and normal chromosomes V and XVI that are presented in Fig. 2. These genetic analyses established that one of the translocated points is at the CYC7 locus on chromosome V and that the other is proximal to mak3 on chromsome XVI (20). Dominant and recessive properties of the CYC7-H1 mutations were investigated with several mutants lacking iso-2-cytochrome c, including a UAG mutation cyc7-H1-1 corresponding to amino acid position 24. Because cyc7-H1-1/CYC7+ diploid strains contain approximately half of the amount of iso-2-cytochrome c found in CYC7-H1 haploid strains and because cyc7-H1-1/CYC7+ diploid strains contain approximately the CYC7+ level, the overproduction of iso-2-cytochrome c caused by the CYC7-H1 translocation is cis dominant and trans recessive.

The normal amino acid composition and peptide map of iso-2-cytochrome c from the CYC7-H1 mutant indicated that the breakpoint of the translocation is outside of the translated portion of the gene (4). Furthermore, the low frequency of CYC7-H1 recombinants among the meiotic progeny of cyc7-H1-1 x CYC7+ crosses is indicative of a short genetic distance between the CYC7-H1 breakpoint and structural gene (4). The abnormal size of the EcoRI and XhoI restriction fragments from the CYC7-H1 DNA shown in Figures 1 and 3 suggests that the translocation altered the EcoRI and XhoI sites that are, respectively, approximately one kbp and 140 bp in front of the translated region of the gene. These altered restriction fragments, along with finding a normal iso-2-cytochrome c, indicate that the breakpoint is between the structural gene and the XhoI site 140 base pairs in front of the AUG initiation codon. Thus it is apparent that the translocation disrupted the CYC7 regulatory region in a way that caused overproduction of iso-2-cytochrome c. The formation of an abnormal regulatory region contiguous to the CYC7 structural gene in the CYC7-H1 mutation is similar to the alteration in the other mutations CYC7-H2 and CYC7-H3, but, as described below, the regulatory region is disrupted at different sites and all three of the abnormal sequences were different from each other.

The CYC7-H2 Mutation

Initial genetic analysis established that the CYC7-H2 mutation contains a lesion at or near the CYC7 structural gene and that the lesions causing overproduction could be separated by recombination from a site in the translated portion of the gene (22). Unlike the CYC7-H1 mutation, the CYC7-H2 mutation was not associated with a gross chromosomal rearrangement that produces deficiencies in the meiotic progeny from heterozygous crosses.

The DNA alteration of the CYC7-H2 mutation was characterized by restriction endonuclease analysis of the cloned segment, by heteroduplex analysis of CYC7+ and CYC7-H2 segments and by DNA sequencing of the pertinent regions (8,9 and B. Kosiba et al., unpublished results). The results, summarized in Figure 4, demonstrate that an approximately 5.5 kbp segment was inserted at a site 184 bp in front of the translated portion of the gene. The inserted segment was found to hybridize to a probe containing the central

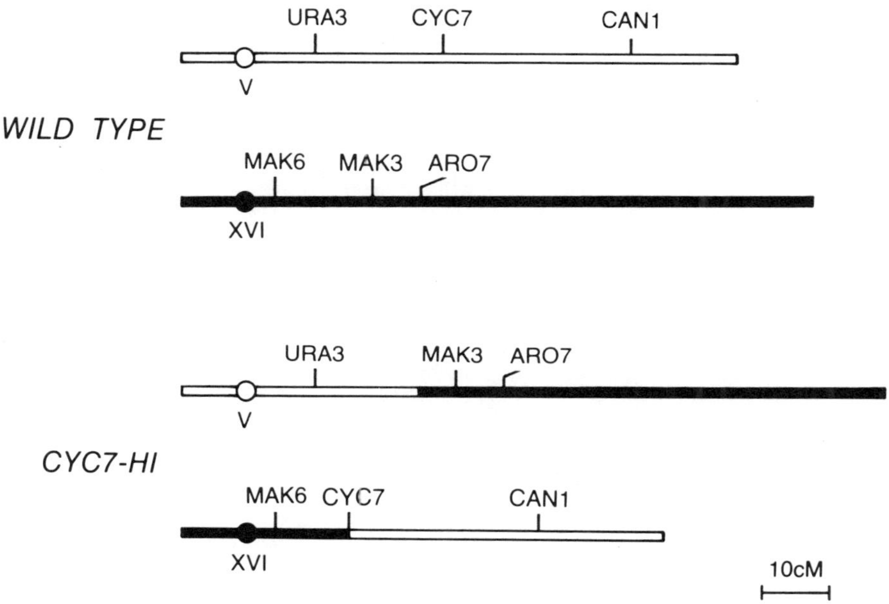

Figure 2. The *CYC7-H1* Mutaticn. A genetic map is shown of the left arm of chromosome V and the right arm of chromosome XVI. Overproduction of iso-2-cytochrome *c* occurs in *CYC7-H1* strains containing a reciprocal translocation between these two chromosomes (20). The breakpoint on chromosome V is adjacent to the *CYC7* gene and, on chromosome XVI is between *MAK6* and *MAK3*.

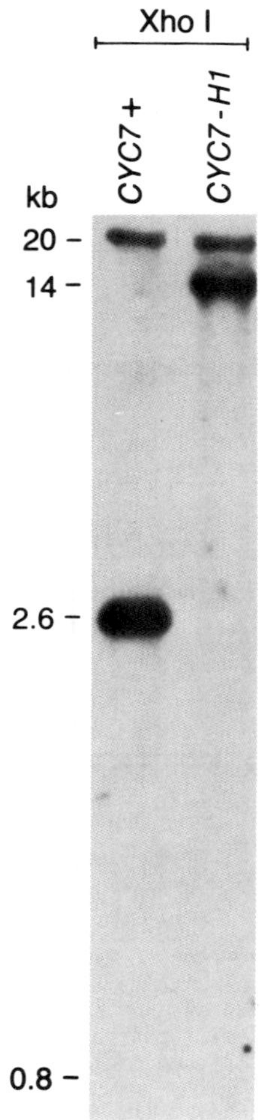

Figure 3. Southern Blot of the CYC7-H1 Mutant. Genomic DNA from CYC+ and CYC7-H1 DNA was digested with XhoI, electrophoresed in 1 percent agarose, transfered to a nitrocellulose sheet and hybridized to a nick translated probe consisted of an EcoR1 fragment containing the CYC7+ region. The autoradiogram shows a different XhoI restriction pattern for the CYC7-H1 and CYC+ region.

portion of a Tyl element which Cameron *et al*. (1) identified and characterized as a family of dispersed repetitive elements in yeast. The DNA sequence of the termini of the inserted segment were found to be closely homologous to the terminal δ sequences of known Ty elements and a 5 bp duplication was created at the site of insertion, similar to duplications created by Ty elements inserted at other sites (10,11,31).

Unlike the *CYC7-H1* mutations discussed above and the *CYC7-H3* mutation discussed below, the *CYC7-H2* mutation has special properties that belong to a class of mutations denoted as ROAM (Regulated Overproducing Alleles responding to Mating signals) by Errede *et al*. (8,9).

The ROAM mutants are defined by the characteristic property that overproduction or constitutivity occurs in *MATa* and *MATα* haploid strains but less so or not at all in *MATa/MATα* diploid strains homozygous for the regulatory mutation. Furthermore, overproduction occurs in *MATa/MATa* and *MATα/MATα* diploid strains capable of mating as well as in specially constructed diploid strains capable of both mating and sporulating. Overproduction is decreased when haploid ROAM mutants also contain the *ste7*, *ste11*, *ste12* or *ste4* mutations which prevent conjugation; however, other *ste* mutations such as *ste3*, *ste5* and *ste13*, do not affect the expression of ROAM mutations. Genetic analysis of diploid strains heterozygous for ROAM mutations and homozygous for *MAT* alleles indicated that the overproduction mutations are caused by *cis* dominant alterations adjacent to the structural genes.

In addition to *CYC7-H2*, the ROAM mutants include *cargA$^+$O^h*, which constitutively overproduces arginase, *cargB$^+$O^h* which constitutively overproduces ornithine transaminase, *durO^h* which constitutively overproduces the urea carboxylase allophanate hydrolase enzyme complex (8,9) and *ADR3^c* which constitutively produces slightly lower amounts of alcohol dehydrogenase II (30,31). The similar genetic properties of the *cargA$^+$O^h*, *cargB$^+$O^h* and *durO^h* suggest that they each contain an inserted Ty element. DNA sequencing of pertinent regions of the four mutants *ADR3-2^c*, *ADR3-3^c*, *ADR3-6^c*, *ADR3-7^c* demonstrate that they contain Ty insertions, that the points of insertion are different for each of the mutations, that the Ty elements are oriented in the same direction and that this orientation is the same as the orientation of the Ty element in the *CYC7-H2* mutant. Thus, insertion of Ty elements in front of a number of unrelated genes can lead to overproduction, constitutivity or both; although the site of insertion in front of the gene may vary, it appears as if the same orientation is required and in this orientation the Ty transcript is synthesized in the opposite direction as the transcript of the affected gene (6,7).

The *CYC7-H3* Mutation

Genetic and molecular analyses have demonstrated that the *CYC7-H3* mutant is a deletion of about 5 kbp from the region immediately 5' to the *CYC7* coding region (14). Subsequent DNA sequence analysis has shown that the deletion breakpoint proximal to the *CYC7* gene is 222 bp 5' to the ATG initiation codon (Fig. 4) (Kosiba and Sherman, unpublished results). In addition to causing a cis-dominant 20-fold overproduction of iso-2-cytochrome *c*, the *CYC7-H3* mutation was found (14) to have the following four recessive phenotypes:

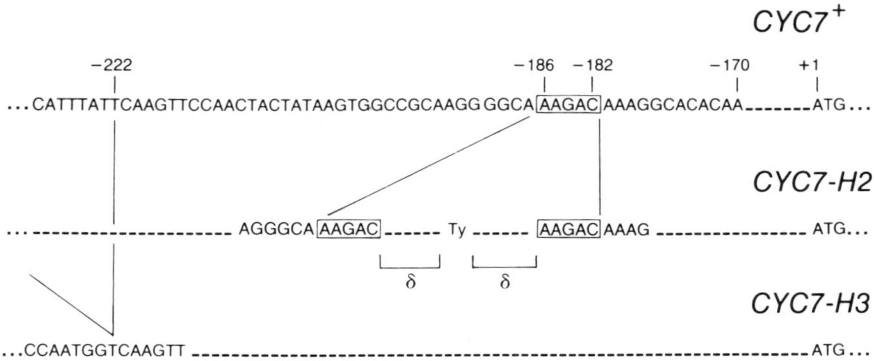

Figure 4. The CYC7-H2 and CYC7-H3 Mutations. The DNA sequence of the CYC7+ (15), CYC7-H2 and CYC7-H3 regions. The CYC7-H2 sequence shows a Ty insertion between -186 and -182 bp from the ATG initiation codon of the iso-2-cytochrome c coding region. This Ty insertion generates the customary 5 bp repeat at each end of the insertion. The CYC7-H3 sequence shows a breakpoint in the wild type CYC7+ DNA sequence 5' to position -222.

sensitivity to ultraviolet radiation; inability to grow on hypertonic medium containing ethylene glycol or KCl; sensitivity to a breakdown product of chloramphenicol, amino nitrophenyl propanediol (ANP); and flocculent cells.

McKnight *et al.* (14) showed that the four recessive phenotypes segregated with the *CYC7-H3* mutation in all of the 294 pedigrees examined and that they were controlled by two new genes, *RAD23* and *ANP1*. Point mutations at the *RAD23* locus failed to complement the UV sensitivity while point mutations at *ANP1* failed to complement any of the other three phenotypes: ANP sensitivity, osmotic sensitivity or cell flocculence. Genetic mapping of two point mutants, *rad23-1* and *anp1-1*, with respect to the *CYC7-H2* mutation showed the gene order to be centromere-*ANP1*-*RAD23*-*CYC1*. From the pattern of gene conversion in this analysis the size of the *ANP1*-*CYC7* interval was estimated at about 2.1 centiMorgans. Thus the *CYC7-H3* mutation is a 5 kbp deletion extending from the *ANP1* region to 222 bp in front of the *CYC7* coding region which results in osmotic sensitivity and associated phenotypes, UV sensitivity and the overproduction of iso-2-cytochrome *c*. This overproduction of cytochrome *c* results from the deletion which removed 5 kbp upstream of the *CYC7* gene, fusing the *CYC7* gene to novel upstream DNA sequences.

RECESSIVE OVERPRODUCERS

By far the most frequent mutations causing overproduction of iso-2-cytochrome *c* were found to be recessive and unlinked to the *CYC7* locus. Genetic complementation tests revealed that these mutations fall into four loci which have been designated *CYC8*, *CYC9*, *CYC10* and *CYC11* (17, Kosiba, Errede and Sherman, unpublished results). These loci may correspond to one or more of the loci *CYP1*, *CYP2*, *CYP4* and *CYP5* described by Verdiere and Petrochila (29) and others (3,16,28).

The *cyc8* mutant

The *cyc8* mutation leads to an increase in iso-2-cytochrome *c* production of approximately three-fold, to a lacy colony morphology and in some genetic background with abnormal cell shape (17).

The diploid cross *cyc1 CYC7+ cyc8*/*cyc1 CYC7+ CYC8+* contains the same normal amount of iso-2-cytochrome *c* as the diploid cross *cyc1 CYC7+ CYC8+*/*cyc1 CYC7+ CYC8+*, thus indicating that the *cyc8* mutation is recessive. Similar tests also revealed that the *cyc9*, *cyc10* and *cyc11* mutations were recessive.

The *cyc8* mutation leads to an increase in iso-2-cytochrome *c* production of approximately three-fold, to a lacy colony morphology and abnormal cell shape in some genetic backgrounds (17).

The *cyc9* Mutant

Similar to *cyc8* mutants, the *cyc9* mutants contain approximately three times then normal amounts of iso-2-cytochrome *c*. The *cyc9* mutation cause a large number of apparently unrelated pleiotropic phenotypes and is, in fact, equivalent to the *tup1*, *umr7* and *flk1* mutations that were uncovered in other studies. The following pleiotropic phenotypes are associated with the *cyc9* etc. mutations: the ability to utilize exogenous dTMP; a characteristic flocculent morphology; the lack of sporulation of homozygous diploids; low frequency of mating and abnormally shaped cells of α strains; the lack of detection of UV-induced mutations at the CAN1 locus, and the lack of catabolite repression of numerous enzymes including maltase, α-methyl-glucosidase invertase and succinate dehydrogenase (see (17) for original citations). However, the cyc9 mutation appears not to appreciably affect iso-1-cytochrome c or at most only slightly lower its concentration. The CYC9 locus, as well as the equivalent loci TUP1, UMR7 and FLK1 is located on the right arm of the chromsome III.

The cyc10 and cyc11 Mutants

Cyc10 and cyc11 mutants were recently uncovered in systematic searches for mutations causing overproduction of iso-2-cytochrome c (B. Errede, B. Kosiba, and F. Sherman, unpublished results). The examination of cytochrome c in diploid crosses established that the cyc10 and cyc11 mutatios are recessive and that they complement each other as well as the cyc8 and cyc9 mutation. In addition, as shown in Table 2, the cyc10 and cyc11 mutations are unlinked to each other and to the cyc8 and cyc9 mutations. The cyc10 and cyc11 mutants overproduce, respectively, approximately 7 and 3 times the normal amount of iso-2-cytochrome c, similar to the overproduction of observed in cyc8 and cyc9 mutants. However, unlike the cyc8 and cyc9 mutations, the cyc10 and cyc11 mutations do not produce any additional effects such as aberrant colony morphology.

Table 2. Independent Segregation of cyc10 and cyc11 with Each Other and with cyc8 and cyc9

	PD	NPD	T		PD	NPD	T
cyc10-cyc8	2	2	10	cyc11-cyc8	2	2	5
cyc10-cyc9	1	2	4	cyc11-cyc9	1	1	9
cyc10-cyc11	0	3	11				

INTERACTION OF OVERPRODUCERS

Haploid strains containing various pairwise combinations of the cyc8, cyc9, cyc10 and cyc11 mutations produced higher levels of iso-2-cytochrome c than haploid strains containing only one of the mutations (Table 3). The overproduction of iso-2-cytochrome c appears approximately additive although combinations with cyc10 produce unexpectedly larger amounts of iso-2-cytochrome c. In contrast, previous results with the CYC7-H2 mutation indicated that the CYC7-H2 overproduction was not compounded with the cyc8 or cyc9 overproduction; cyc1 CYC7-H2 cyc8 and cyc1 CYC7-H2 cyc9 segregants did not appear to contain higher levels of iso-2-cytochrome c than the cyc1 CYC7-H2 segregants (17).

Table 3. Overproduction of Iso-2-cytochrome c in Haploids Containing Combinations of Recessive Mutations

Single Mutations		Double Mutations	
Genotype	Fold overproduction	Genotype	Fold overproduction
cyc8	3	cyc8 cyc9	6
cyc9	3	cyc8 cyc10	15
cyc10	7	cyc8 cyc11	6
cyc11	3	cyc9 cyc10	15
		cyc9 cyc11	6
		cyc10 cyc11	15

OVERPRODUCTION OF THE CYC7 TRANSCRIPT

Overproduction of iso-2-cytochrome c in the three CYC7-H mutants and the four cyc8, cyc9, cyc10 and cyc11 recessive mutants approximately parallels overproduction of CYC7 transcription. Total cellular RNA, extracted from each of the overproducers, was separated on a denaturing agarose gel, transferred to a nitrocellulose membrane and hybridized with a specific CYC7 probe. The results, shown in Figure 5, demonstrate that little CYC7 mRNA is produced by the CYC7+ strain compared to the strains overproducing iso-2-cytochrome c. Although precise quantitative determinations were not made, the level of CYC7 transcription was higher in the three CYC7-H mutants than the four recessive mutants, suggesting that the amount of the iso-2-cytochrome c protein corresponds to the amount of the CYC7 transcript.

CONCLUSION

Mutations overproducing iso-2-cytochrome c consist of either gross structural changes upstream of the coding region or extragenic mutations at several unlinked loci. The former class of mutations includes a 5 kbp deletion, a 5.6 kbp insertion and a reciprocal translocation between chromosomes V and XVI. Each of these mutations introduces entirely new DNA sequences within the first 230 bp upstream of the iso-2-cytochrome c ATG translation start site. These abnormal DNA sequences may function directly, by providing the basic DNA sequence necessary to bind RNA polymerase and initiate transcription at abnormally high rates. Alternatively, these sequences may function indirectly by enhancing the utilization of pre-existing signals by some unknown means such as altering the chromatin structure.

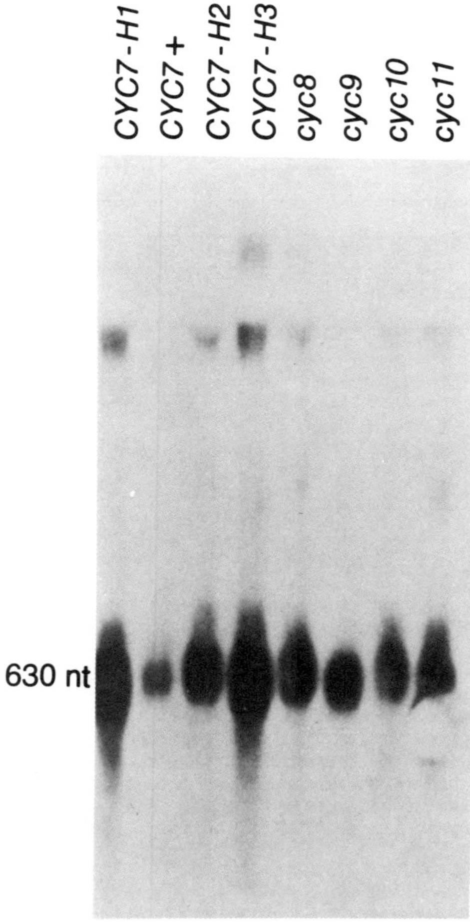

Figure 5. CYC7 Specific mRNA in Iso-2-cytochrome c Overproducer Mutants. Total cellular RNA was extracted from CYC7+ and iso-2-cytochrome c overproducer strains. Approximately 20 μg of each sample was denatured, run in 1.8 percent agarose, 50 percent formamide, transfered to a nitrocellulose sheet and hybridized to single strainded radiolabeled probe consisting of a Sau3AI/HpaII fragment containing most of the CYC7 cistron. The autoradiogram shows a low level of 630 nt CYC7-specific message from CYC7+ cells, moderately increased levels of CYC7-specific message from cyc8, cyc9, cyc10 and cyc11 RNA and greatly increased levels of CYC7-specific message from CYC7-H1, CYC7-H2 and CYC7-H3 cells.

The CYC7-H1 mutation contains new DNA sequences less than 140 bp from the CYC7 ATG initiation codon and produces a transcript which appears to be slightly shorter than that of the wild type CYC7+ gene. Thus the overproduction of iso-2-cytochrome c seen in CYC7-H1 mutants may result from an abnormal 5' region determining the rate and origin of CYC7 transcription. However, the CYC7-H2 and CYC7-H3 transcripts both appear to be the same size as the CYC7+ transcript. Although the mechanisms of overproduction in CYC7-H2 and CYC7-H3 are unknown, a common mode of action can be suggested by considering the properties of CYC7-H2 and other ROAM mutations. In the case of the CYC7-H2 and four ADR3^{c} mutants, the transcripts of the affected gene and the adjacent Ty element are synthesized divergently. The size of the CYC7-H2 transcript (Fig. 5) and the 5' ends of seven ADR3^{c} transcripts (31) indicate that the mRNAs in ROAM mutants are initiating at the normal site. In addition, the amount of Ty RNA is dependent on the mating type system, there being 20 times more in MATa and MATα haploid cells than in MATa/MATα cells (7). Thus it appears that transcription away from the 5' end of certain genes causes overproduction and constitutivity. Also, the insertion of just any DNA sequence does not cause equivalent effects; there is a lack of constitutive expression of alcohol dehydrogenase II when an insert of approximately 350 bp solo δ is at the same site as the inserted Ty element in ADR3^{c} mutants (2). In fact, the insertion of a solo δ sequence at various sites causes diminution and complete deficiency of the gene product. These results, together with the CYC7-H2 and other ROAM mutations, lead us to suggest that the overproduction in the CYC7-H3 mutant may also be due to transcription away from the CYC7 gene on the abnormal DNA sequence. The hypothetical transcript should not, however, be diminished in MATa/MATα cells. A more complete analysis of the abnormal DNA region in the CYC7-H3 mutant and an analysis of the corresponding transcripts may reveal whether or not there is a similar mode of action between overproduction in the CYC7-H2 and CYC7-H3 mutants.

The second class of mutations includes alterations at four unlinked loci which cause increases of 3 to 7 fold in iso-2-cytochrome c production. While it is possible that one or more of these mutations is within the gene of a specific regulator of iso-2-cytochrome c production, the pleiotropic nature of two of them (cyc8 and cyc9) suggests that this may not be the case. In fact, the extreme pleiotropy of the cyc9 mutant suggests that it plays a fairly nonspecific, though important role in determining levels of gene transcription. It is tempting to speculate that the recessive mutations destroy or modify components required for normal transcription or for chromatin integrity which indirectly modify transcription.

LITERATURE CITED

1. Cameron, J.R., Loh, E.Y., Davis, R.W. 1979. Evidence for transposition of dispersed repetitive DNA families in yeast. Cell 16:739-751.

2. Ciriacy, M. and Williamson, V.M. 1981. Analysis of mutations affecting Ty-mediated gene expression in Saccharomyces cerevisiae. Molec. Gen. Genet. 182:159-163.

3. Clavilier, L., Pere, G. and Slonimski, P.P. 1969. Mise en evidence de plusieurs loci independants impliques dans la synthese de l'iso-2-cytochrome c chez la levure. Molec. Gen. Genet. 104:195-218.

4. Downie, J.A., Stewart, J.W., Brockman, N., Schweingruber, A.M. and Sherman, F. 1977. Structural gene for yeast iso-2-cytochrome c. J. Molec. Biol. 113:369-384.

5. Downie, J.A., Stewart, J.W., and Sherman, F. 1977. Yeast mutants defective in iso-2-cytochrome c. J. Molec. Biol. 117:369-386.

6. Elder, R.T. 1981. (abst.) Both the 5' and 3' ends of Ty1 RNA are in the delta sequences. Abstracts of Molecular Biology of Yeast. Cold Spring Harbor Laboratory, Cold Spring Harbor, NY.

7. Elder, R.T., St. John, T.P., Stinchcomb, D.T. and Davis, R.W. 1981. Studies on the transposable element Ty1 of yeast. I. RNA homologous to Ty1. Cold Spring Harbor Laboratory Sumposium of Quantitative Biology 45:581-584.

8. Errede, B., Cardillo, T.S., Sherman, F., Dubois, E., Deschamps, J. and Wiame, J.-M. 1980. Mating signals control expression of mutations resulting from insertion of a transposable repetitive element adjacent to diverse yeast genes. Cell 22:427-436.

9. Errede, B., Cardillo, T.S., Wever, G. and Sherman, F. 1981. Studies on transposable elements in yeast. I. ROAM mutations causing increased expression of yeast genes: Their activation by signals directed toward conjugation functions and their formation by insertion of Ty1 repetitive elements. Cold Spring Harbor Symposium of Quantitative Biology 45:593-602.

10. Farabaugh, P.J. and Fink, G.R. 1980. Insertion of the eukaryotic transposable element Ty1 creates a 5-base pair duplication. Nature 286:352-356.

11. Gafner, J. and Philippsen, P. 1980. The yeast transposon Ty1 generates duplications of target DNA upon insertion. Nature 286:414-418.

12. Lawrence, C.W., Sherman, F., Jackson, M. and Gilmore, R.A. 1975. Mapping and gene conversion studies with the structural gene for iso-1-cytochrome c in yeast. Genetics 81:615-629.

13. Mattoon, J.R. and Sherman, F. 1966. Reconstitution of phosphorylating electron transport in mitochondria from a cytochrome c-deficient yeast mutant. J. Biol. Chem. 241:4330-4338.

14. McKnight, G.L., Cardillo, T.S. and Sherman, F. 1981. An extensive deletion causing overproduction of yeast iso-2-cytochrome c. Cell 25:409-419.

15. Montgomery, D.L., Leung, D.W., Smith, M., Shalit, P., Faye, G. and Hall, B.D. 1980. Isolation and sequence of the gene for iso-2-cytochrome c in Saccharomyces cerevisiae. Proc. Natl. Acad. Sci. 77:541-545.

16. Petrochilo, E. and Verdiere, J. 1977. A glycine to serine substitution identifies the CYP3 locus as the structural gene of iso-2-cytochrome c in Saccharomyces cerevisiae. Biochem. Biophys.Res. Commun. 79:364-371.

17. Rothstein, R.J. and Sherman, F. 1980. Genes affecting the expression of cytochrome c in yeast: Genetic mapping and genetic interactions. Genetics 94:871-889.

18. Sels, A.A., Fukuhara, H., Pere, G. and Slonimski, P.P. 1965. Cinetique de la biosynthese induite de l'iso-1-cytochrome c et de l'iso-2-cytochrome c au cours de l'adaptation a l'oxygene. Biochim. Biophys. Acta 95:486-502.

19. Sherman, F., Cardillo, T.S., Errede, B., Friedman, L., McKnight, G. and Stiles, J.I. 1981. Yeast mutants overproducing iso-cytochromes c. In: D. von Wettstein, J. Friis, M. Kielland-Brandt and A. Stenderup (eds.), Molecular Genetics in Yeast, Alfred Benzon Symposium 16. Munksgaard, Copenhagen.

20. Sherman, F. and Helms, C. 1978. A chromosomal translocation causing overproduction of iso-2-cytochrome c in yeast. Genetics 88:689-707.

21. Sherman, F. and Stewart, J.W. 1971. Genetics and biosynthesis of cytochrome c. Ann. Rev. Genet. 5:257-296.

22. Sherman, F., Stewart, J.W., Helms, C. and Downie, J.A. 1978. Chromosome mapping of the CYC7 gene determining yeast iso-2-cytochrome c: Structural and regulatory regions. Proc. Natl. Acad. Sci. 75:1437-1441.

23. Sherman, F., Stewart, J.W., Jackson, M., Gilmore, R.A. and Parker, J.H. 1974. Mutants of yeast defective in iso-1-cytochrome c. Genetics 77:255-284.

24. Sherman, F., Stewart, J.W., Margoliash, E., Parker, J. and Campbell, W. 1966. The structural gene for yeast cytochrome c. Proc. Natl. Acad. Sci. 55:1498-1504.

25. Sherman, F., Taber, H. and Campbell, W. 1965. Genetic determination of iso-1-cytochrome c in yeast. J. Molec. Biol. 13:21-39.

26. Smith, M., Leung, D.W., Gillam, S., Astell, C.R., Montgomery, D.L. and Hall, B.D. 1979. Sequence of the gene for iso-1-cytochrome c in Saccharomyces cerevisiae. Cell 16:753-761.

27. Stewart, J.W., Sherman, F., Jackson, M., Thomas, F.L.X. and Shipman, N. 1972. Demonstration of the UAA ochre codon in bakers' yeast by amino acid replacements in iso-1-cytochrome c. J. Molec. Biol. 68:83-96.

28. Verdiere, J. and Petrochilo, E. 1975. CYP3 a likely candidate for the structural gene of iso-2-cytochrome c in Saccharomyces cerevisiae. Biochem. Biophys. Res. Commun. 67:1451-1458.

29. Verdiere, J. and Petrochilo, E. 1979. CYP3-15: an up-promotor mutation at the iso-2-cytochrome c structural gene locus and its interaction with other independent regulatory mutations of cytochrome c synthesis in Saccharomyces cerevisiae. Molec. Gen. Genet. 175:209-216.

30. Williamson, V., Young, E.T. and Ciriacy, M. 1981. Transposable elements associated with constitutive expression of yeast alcohol dehydrogenase-II. Cell 23:605-614.

31. Young, T., Williamson, V., Taguchi, A., Smith, M., Sledziewski, A., Russell, D., Osterman, J., Denis, C., Cox, D. and Beier, D. 1982. The alcohol dehydrogenase genes of the yeast, Saccharomyces cerevisiae: Isolation, structure and regulation, pp. 335-361. In: A. Hollaender, R.D. DeMoss, S. Kaplan, J. Konisky, D. Savage and R.S. Wolfe (eds.), Genetic Engineering of Microorganisms for Chemicals. Plenum Publ. Corp., NY.

13. EXPRESSION, PROCESSING, AND SECRETION OF HETEROLOGOUS GENE PRODUCTS BY YEAST

R.A. Hitzeman,* D.W. Leung,* L.J. Perry,** W.J. Kohr,** F.E. Hagie,* C.Y. Chen,* J.M. Lugovoy,* A. Singh,* H.L. Levine,** R. Wetzel,** and D.V. Goeddel*

Departments of Molecular Biology* and Protein Biochemistry**
Genentech, Inc., South San Francisco, California 94080

Rec. Adv. Yeast Mol. Biol. (1982) 1:173-190

A DNA sequence, capable of transcription starting specificity in *Saccharomyces cerevisiae* (yeast), has been isolated from the 5'-flanking sequence of the yeast 3-phosphoglycerate kinase (PGK) gene by introduction of specific restriction sites between this flanking sequence and the ATG-translational start. This portable PGK promoter fragment has been used to express the genes for both mature and pre-forms (secretion signal sequence present) of human interferons. All of the expressed mature forms remain in the yeast cell, while the pre-forms are found outside as well as inside the cell with some of the interferon activity also present in the cell wall. The proportions of interferon secretion vary somewhat depending on which interferon gene is expressed and at what stage of cell growth the interferon is assayed. One gene product, expressed from the leukocyte A (LeIF A) gene containing a hybrid secretion signal derived from LeIF D and LeIF A pre-sequences, has been purified from both outside and inside the cell. The interferon found free in the medium is mostly properly processed LeIF A by amino-terminal analysis, while another form containing 3 additional amino acids of pre-sequence is also present. The interferon purified from the cell is also all processed; however, three forms are present with proportions that are different from extracellular forms.

INTRODUCTION

We have previously described a plasmid which is capable of expression in yeast of the mammalian gene product, human leukocyte interferon (1,22). The expression of this heterologous gene in yeast was obtained by the proper insertion of the structural gene between two yeast derived sequences in a plasmid capable of replication and selection in both yeast and *E. coli*. The 5'-flanking DNA sequence (promoter fragment) used to obtain transcription was derived from the 5'-flanking sequence of the yeast alcohol dehydrogenase I (ADH1) gene and the 3'-flanking sequence (transcription terminator) was the 3'-end of the *TRP1* gene (31,33) from yeast. The structural gene was a modified cDNA made by replacing the 23-amino acid signal peptide sequence with an ATG-translational initiation codon preceded by an *Eco*RI linker (P. Gray and D.G., unpublished results).

The ADH1 promoter fragments used for initiation of transcription were isolated by *Bal*31 exonuclease digestions from a restriction site within the ADH structural gene through the ATG translation start, followed by the addition of an *Eco*RI linker at the 3'-ends of fragments of various lengths.

We describe here the construction of a new yeast promoter fragment from another highly expressed yeast gene; that is the gene for 3-phosphoglycerate kinase. The gene that codes for this enzyme in *Saccharomyces cerevisiae* has been isolated, described, and characterized (21). Furthermore, evidence suggests that there is a single copy of this gene per haploid cell that is responsible for the high expression (20). We have obtained additional evidence that the gene isolated above is that for 3-phosphoglycerate kinase; in that the DNA sequence shows an amino acid sequence homologous with human (23) and horse (3) PGKs. We here describe a portion of the DNA sequence from this gene and how this sequence information was used to add specific restriction sites, so that this promoter fragment can be used as a portable restriction fragment for the construction of plasmids to express heterologous gene products in yeast.

Using the PGK promoter fragment in a 2μm origin-based (7) plasmid, we have expressed the cDNA genes for both mature and pre-interferons. Pre-interferon genes code for mature interferons as well as the natural human amino acid sequences which signal secretion of these proteins from the human cell and which are cleaved during this process to give mature interferons (13). Mature interferon genes have been constructed *in vitro* from cDNA sequences by removal of the signal coding DNA sequence and replacement of this sequence with an ATG translational start preceded by a restriction site (15). Expressing such a modified gene in *E. coli* (15) or yeast (22) results in the formation of mature interferon activity within the cell without secretion. However, we now report that when the human secretion signal sequences are not removed but expressed using the PGK promoter, interferons are secreted into the yeast cell wall and into the medium. Although this process varies in efficiency from one interferon gene to another, it is a true secretion process; that is, it is not the result of partial cell lysis. One of the pre-interferon gene products (LeIF A activity) has been purified from both cell extract and medium. The interferon protein from the medium is composed of two forms. The major form (60-70 percent) is properly processed mature LeIF A, identical to the mature interferon that results after secretion from human cells. This demonstrates that the low eukaryote yeast is capable of recognizing and processing the secretion signal of the highest eukaryotic organism. Another form (30-40 percent), having 3 extra amino acids of pre-sequence at the amino terminus suggests that yeast does not process the protein with perfect fidelity. The nature of the interferon remaining inside the cell is even more complex since three forms are present. This variation in processing may be related to the efficiency of secretion to the outside of the cell.

MATERIALS AND METHODS

Strains and Media

E. coli K-12 strain 294 (2) was used for all bacterial transformations. Yeast strains 20B-12 (α *trp1* *pep4-3*) (24) and GM3C-2 (a, *leu2-3* *leu2-112* *trp1-1* *his4-519* *cyc1-1* *cyp3-1*) (12) were used.

Luria broth was as described by Miller (27) with the addition of 20 μg/ml ampicillin. Yeast were grown on the following media: YEPD contained 1 percent yeast extract, 2 percent peptone and 2 percent glucose. YNB+CAA (used for Trp^+ selection) contained 6.7 grams of yeast nitrogen base (without amino acids) (YNB) (Difco), 10 mg of adenine, 10 mg of uracil, 5 grams Difco casamino acids (CAA), and 20 grams glucose per liter. Solid medium contained 3 percent agar.

Plasmid DNA Preparation, Transformations, and Stabilities

Purification of covalently closed circular plasmid DNAs from *E. coli* (9) and transformation of *E. coli* (8) were done in accordance with previously described procedures. *E. coli* miniscreens were as described by Birmboim and Doly (5). Transformation of yeast was done essentially as previously described (19). Stability of plasmids in yeast was determined by replica plating colonies from YEPD (nonselective) to YNB+CAA (selective) media.

Extract Preparation and Interferon Assays

Extracts and media from yeast were assayed for interferon by comparison with interferon standards using the cytopathic effect (CPE) inhibition assay (30). Media were assayed directly after cell removal while yeast extracts were prepared as follows: Cultures were grown in YNB+CAA and 10 ml aliquots of cells were collected by centrifugation then resuspended in 3 ml of 1.2 M sorbitol, 10 mM KH_2PO_4 (pH 6.8) and 1 percent zymolyase 60,000 followed by incubation at 30°C for 30 min (to 90 percent spheroplasting). Spheroplasts were pelleted at 3000 xg for 10 min., then resuspended in 150 µl of 7M guanidine hydrochloride (GHCl). Extracts were diluted in PBS/BSA buffer (20 mM NaH_2PO_4 (pH 7.4), 150 mM NaCl, and 0.5 percent BSA). Alternatively, 10 ml of cells at the same A_{660} were pelleted and resuspended in 0.4 ml of 7M GHCl in an Eppendorf (1.5 ml) tube containing about 0.4 ml of glass beads (0.45 to 0.5 mm, B. Braun Melsurgen AG). These tubes were vortexed twice for 2 min at the highest vortex setting, keeping on ice in between. The extracts were centrifuged 0.5 min in an Eppendorf centrifuge and diluted in PBS/BSA buffer as above. Bioassays were performed with MDBK cells (30) for LeIF A, LeIF D, and the pre-forms, but with HeLa cells (30) for IFN-γ and preIFN-γ.

Purification of (pre D/A) LeIF A from the Medium and Cell Extract

Yeast strain YEp1PT-preLeIF A53t/20B-12 was grown at 30°C to A_{660} of 4. At this time the 5 liter culture was harvested by centrifugation. Ten milliliter aliquots were withdrawn periodically during the fermentation to measure optical density, interferon production, and secretion. The medium and cells were assayed as described above.

The medium was concentrated and diafiltered in an Amicon thin channel apparatus or a 2.5 liter stirred cell. The retentate was further purified by ion exchange chromatography on CM-52 followed by immunoaffinity column chromatography as described for LeIF A (29). The peak of activity was lyophilized and the residue redissolved in 0.1 percent trifluoroacetic acid (TFA), pH 2.5. The sample was then further purified by high pressure liquid chromatography (HPLC) on a Synchropak RP-P column. The column was eluted at a flow rate of 1 ml per minute with a linear gradient of 0 to 100 percent acetonitrile, 0.1 percent TFA, pH 2.5 in 60 minutes. The protein in the peak fractions containing interferon activity was then sequenced at the amino terminal end.

The cells harvested above for medium isolation were disrupted in a Bead Beater (Biospec Products). The lysate was centrifuged, the pellet washed, and the supernatants combined. The supernatants were dialyzed and the interferon purified as described above for medium material. A portion of the interferon was further purified by HPLC with a linear gradient of 0 to 100 percent acetonitrile. The protein from fractions containing interferon activity was then sequenced at the amino terminal end.

Protein Sequence Analysis

Sequence analysis was based on the Edman degradation (10). The sample was introduced into the cup of a Beckman 890B spinning cup sequencer. Polybrene™ (1,5-dimethyl-1,5-diazaundecamethylene polymethobromide) was used as a carrier in the cup (32). The sequencer was modified with a cold trap and some program changes to reduce background peaks. The reagents were Beckman's sequence grade 0.1 molar Quadrol buffer, phenylisothiocyanate, and heptafluorabutyric acid.

A modification also included automatic conversion of the 2-anilino-5-thiazolinone derivatives as they were extracted from the sequencer. The 1-chlorobutane was collected in a Pierce Reacti-Vial™ and dried under nitrogen. Then 25 percent TFA in water was added to the 2-anilino-5-thiazolinone and heated to 70°C for 10 min to convert it into the 3-phenyl-2-thiohydantoin (PTH derivative) (35). The PTH-amino-acid residue was then automatically dissolved in 50 percent acetonitrile and water and injected into a reverse-phase high-pressure liquid chromatograph. Each PTH-amino acid was then identified by comparison to the retention times of a standard mixture of PTH-amino acids.

RESULTS

Identification of PGK 5'-flanking DNA Sequence

Fig. 1 illustrates a partial restriction map of a cloned 3.1 kbp HindIII fragment from chromosome III of yeast, which contains the gene for yeast 3-phosphoglycerate kinase (20,21). We have determined the exact location of the structural gene by DNA sequencing from the PvuI site to the HindIII site. The relevant portion of the sequence is shown in Fig. 1 as an insert between the Sau3A through the ATG translational start and 8 amino acids of additional coding sequence. This is thought to be the start of the structural gene due to the length of the open reading frame (consistent with polypeptide size), the DNA sequence 5' to the ATG, and extensive homology of protein sequence between yeast PGK and human (23) or horse (3) sequence.

Construction of a PGK Promotor Fragment

Having identified the start of the structural gene, a design was required for the isolation of a promoter fragment (5'-flanking sequence of the PGK gene) on a specific restriction fragment for expression of fused heterologous genes (portable promoter fragment). This was done for the alcohol dehydrogenase I (ADH1) gene by Bal31 digestion through the ATG into the 5'-flanking sequence followed by the addition of an EcoRI linker (22). Since deletion of up to 32 base pairs of 5'-flanking sequence adjacent to the ATG had no gross effects on the expression of a heterologous gene

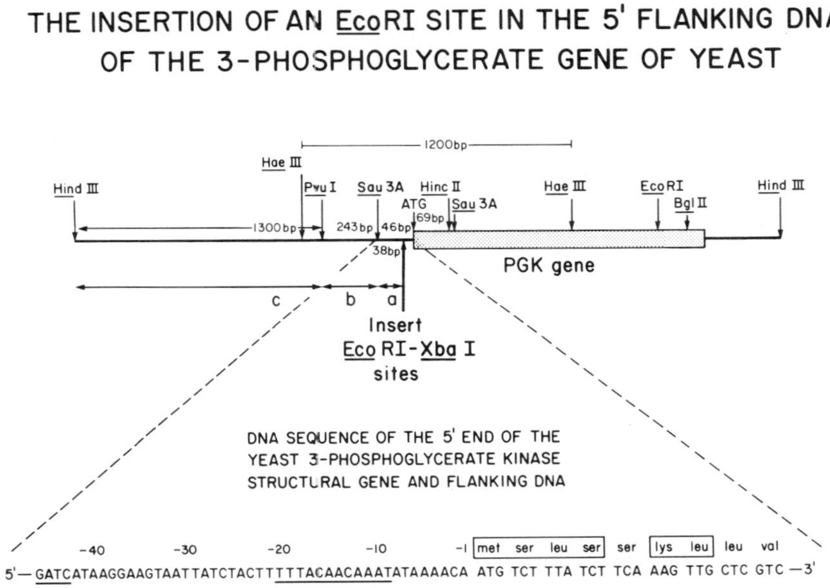

Fig. 1. Partial restriction and sequence map of the PGK gene. The PGK structural gene is shown as a bar region in the partial restriction map of a 3.1 kbp HindIII fragment. Note that distances are not drawn to scale; however, distances of fragments relevant to the discussed constructions are designated. The inserted 5' to 3' sequence (left to right) is from the Sau3A site through 9 amino acids of coding sequence. The underlined sequence was used to make a complementary oligonucleotide for the primer repair reaction. Boxed in amino acids are identical to amino acids in the amino-terminal end of horse and human PGKs.

(leukocyte interferon D), Bal31 digestions were not used for construction of the PGK promoter fragment.

Instead, we used the primer repair technique described by Goeddel et al. (14). An oligonucleotide (12 nucleotides long) was synthesized which is complementary to the -9 through -20 nucleotide sequence of the PGK strand shown in the insert of Fig. 1. When such an oligonucleotide is hybridized with denatured DNA from this region of the PGK gene in the presence of Klenow DNA polymerase I, polymerization occurs 5' -> 3' through the upstream Sau3A site while the 3' -> 5' exonuclease degrades phosphodiester bonds from the 3' end of the fragment until the double stranded region is reached. By cutting the repair product with Sau3A a small region of the PGK promoter fragment can be isolated as a Sau3A to blunt end DNA (-46 to -9) fragment. The loss of -8 through -1 of the 5'-flanking sequence was thought not to be a problem due to the ADH1 promoter fragment results already discussed.

The PGK fragment which contains a blunt end and a Sau3A sticky end (sequence "a" in Fig. 1) can then be ligated into a plasmid vector containing a BamHI sticky end and a filled-in XbaI blunt end. This allows a portion of the 5'-flanking sequence of the PGK gene to then be isolated from this vector as a Sau3A to XbaI restriction fragment. However, since this fragment is small and since transcriptional starting efficiency and control of yeast genes is sometimes affected by yeast DNA far upstream from the ATG translational start (12,17), the plan to make a PGK promoter fragment required the reconstruction of 5'-flanking sequence upstream from the Sau3A site to the HindIII site (thus sequences "b" and "c" in Fig. 1 were added back to sequence "a"). During this reconstruction an EcoRI site was also added adjacent to the XbaI site so that the 1600 bp promoter fragment can be isolated as a HindIII-to-XbaI or as a HindIII-to-EcoRI fragment.

A Yeast Plasmid for Expression of Heterologous Genes

A plasmid containing the PGK promoter fragment and other necessary components is shown in Fig. 2. This plasmid contains a portion of pBR322 (6) with the ampicillin resistance (Ap^R) gene and the *E. coli* origin of replication for selection and stable growth in *E. coli*. The plasmid also contains the TRP1 gene on an EcoRI to Pstl fragment which originates from chromosome IV of yeast (31,33). This gene allows for selection and maintenance of the plasmid in trpl yeast. The plasmid also contains a yeast origin of replication on a 2.0 kbp fragment from endogenous yeast 2µm plasmid DNA (7). This origin allows the DNA to replicate autonomously in yeast and be maintained as a plasmid.

Furthermore, the plasmid system contains the yeast PGK promoter fragment which originates transcription near the only EcoRI site in the plasmid (the other EcoRI site was removed by standard procedures). A HindIII/BglII fragment from the yeast TRP1 gene region (31,33) of chromosome IV was used as a convertor of HindIII to BglII for ligation with the BamHI site of pBR322. The 2.0 kbp fragment from 2µ DNA contains the transcription termination/polyadenylation signals which are normally the signals for the "Able" gene in 2µm plasmid (18). Such a region appears to be essential for such a yeast expression system (data not shown) as well as for a natural yeast gene (36). Gene inserts as EcoRI fragments in the proper orientation can thus be expressed as protein when the vector is put into yeast. Many

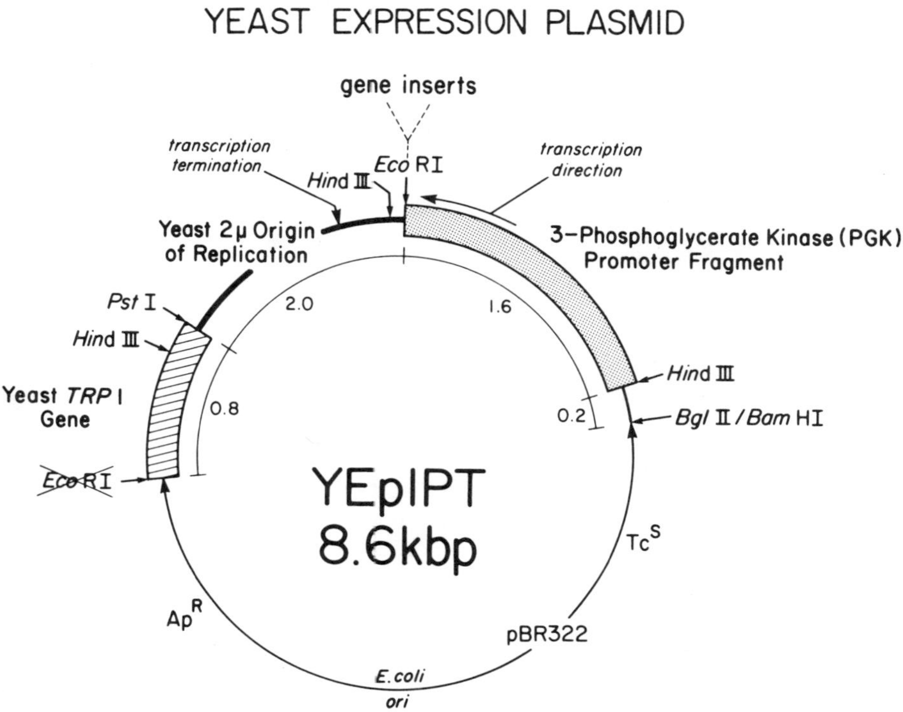

Fig. 2. Yeast expression plasmid. The partial restriction map of YEp1PT is shown with components designated that are required for transcription and translation of a heterologous gene inserted at the single EcoRI site.

heterologous genes have been expressed in this system. Some examples besides the interferon genes mentioned here are human serum albumin (26) and hepatitis surface antigen (34).

Interferon Genes

Fig. 3 illustrates some EcoRI fragment constructions containing interferon cDNAs that were used in the YEp1PT expression plasmid. The details of these constructions will only briefly be discussed here. The modifications involved in the construction of mature LeIF D (22), LeIF A (15), and IFN-γ (16) have already been published. Mature LeIF A required a final modification which we refer to as a converter. This 245 bp DNA fragment was obtained from yeast 2μm plasmid DNA (18) and due to its location was thought to be devoid of transcripion terminators. However, recent data have shown that there is a terminator on this fragment which functions as well as the "Able" terminator. Pre-interferon 5'-ends were converted to EcoRI by the primer repair technique (14) or by conversion of a convenient restriction sites.

Expression and Secretion of Interferons

The genes shown in Fig. 3 were inserted into YEp1PT (Fig. 2), and yeast were transformed with these plasmids. Transformants were assayed for interferon using the cytophathic effect bioassay (30). Interferon assays were done on three distinct compartmental locations in the yeast culture. The results of such assays are shown in Table 1.

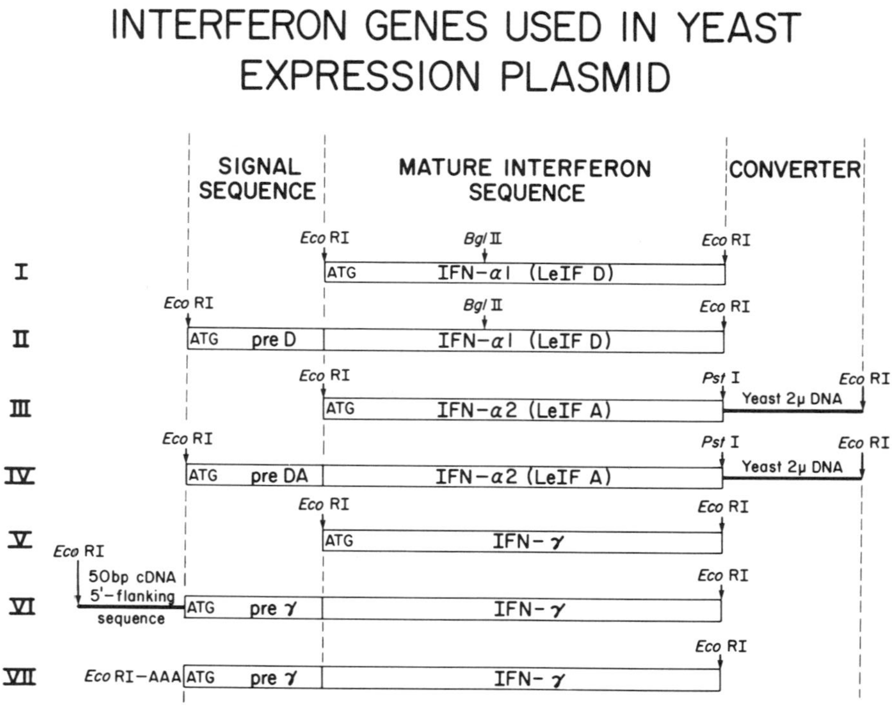

Fig. 3. Modifications of cDNAs for insertion into the EcoRI site of YEp1PT.

Table 1. Interferon Expression Levels

Gene Construction No.	YEp1PT plasmid containing these EcoRI fragments	Yeast	Inside[a] cell U/l/Abs_{660}=1	Pct.[b] cell protein	Released after cell wall removal[c] (U/l/Abs_{660})	Pct. cell protein	Outside cell (media) U/l/Abs_{660}	Pct. cell protein	Final[d] Abs_{660}	Pct. of activity secreted[e]
I	LeIF D	GM3C-2	$130x10^6$	1.0	0	0	0	0	1.0	0
II	pre LeIF D	GM3C-2	$27x10^6$	0.3	$0.4x10^6$	.004	$0.8x10^6$	.008	1.4	4
III	LeIF A	20B-12	$130x10^6$	1.0	0	0	0	0	1.0	0
IV	(pre D/A) LeIF A	20B-12	$19x10^6$	0.2	$0.5x10^6$	.005	$0.5x10^6$	.005	1.0	5
IV	(pre D/A) LeIF A	20B-12	$25x10^6$	0.1	N.D.	—	$2x10^6$	.007	3-4	8
IV	(pre D/A) LeIF A	GM3C-2	$28x10^6$	0.3	$0.3x10^6$	.003	$0.5x10^6$	.005	1.3	3
V	IFN-γ	20B-12	$0.6x10^6$	N.D.	N.D.	—	0	0	1.0	0
VI	pre IFN-γ + cDNA 5' flanking sequence	20B-12	$0.2x10^6$	N.D.	N.D.	—	$.03x10^6$	N.D.	1.2	15
VI	pre IFN-γ + cDNA 5' flanking sequence	GM3C-2	$0.38x10^6$	N.D.	N.D.	—	$.06x10^6$	N.D.	0.93	16
VII	pre IFN-γ	20B-12	$0.9x10^6$	N.D.	N.D.	—	$.19x10^6$	N.D.	1.0	21
VII	pre IFN-γ	GM3C-2	$1.9x10^6$	N.D.	N.D.	—	$.19x10^6$	N.D.	0.93	10

a) See Methods for extract preparation. Note that two methods were used for extracts. When cells were spheroplasted the "inside cell" amount was really inside material; however, when N.D. (not determined) is specified the "inside cell" amount and the "released after cell wall removal" were both part of "inside cell" amount--this type of extract involves glass beading cells without cell wall removal. Note that glass bead extracts without spheroplasting were always done for IFN-γ and preIFN-γ and that PBS buffer was used instead of 7M GHCl.

b) The specific activities of LeIFA and LeIFD were both assumed to be 10^8 U/mg protein for the calculations. A yeast culture contains about 100 mg of protein in the culture at an Abs_{660} = 1.

c) See Methods for spheroplasting procedure.

d) Abs of culture at which assay done.

e) The percent secretion is the percent "released after cell wall removal" plus the percent "outside cell". When spheroplasting was not done the "percent of activity secreted" does not include this cell wall secretion activity and the percent is lower (maybe 1/2) than it actually should be.

The first compartment is inside the cell. This fraction is measured by making a cell extract after the cell wall is removed and is defined as interferon activity that is not secreted. The other two compartments are the medium (material completely separate from yeast cell) and the activity released from the cells after cell wall removal using zymolyase. Both the medium fraction and fraction released after cell wall removal represent the total secreted material. Alternatively when cell walls were not removed (see Methods), "inside cell" activity also includes the secreted activity present in the cell wall.

It should be noted that (pre D/A) LeIF A is a mature LeIF A gene with a hybrid signal peptide sequence (see Figs. 3 and 4). This construction was made using the DdeI restriction site common to both preLeIF D and preLeIF A. Fig. 4 shows this DdeI site at amino acid -10. The underlined amino acids represent differences between the amino acid sequence of preLeIF D and preLeIF A. The hybrid (pre D/A) LeIF A is more like preD than preA. Both mature LeIF A and LeIF D genes (constructions I and III) were expressed in the yeast as 1.0 percent of the total cellular protein. The wrong orientations of all genes inserted into YEp1PT did not express interferon activity. For these two mature genes as well as the mature IFN-γ gene, no secretion occurred. However, when presequences were used on these genes, all three protein products were found in the media as secreted products.

As shown in Table 1, levels of interferon secretion vary from one gene to another with immune interferon giving the highest level of secretion. Secretion into the medium of this interferon varies from 10-21 percent; however, another 10-21 percent may be in the cell wall (not determined). By comparison of the production of interferon for constructions VI and VII, it is interesting that yeast tolerates 50 bp of preIFN-γ 5'-flanking cDNA sequence; even though such an intrusion between the yeast promoter and structural gene results in about a 5-fold decrease in expression. Pre LeIF D and (pre D/A) LeIF A secrete from 3-5 percent into the cell wall and medium combined at an $A_{660} \approx 1$; however, when (pre D/A) LeIF A/20B-12 was grown to A_{660} of 3-4, 8 percent secretion into the medium was observed. Although the percent in the cell wall was not determined at this higher A_{660}, as much as 16 percent of the activity was probably secreted.

Growth Curve and Interferon Production in the Medium from a Yeast Containing the (pre D/A) LeIF A Gene

Two interferon-producing yeast were investigated by further characterization. These were YEp1PT-preLeIF A53t/20B-12 and YEp1PT-LeIF A1/20B-12. The former contains two copies of construction IV (Fig. 3) in the EcoRI site of YEp1PT and results in activity being inside the cell, in the cell wall, and outside the cell (medium). This two copy gene (in tandem - both in orientation for proper expression) containing plasmid was used instead of the single copy construction since it sometimes gave higher levels of interferon activity in the medium. However, more careful comparisons have shown this not to be the case (both express about the same). Furthermore, the transcript stops at the end of the first structural gene in the converter as previously discussed. Yeast strain 20B-12 was used since it has greatly reduced protease levels (24), which might be an advantage for obtaining undegraded protein (interferon) from the

DdeI site in DNA (between −11 and −10) — Cleavage site (between −1 and +1)

	−23	−22	−21	−20	−19	−18	−17	−16	−15	−14	−13	−12	−11	−10	−9	−8	−7	−6	−5	−4	−3	−2	−1	+1	+2
preA	Met	Ala	Leu	Thr	Phe	Ala	Leu	Leu	Val	Ala	Leu	Leu	Val	Leu	Ser	Cys	Lys	Ser	Ser	Cys	Ser	Val	Gly	Cys	Asp
preD	Met	Ala	Ser	Pro	Phe	Ala	Leu	Leu	Met	Val	Leu	Val	Val	Leu	Ser	Cys	Lys	Ser	Ser	Cys	Ser	Leu	Gly	Cys	Asp
preD/A	Met	Ala	Ser	Pro	Phe	Ala	Leu	Leu	Met	Val	Leu	Val	Val	Leu	Ser	Cys	Lys	Ser	Ser	Cys	Ser	Val	Gly	Cys	Asp
preγ				Met	Lys	Tyr	Thr	Ser	Tyr	Ile	Leu	Ala	Phe	Gln	Leu	Cys	Ile	Val	Leu	Gly	Ser	Leu	Gly	Cys	Tyr
preβ				Met	Thr	Asn	Lys	Cys	Leu	Gln	Ile	Ala	Leu	Leu	Leu	Cys	Phe	Ser	Thr	Thr	Ala	Leu	Ser	Met	Ser

Fig. 4. Amino acid sequences of secretion signals. A comparison of the amino acid sequence of the signal prepeptide regions of human IFN-α1 (pre D), IFN-α2 (pre A), IFN-α1,2 (pre D/A), IFN-γ (pre γ), and IFN-β (pre β) is shown. The amino acids underlined represent differences between the amino acid sequences of pre A and pre D. The DdeI site indicates the junction of the D and A pre-sequences in preparation of the hybrid pre D/A pre-sequence. The cleavage site that is used in human cells is shown.

medium. The latter yeast contain construction III (Fig. 3) in YEp1PT and express mature LeIF A inside the cell but do not secrete.

Fig. 5 illustrates an essentially identical growth curve for these two yeast strains in YNB+CAA (Trp^+ selective growth). Bioassays, done on the medium at various times during cell growth, demonstrate clearly that the pre-sequence on LeIF A is causing a release of interferon activity into the media. Without this pre-sequence essentially no activity is released. This has also been confirmed by "Western" gel analysis (11) of proteins from concentrated media. Labelled antibody reacts only with a LeIF A size protein in medium from a pre-LeIF A expressor and not from a mature LeIF A expressor (data not shown). It is also evident that levels of activity in the medium reach a maximum near stationary phase.

One possible explanation for this apparent secretion process is that the pre-sequence of interferon somehow makes the cells more susceptible to lysis during cell growth. This possibility was carefully examined by measuring levels of protein in the media at stationary phase. Both yeast media contained essentially identical concentrations of protein which showed identical patterns on SDS-gel electrophoresis. Thus a true secretion process must be occurring.

Purification of (pre D/A) LeIFA from the Medium and Cells

In order to determine the nature of the secretion process for (pre D/A) LeIF A into yeast medium, it was necessary to purify the protein product from the medium and cells separately. Since yeast are able to secrete this protein, processing of the amino-terminal end may occur in some manner during the process as occurs in mammalian cells. The object of further experiments was to determine the nature of this processing.

Cell extracts and media were obtained from 5 l fermentations and the cellular interferon was purified as described in Methods. It should be noted that cell extracts also contain cell wall secreted material since the cell wall was not removed prior to extract preparation. The medium interferon was first concentrated by diafiltration, purified by CM-52 ion exchange chromatography, followed by the last two steps for cellular interferon purification.

NH_2-terminal Sequence of (pre D/A) LeIF A Purified from Yeast Medium and Cells

Fig. 6 shows the results of sequencing the purified interferon. The sequence shown is that expected for (pre D/A) LeIF A if no processing occurs. The normal cleavage point of this interferon that is recognized by mammalian cells is shown. Two sequence runs were performed on two different purified samples from cells and media.

The protein sequence was interpreted by noting which PTH amino acid increased in each corresponding Edman cycle and then decreased in the following cycle. PTH amino acids that normally give low recoveries (cys, ser, thr, arg, his) were assumed when no increase in any other PTH amino acid was seen. The percentage of each form was estimated by comparing the areas of the interpreted residue with area from a standard mixture of PTH

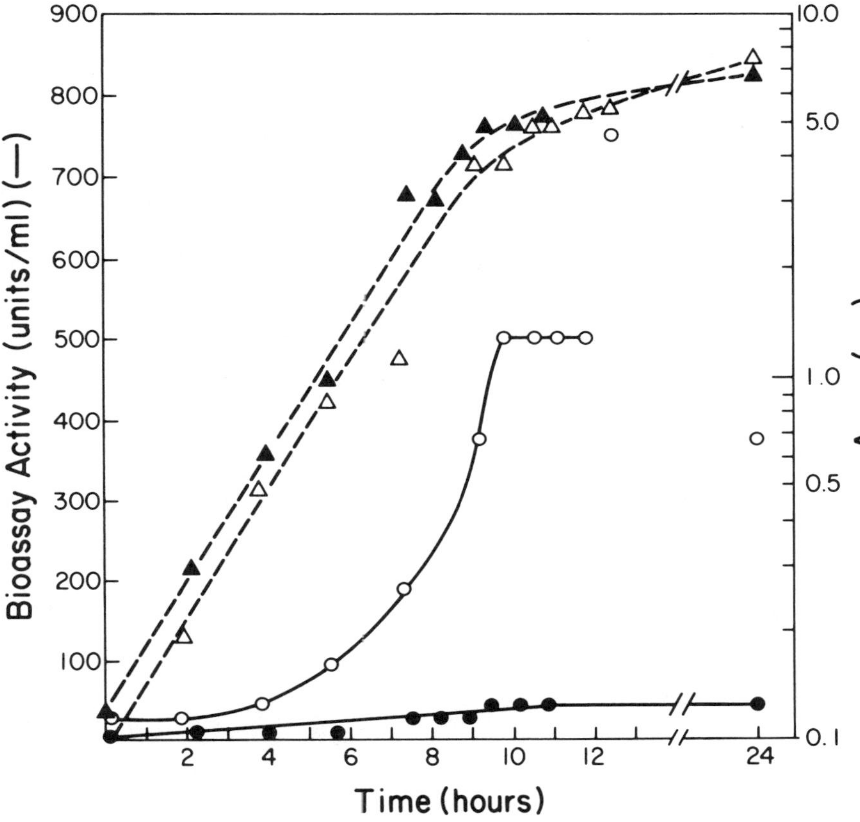

Fig. 5. Growth curve showing secretion into the medium. YEpIPT-LeIF A1/20B-12 (▵ , O) expresses mature interferon which remains within the cell; while YEpIPT-LeIF A53t/20B-12 (▵ , O) expresses the pre-sequence containing gene and produces interferon activity in the medium.

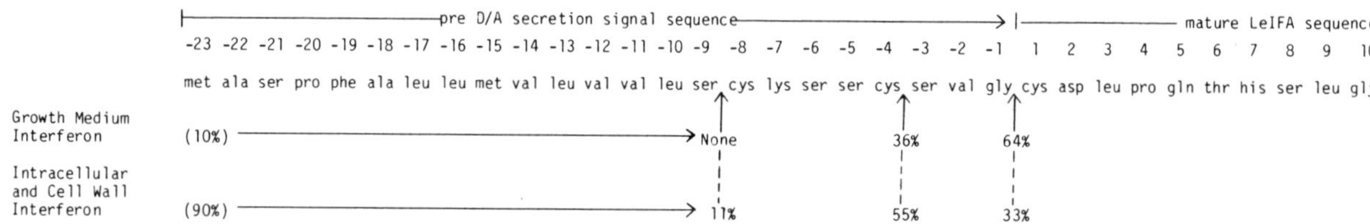

Fig. 6. Processing of (pre D/A) LeIF A by yeast. The 23 amino acid pre-sequence of (pre D/A) LeIF A interferon is shown as well as 10 amino acids of mature interferon sequence. Ten percent of the interferon made was found in the medium and 90 percent was cell associated (about 80 percent in the cell and 10 percent associated with the cell wall). So a small percentage (10 percent) of the 90 percent was also secreted interferon. The interferon from these two growth fractions (medium and cells) was processed as shown.

amino acids run on the same HPLC. Most of the forms were sequenced for 21 cycles or much further into mature LeIF A sequence than shown. An internal standard of Nor-Leucine was introduced in each chromatogram to assure that retention times were reproducible.

Figure 6 shows that most of the interferon in the medium was properly processed (64 percent) as done by human cells (13); however, another form (36 percent) containing three additional amino acids of pre-sequence was also present. The cellular interferon also contained these two forms, but in slightly different proportions, as well as a third form containing 8 amino acids of pre-sequence that was not detected in the medium. In no sequencing experiment was a full length pre-sequence ever seen suggesting that yeast process all of the pre-interferon in some manner. It is possible that sorting is a problem and that transport into other organelles occurs. Examination of the cellular location of these forms in the yeast cell versus the expressed mature form should give some insight into this possibility.

DISCUSSION

Using the isolated PGK gene (21), we have again demonstrated that the 5'-flanking DNA sequence of a highly expressed gene from yeast can be isolated and used as a portable restriction fragment for the expression of heterologous gene products in yeast. Careful comparisons of this promoter fragment and an ADH1 promoter fragment (fragment no. 906, 22) have shown essentially identical expression levels of leukocyte interferon D (data not shown).

In previous experiments, we suggested the need for a yeast 3'-flanking sequence for termination of transcription (22). By the size of the transcripts, it appeared that the mRNA of the hybrid expression unit, 5'-ADH1 promoter-LeIF D-TRP1-3', was terminating in the yeast TRP1 gene terminator. Indeed further experiments demonstrate the need for this signal with little or no expression occurring if such a signal is not present adjacent to the gene (data not shown). Use of such a hybrid expression system should allow further insights into terminator function, factors involved in mRNA synthesis and stability, mRNA translatability, and possible implications of codon usage (4) on expression of heterologous genes.

Using a plasmid containing the PGK-derived hybrid expression system, we have demonstrated the expression of both mature and pre-forms of human interferon cDNAs. The mature forms lack the secretion signal sequences and produce interferon within the yeast cell. The pre-forms contain the secretion signal coding sequences and produce interferon activities both inside and outside the cell. The secreted interferons were found both in the cell wall and unattached to the cell in the medium. The interferon, from yeast expressing the (pre D/A) LeIF A cDNA, was purified from cells (containing both intracellular and cell wall secreted material) and from medium. Two forms of interferon were found in the medium (10 percent of the expressed interferon), one properly processed (+1, Fig. 6) and the other with three additional amino acids of pre-sequence (-3, Fig. 6). The second form may be demonstrating a difference in yeast versus human cell processing recognition or it may be the result of the hybrid signal sequence. Further purification and sequencing of pre LeIF D should address this possibility.

It might also be possible that human cells make some of this -3 form that has not been noticed in interferon preparations.

Three forms of interferon were isolated from cells representing 90 percent of the expressed interferon. One form (33 percent) was properly processed (+1, Fig. 6), a second form (55 percent) contained 3 additional amino acids (-3, Fig. 6), and a third form (11 percent) contained 8 additional amino acids (-8). The last form was not seen in medium and a full length pre-sequence was never observed. However, it is possible that this peptide lacks interferon activity and thus was lost during purification. Therefore yeast appear to process both the secreted and nonsecreted interferon. The reason for only 10-20 percent secretion (20 percent if cell wall bound material included) instead of higher secretion levels may be due to a sorting problem associated with the heterologous signal. Interferon may be going into various organelles instead of through the plasma membrane. To examine this, we plan to compare cellular locations of interferon from both mature and pre-interferon producing yeast.

It would also be interesting to see if the interferons are secreted by yeast using the exocrine-like pathway which has been defined in yeast by sec mutants (28). Unlike other enzymes secreted by this pathway, the mature leukocyte interferons lack N-glycosylation and are probably not glycosylated when secreted by yeast. The effect of tunicamycin, which inhibits acid phosphatase and invertase secretion (25), would also be of interest. Perhaps a new secretion pathway might be defined by leukocyte interferon secretion from yeast.

REFERENCES

1. Ammerer, G., Hitzeman, R., Hagie, F., Barta, A. and Hall, B.D. 1981. In: Recombinant DNA, Proceedings of the Third Cleveland Symposium on Macromolecules, A.G. Walton, ed. (Amsterdam, Netherlands, Elsevier Scientific Publishing Co.) pp. 185-197.
2. Backman, K., Ptashne, M. and Gilbert, W. 1976. Construction of plasmids carrying the cI gene of bacteriophage λ. Proc. Natl. Acad. Sci. U.S.A. 73:4174-4178.
3. Banks, R.D., Blake, C.C.F., Evans, P.R., Haser, R., Rice, D.W., Hardy, G.W., Merrett, M. and Phillips, A.W. 1979. Sequence structure and activity of phosphoglycerate kinase: a possible hinge-bending enzyme. Nature 279:773-777.
4. Bennetzen, J.L. and Hall, B.D. 1982. Codon selection in yeast. J. Biol. Chem. 257:3026-3031.
5. Birnboim, H.C. and Doly, J. 1979. A rapid alkaline extraction procedure for screening recombinant plasmid DNA. Nucleic Acids Res. 7:1513-1523.
6. Bolivar, F., Rodriguez, R.L., Green, P.Y., Betlach, M.C., Heyneker, H.L., Boyer, H.W., Crosa, Y.H. and Falkow, S. 1977. Construction and characterization of new cloning vehicles. II. A multipurpose cloning system. Gene 2:95-113.
7. Broach, J.R., Strathern, J.N. and Hicks, J.B. 1979. Transformation in yeast: Development of a hybrid cloning vector and isolation of the CAN1 gene. Gene 8:121-133.
8. Clarke, L. and Carbon, J. 1975. Biochemical construction and selection of hybrid plasmids containing specific segments of the

Escherichia coli genome. Proc. Natl. Acad. Sci. USA 72:4361-4365.
9. Clarke, L. and Carbon, J. 1976. A colony bank containing synthetic Col EI hybrid plasmids representative of the entire E. coli genome. Cell 9:91-99.
10. Edman, P. and Begg, G. 1967. A Protein Sequencer. Eur. J. Biochem. 1:80-91.
11. Erlich, H.A., Levinson, J.R., Cohen, S.N., and McDevitt, H.O. 1979. Filter affinity transfer. J. Biol. Chem. 254:12240-12247.
12. Faye, G., Leung, D.W., Tatchell, K., Hall, B.D. and Smith, M. 1981. Deletion mapping of sequences essential for in vivo transcription of the iso-1-cytochrome C gene. Proc. Natl. Acad. Sci. USA 78:2258-2262.
13. Goeddel, D.V., Leung, D.W., Dull, T.J., Gross, M., Lawn, R.M., McCandliss, R., Seeburg, P.H., Ullrich, A., Yelverton, E. and Gray, P.W. 1981. The structure of eight distinct cloned human leukocyte interferon cDNAs. Nature 290:20-26.
14. Goeddel, D.V., Shepard, H.M., Yelverton, E., Leung, D. and Crea, R. 1980. Synthesis of human fibroblast interferon by E. coli. Nucleic Acids Res. 8:4057-4073.
15. Goeddel, D.V., Yelverton, E., Ullrich, A., Heyneker, H.L., Miozzari, G., Holmes, W., Seeburg, P.H., Dull, T., May, L., Stebbing, N., Crea, R., Maeda, S., McCandliss, R., Sloma, A., Tabor, J.M., Gross, M., Familletti, P.C. and Pestka, S. 1980. Human leukocyte interferon produced by E. coli is biologically active. Nature 287:411-416.
16. Gray, P.W., Leung, D.W., Pennica, D., Yelverton, E., Najarian, R., Simonsen, C.C., Derynck, R., Sherwood, P.J., Wallace, D.M., Berger, S.L., Levinson, A.D. and Goeddel, D.V. 1982. Expression of human immune interferon cDNA in E. coli and monkey cells. Nature 295:503-508.
17. Guarente, L. and Ptashne, M. 1981. Fusion of Escherichia coli lacZto the cytochrome C gene of Saccharomyces cerevisiae. Proc. Natl. Acad. Sci. USA 78:2199-2203.
18. Hartley, J.L. and Donelson, J.E. 1980. Nucleotide sequence of the yeast plasmid. Nature 286:860-865.
19. Hinnen, A., Hicks, J.B. and Fink, G.R. 1978. Transformation of yeast.Proc. Natl. Acad. Sci. USA 75:1929-1933.
20. Hitzeman, R.A., Chinault, A.C., Kingsman, A.J. and Carbon, J. 1979. Detection of E. coli clones containing specific yeast genes by immunological screening, p. 58-68. In: R. Axel and T. Maniatis (eds.), ICN-UCLA Symposia on Molecular and Cellular Biology. Vol. 14. Academic Press, New York.
21. Hitzeman, R.A., Clarke, L. and Carbon, J. 1980. Isolation and characterization of the yeast 3-phosphoglycerokinase gene (PGK) by an immunological screening technique. J. Biol. Chem. 255:12073.
22. Hitzeman, R.A., Hagie, F.E., Levine, H.L., Goeddel, D.V., Ammerer, G. and Hall, B.D. 1981. Expression of a human gene for interferon in yeast. Nature 293:717-722.
23. Huang, I., Welch, C.D. and Yoshida, A. 1980. Complete amino acid sequence of human phosphoglycerate kinase. J. Biol. Chem. 255:6412-6420.
24. Jones, E. 1976. Proteinase mutants of Saccharomyces cerevisiae. Genetics 85:23-33.
25. Kuo, S.-C. and Lampen, J.O. 1974. Tunicamycin--an inhibitor of yeast glycoprotein synthesis. Biochem. Biophys. Res. Commun. 58:287-295.
26. Lawn, R.M., Adelman, J., Bock, S.C., Franke, A.E., Houck, C.M., Najarian, R.C., Seeburg, P.H. and Wion, K.L. 1981. The sequence of

human serum albumin cDNA and its expression in E. coli. Nucleic Acids Res. 9:6103-6113.

27. Miller, J.H. 1972. Experiments in Molecular Genetics, Cold Spring Harbor Laboratory, Cold Spring Harbor, NY.
28. Novick, P., Field, C., and Schekman, R. 1980. Identification of 23 complementation groups required for post-translational events in the yeast secretory pathway. Cell 21:205-215.
29. Staehelin, T., Hobbs, D.S., Kung, H., Lai, C.Y. and Pestka, S. 1981. Purification and characterization of recombinant leukocyte interferon with monoclonal antibodies. J. Biol. Chem. 256:9750-9754.
30. Stewart II, W.E. 1979. Interferon Assays In: The Interferon System, Springer, New York.
31. Stinchcomb, D.T., Struhl, K. and Davis, R.W. 1979. Isolation and characterization of a yeast chromosomal replicator. Nature 282:39-43.
32. Tarr, G.E., Beecher, J.F., Bell, M. and McKean, D.J. 1978. Polyquarternary amines prevent peptide loss from sequenators. Anal. Biochem. 84:622-627.
33. Tschumper, G. and Carbon, J. 1980. Sequence of a yeast DNA fragment containing a chromosomal replicator and the TRP1 gene. Gene 10:157-166.
34. Valenzuela, P., Gray, P., Ouiroga, M., Zaldivar, J., Goodman, H.M. and Rutter, W.J. 1979. Nucleotide sequence of the gene coding for the major protein of hepatitis B virus surface antigen. Nature 280:815-819.
35. Wittmann-Liebold, B., Graffunder, H. and Kohls, H. 1976. A device coupled to a modified sequenator for the automated conversion of anilinothiazolinones into PTH amino acids. Anal. Biochem. 75:621-633.
36. Zaret, K.S. and Sherman, F. 1982. DNA sequence required for efficient transcription termination in yeast. Cell 28:563-573.

14. DOUBLE-STRAND BREAKS AND GENETIC RECOMBINATION

Rodney J. Rothstein,* Terry L. Orr-Weaver† and Jack W. Szostak†

*Department of Microbiology, New Jersey Medical School, Newark, New Jersey 07103
†Sidney Farber Cancer Institute, Department of Biological Chemistry, Harvard Medical School, Boston, Massachusetts 02115

Rec. Adv. Yeast Mol. Biol. (1982) 1:191-199

Recombination of a plasmid with the yeast genome after transformation is stimulated by introducing a double-strand break in the plasmid molecule within a DNA fragment that is homologous to the chromosome. Transformation with a plasmid containing two double-strand breaks (a gap) within a single yeast DNA fragment results in a recombination event in which the gap is repaired using chromosomal information as the template. We have shown that gapped plasmids that contain an autonomous replicating sequence are also repaired and either a replicating or integrated plasmid is produced. We propose a model based on these observations that explains the repair of double-strand breaks and that postulates that double-strand breaks are important initiators of genetic recombination in yeast.

INTRODUCTION

The process of homologous genetic recombination requires 1) synapsis, 2) strand exchange and 3) resolution. We are using DNA transformation of yeast cells as a model system to study the various processes involved in recombination (9). The ability to experimentally manipulate plasmid DNA facilitates these studies. We have examined the interactions of both replicating and non-replicating plasmids with homologous chromosomal sequences. Non-replicating circular plasmids require integration into the chromosome to be stably inherited and the plasmids integrate by homologous recombination into the genome (3,10). In about one-third of the transformants, yeast sequences on the plasmid are substituted for the chromosomal sequences without integration of the entire plasmid (3). This class is functionally a gene conversion event. Replicating plasmids are maintained exogenously and consequently show an unstable phenotype for selectable markers on the plasmid.

We have previously shown that circular plasmids linearized by a restriction enzyme cleavage <u>within</u> a yeast sequence recombine into the homologous region of the genome (9). The double-strand break produced by the restriction enzyme cleavage stimulates recombination between 10- and 1000-fold over that found for circular integrating plasmids. We have also shown that a gap (two double-strand breaks) within a yeast segment is repaired using chromosomal sequences, and the resultant integrated plasmid is flanked by two complete direct repeats of the target sequence. These observations have led to methodologies for the directed integration of yeast plasmids into desired regions of the genome, for the recovery of chromosomal alleles, and for mapping the position of alleles within cloned segments (10). We propose that double-strand break repair provides a general model for recombination in fungi.

RESULTS

Circular molecules integrate by homology. Integrations into the yeast genome occur by homologous recombination. The plasmid pWJ12 contains a. 1.8kb EcoRI - BamHI fragment from the sup3+ region of chromosome XV as well as a 1.7kb BamHI fragment of the HIS3+ region from the other side of the centromere on chromosome XV (Figure 1). After transformation of 10^8 cells with 10ug of DNA we observed 16 HIS3+ transformants. Colonies with integrated plasmid were detected by standard yeast filter colony hybridization using bacterial plasmid sequences as a probe (3). HIS3+ colonies that failed to hybridize with the probe presumably became HIS3+ by a gene conversion event. These colonies (4 of the 16) were not further analyzed. Southern blot analysis (14) was performed after isolating DNA from twelve transformants to determine the site of plasmid integration. Integration into either the sup3 or the HIS3 region produced a characteristic restriction fragment after digestion of the genomic DNA with the restriction enzyme SalI (see Figure 1). The fragments were detected by hybridization of genomic blots with labeled plasmid as a probe. Of 12 transformants examined, 4 contained the plasmid integrated at the HIS3 region and 8 at the sup3 region (Figure 2). The larger number of integrations into the sup3 region reflects an as yet undetermined but reproducible property of the sup3 region.

Double-strand breaks stimulate recombination. The enzyme, XhoI, creates a double-strand break outside the coding sequence of the HIS3+ gene but within the HIS3+ fragment. After transformation with 10ug of XhoI-cut plasmid DNA 140 transformants/10^8 cells were detected. Approximately 90% contained pBR322 sequences and 12 were analyzed by Southern blots. The appearance of a genomic 24kb SalI fragment in each transformant after probing with radioactive pBR322 sequences indicated that the plasmid had integrated into the HIS3 region (data not shown). The creation of the double-strand break both stimulated recombination (10-fold more than circles in this example) as well as directed integrations to the HIS3 region.

Gapped molecules are repaired. The plasmid pWJ12 was digested with SmaI to create a 1200bp gap within the sup3+ region. Forty-three transformants were recovered after transforming with 10ug of DNA/10^8 cells. The 3-fold stimulation of transformation is low presumably due to the small amount of homology left near the border of the sup3+ DNA and pBR322 sequences (170bp). All transformants that contained plasmid sequences were integrated at the sup3 region as demonstrated by Southern blots. The gap was repaired and the sup3 regions flanking the plasmid sequence were both full length. We have previously shown that chromosomal information was used as template for the gap repair (10).

We constructed a plasmid that contains the HIS3 region as well as the 2kb SalI-XhoI LEU2+ gene and the autonomous replication sequence arsl. The plasmid was made $his3^-$ by creating a 60bp deletion between the two BglII sites within the HIS3+ gene. Therefore, digestion with BglII generates a 60bp gap. When yeast is transformed with BglII-cut pSZS11 and selection is made for HIS3+, transformants arise only if the gap is repaired from the

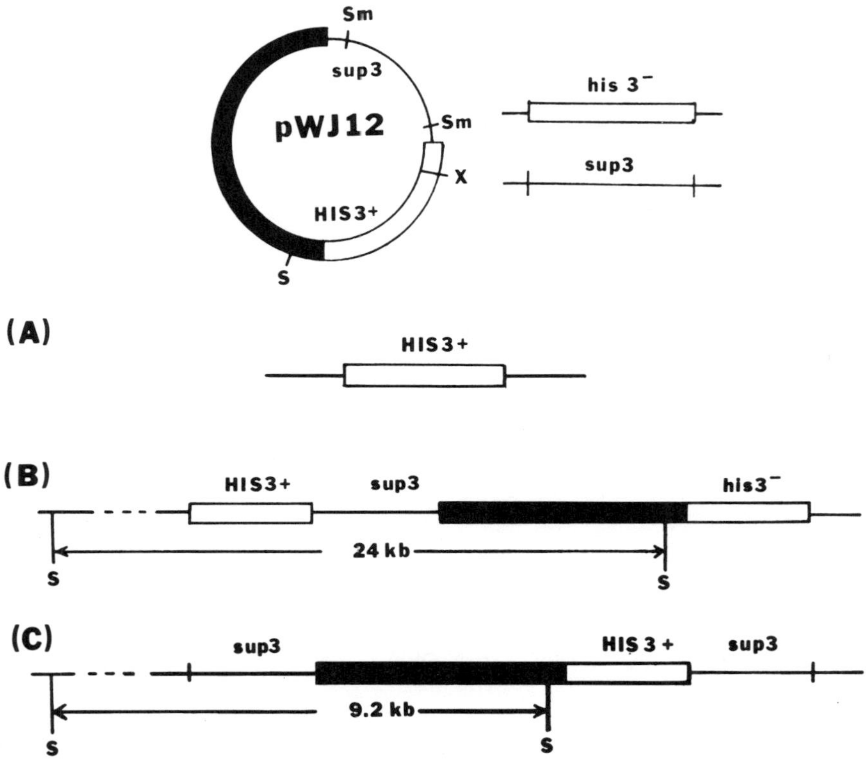

Figure 1. Integration of pWJ12 into the yeast genome. The 7.4kb plasmid, pWJ12, contains DNA homologous to the two yeast regions HIS3 and sup3 (shown to the right) cloned in pBR322. HIS3+ transformants can arise in three ways: (A) gene conversion of HIS3+ information from pWJ12 into the genome without integration of the vector sequences, (B) integration of pWJ12 by a single crossover at his3⁻ or (C) integration of pWJ12 by a single crossover at sup3. Restriction sites SalI, SmaI and XhoI are designated as S, Sm and X respectively.

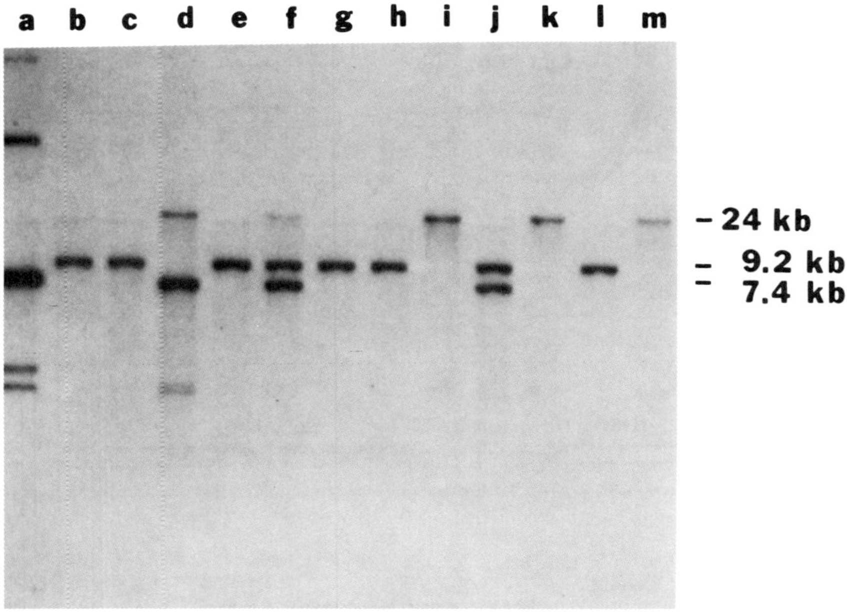

Figure 2. Genomic blot of pWJ12 transformants. DNA was isolated from 12 strains that had been transformed with circular plasmid DNA from pWJ12. Approximately 0.5ug of total genomic DNA was digested with the restriction enzyme SalI and electrophoresed for 16 hours in a 0.7% agarose gel at 75 volts in tris borate buffer as described previously (12). The DNA was blotted and hybridized with radioactively labelled pBR322 (12). Lane a shows the position of linear pWJ12 digested with SalI. The plasmid has integrated at sup3 in strains represented by lanes b,c,e,f,g,h,j and l as indicated by the 9.2 kb band. Integrations at his3 are indicated by the 24 kb band (lanes d,i,k and m). The presence of the additional band at 7.4 kb in lanes d,f and j is indicative of multiple tandem integrations (9).

chromosomal HIS3 region. Re-ligation of the gapped plasmid would result in a his3⁻ transformant (Figure 3A). After transformation with BglII-cut linear molecules, both stable and unstable transformants are observed at approximately equal frequency. The stable transformants contain the repaired plasmid integrated at the HIS3 region (Figure 3B). The HIS3+ transformants exhibiting the unstable phenotype are due to molecules that interacted with the chromosome and repaired the gap without subsequent integration (Figure 3C). The generation of unstable gap-repaired molecules at such a high frequency indicates that they are one of the normal products of the molecular interaction between the incoming plasmid and the chromosome - the other potential product being an integrated plasmid.

DISCUSSION

It is clear from our studies that free DNA ends are recombinogenic. The properties of gapped plasmid integration demonstrate that an efficient repair mechanism exists for resolving these structures. These results suggest new methods for the manipulation of plasmids during yeast transformation (10). We propose a mechanism for genetic recombination based on the idea that double-strand breaks and their subsequent repair can explain the observed fungal data for gene conversion and crossing over. As a detailed model will appear in the future, in this communication we briefly outline some aspects of this proposal.

The initial steps in the repair of gapped molecules are presented in Figure 4. In this model both ends invade the target double helix in a concerted fashion. We imagine that the initial invasive event actually occurs at either end of the double-strand gap. After the initial invasion the other free end has a high probability of invading its target site due to the close proximity of its region of homology. Each invading 3' end primes the repair synthesis necessary to fill in the gapped region using the existing chromosomal sequence as template. At the end of the replication-repair event, the repaired region is flanked by two Holliday structures (4). Theoretically, each Holliday structure could be independently resolved to yield two separate duplexes that have either crossed over with respect to each other or have resolved without a crossover. If each of the two Holliday structures in Figure 4 are resolved differently - crossover and non-crossover - the net result is a single crossover and the plasmid integrates. If both are resolved similarly, no plasmid integration occurs. However, the heteroduplex present at the initial site of pairing of the invading strand may result in gene conversion.

We have generalized this view of gapped plasmid integration in a model for genetic recombination based on double-strand break repair. In brief, the initiation of genetic recombination occurs at or near selected sites and is initiated by a double-strand break. The polarity of gene conversion suggests the existence of specific initiation sites. Exonucleolytic degradation exposes single strands which invade homologous chromosomal sequences; repair synthesis ensues and leads to the double Holliday structures shown for plasmid integration in Figure 4. Earlier investigators including Sobell (13) and Holliday (5) have postulated such intermediates for recombination. It is noteworthy that Leslie Bell and Breck Byers (personal communication) have observed DNA structures during meiosis in yeast that are consistant with the view that Holliday structures are frequently found in pairs separated by 300-800 base pairs.

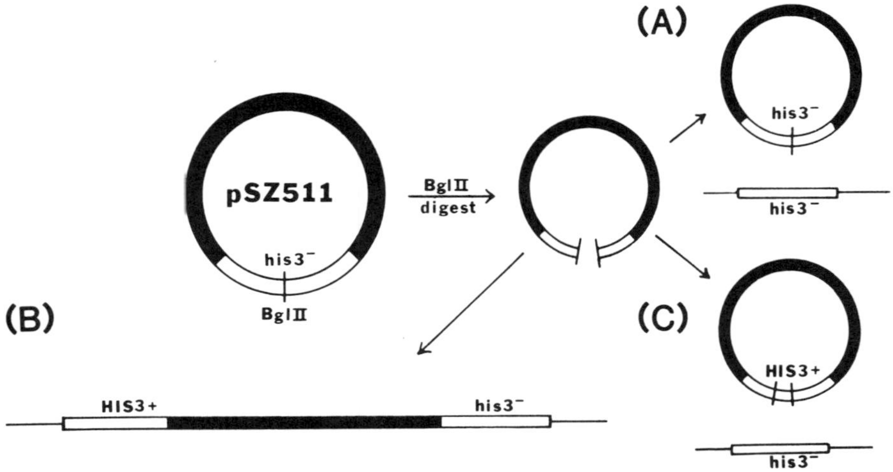

Figure 3. Transformation with gapped pSZ511. The LEU2+ ars1 his3⁻ plasmid pSZ511 contains an approximately 60 base pair deletion within the HIS3+ DNA sequence and is, therefore, his3⁻. The LEU2+ ars1 pBR322 portion of the plasmid is indicated by the thick solid black line. The plasmid was linearized with BglII and transformed into a his3⁻ yeast cell. If re-ligation of the plasmid occurs, the cell, remains his3⁻ due to the deletion of the DNA between the two BglII sites (A). Gap repair and integration as described in the text could occur and a stably integrated HIS3+ plasmid would result (B). Lastly the plasmid could interact with the chromosomal his3 region, repair the gap without subsequent integration, and result in an unstable HIS3+ phenotype due to the repaired autonomously replicating plasmid (C).

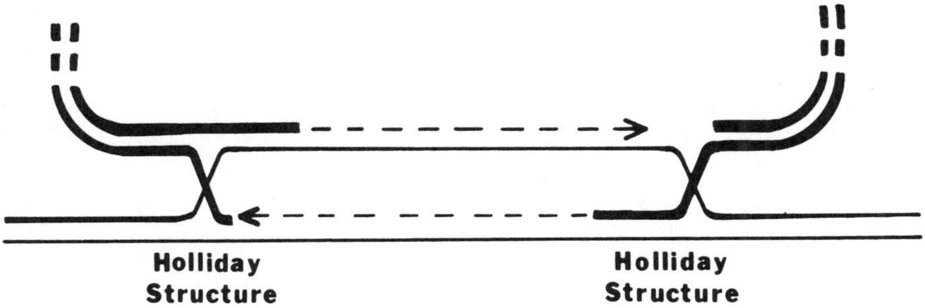

Figure 4. Model for the repair of a double-strand gap. The thin lines represent chromosomal duplex DNA and the thick lines represent plasmid DNA. The dashed lines represent repair synthesis. The incoming plasmid DNA is shown invading the homologous region of the chromosome. The 3' end can act as a primer for repair synthesis. After repair two Holliday structures are formed. Resolution of the Holliday structures results in either integration of the plasmid or a donation of a small portion of plasmid sequences (depicted as a thick line) without the integration of the entire plasmid.

After the formation of the two Holliday structures, rotary diffusion (7) of the structure may occur. This feature accommodates the doubly heteroduplex regions necessary to generate aberrant 4:4s observed in Ascobolus and Sordaria (6,11). For yeast, in which there is a paucity of aberrant 4:4s (1), we postulate that such rotary diffusion cannot occur. The generation of 6:2 and 2:6 tetrads is postulated to occur by the repair of a gapped region. 5:3 and 3:5 segregations are accomodated by failure to repair the heteroduplex region created by strand invasion. There are two notable features of our model that fit with observations of conversion in yeast tetrad data. One is that the 6:2 and 2:6 tetrads, which rarely favor one parent versus the other (1), can arise wholly from degradation of either chromatid and the subsequent repair of the duplex. This event does not require non-random mismatch correction as implied in other models (8). Secondly, when Fogel and his co-workers examined post-meiotic segregations at the arg4 locus, they found that the associated crossover could occur on either side of the conversion event (1). The double-strand break repair model postulates the existence of two Holliday structures and thus accomodate this observation.

The features of the double-strand break repair model for recombination also make it an attractive model for HO-initiated mating type interconversion (2). The mating type system, however, puts an additional constraint on the crossover outcome - there can be no (or extremely rare) exchanges since this would cause lethal deletions of part of chromosome III.

ACKNOWLEDGEMENTS

We thank Frank Stahl for suggesting that our model for plasmid integration may apply generally to meiotic recombination. T.L.O.-W. was supported by NIH Training grant CA 09361. This work was supported by NIH grant GM 27862 to J.W.S. and NSF grant PCM-8003805 to R.J.R.

REFERENCES

1. Fogel, S., Mortimer, R.K. and Lusnak, K. 1981. Mechanisms of meiotic gene conversion, or "wanderings on a foreign strand." pp. 289-339. In: Strathern, J.N., Jones, E.W. and Broach, J.R. (eds), Molecular Biology of the Yeast Saccharomyces: Life Cycle and Inheritance. Cold Spring Harbor Laboratory, Cold Spring Harbor, N.Y.

2. Herskowitz, I and Oshima, Y. 1981. Control of cell type in Saccharomyces cerevisiae: mating type and mating-type interconversion. pp. 181-209. In: Strathern, J.N., Jones, E.W. and Broach, J.R. (eds), Molecular Biology of the Yeast Saccharomyces: Life Cycle and Inheritance. Cold Spring Harbor Laboratory, Cold Spring Harbor, N.Y.

3. Hinnen, A., Hicks, J.B. and Fink, G.R. 1978. Transformation of yeast. Proc. Natl. Acad. Sci. USA 75:1929-1933.

4. Holliday R. 1964. A mechanism for gene conversion in fungi. Genet. Res. 5:282-304.

5. Holliday, R. 1974. Molecular aspects of genetic exchange and gene conversion. Genetics 78:273-287.

6. Kitani, Y. 1978. Aberrant 4:4 segregation. Jpn. J. Genet. 53:301-319.

7. Meselson, M. 1972. Formation of hybrid DNA by rotary diffusion during genetic recombination. J. Mol. Biol. 71:795-798.

8. Meselson, M.S. and Radding, C.M. 1975. A general model for genetic recombination. Proc. Natl. Acad. Sci. USA 72:358-361.

9. Orr-Weaver, T.L., Szostak, J.W. and Rothstein, R.J. 1981. Yeast transformation: a model system for the study of recombination. Proc. Natl. Acad. Sci. USA 78:6354-6358.

10. Orr-Weaver, T.L., Szostak, J.W. and Rothstein, R.J. 1982. Genetic applications of yeast transformation with linear and gapped plasmids. In: L. Grossman, K. Moldave and R. Wu (eds.), Methods in Enzymology: Recombinant DNA. Academic Press, New York, N.Y. (in press).

11. Rossingnol, J.L., Paquette, N. and Nicolas, A. 1978. Aberrant 4:4 asci, disparity in the direction of conversion and the frequency of conversion in Ascobolus immersus. Cold Spring Harbor Symp. Quant. Biol. 43:1343-1352.

12. Rothstein, R. 1979. Deletions of a tyrosine tRNA gene in S. cerevisiae. Cell 17:185-190.

13. Sobell, H.M. 1972. Molecular mechanism for genetic recombination. Proc. Natl. Acad. Sci. USA 69:2483-2487.

14. Southern, E.M. 1975. Detection of specific sequences among DNA fragments separated by gel electrophoresis. J. Mol. Biol. 98:503-517.

Part III

Recombination and Repair

15. IDENTIFICATION OF CLONED GENES THAT COMPLEMENT THE rad50-1, rad51-1, rad54-3 and rad55-3 MUTATIONS IN YEAST

I.L. Calderon, C.R. Contopoulou and R.K. Mortimer

Department of Biophysics and Medical Physics
University of California
Berkeley, California 94720

Rec. Adv. Yeast Mol. Biol. (1982) 1:200-212

Plasmids that complement the rad50-1, rad51-1, rad54-3 and rad55-3 mutations in yeast, have been isolated. They were obtained by transforming strains, carrying the leu2-112 leu2-3 alleles and the particular rad mutation, with YEp13 plasmids containing near random yeast DNA inserts. Rad^+ clones were identified among the Leu^+ transformants. Integration by targeting into the RAD55 locus showed that the rad55-3 complementing plasmid contained the actual RAD55 gene. BamHI fragments from each of the plasmids that complement rad50-1, rad51-1 and rad54-3, all of which lacked Rad^+ activity, were subcloned into the integrating plasmid YIp5 and the hybrid plasmids were used to transform a Rad^- Ura^- strain to Ura^+. By genetic mapping, the rad51 and rad54 subclones were shown to integrate at their respective loci. However, the rad50 subclones integrated at a site unlinked to the RAD50 locus. This suggests that no homology exists between this BamHI fragment and the RAD50 gene. Integration at the RAD54 locus of the rad54 subclone made the host cell Ura^+ but Rad^-; excision of the plasmid was shown to be X-ray inducible and to restore the Ura^- Rad^+ phenotype. These results indicate that the BamHI fragment of the RAD54 plasmid is internal to the RAD54 gene. We can conclude also that the RAD54 gene is not essential as cells bearing a disrupted copy of this gene are able to survive.

Additionally, a plasmid carrying an amber suppressor has been isolated and characterized.

* Support was obtained from the Office of Health and Environmental Research of the U.S. Department of Energy under Contract No. DE-AC03-76SF00098 and a "Fundacion Juan March" fellowship to I.L.C.

INTRODUCTION

The ability to repair DNA that has been damaged seems to be an important characteristic, common to all living organisms. Damage can arise spontaneously or can be induced by both physical and chemical agents; repair of the damage can be accomplished through different pathways, depending on, for example, the kind of damage or the physiological state of the cell.

In the yeast *Saccharomyces cerevisiae* evidence for the repair of potentially lethal damage has been provided by the isolation of a large number of mutant strains that are more sensitive than the wild type to the lethal effects of various mutagens, such as ultraviolet light, X-rays and methyl-methane-sulfonate (see (10), (14) and (15) for recent reviews). These mutants fall into more than 50 different complementation groups: one of them is known to define a photoreactivation gene (19); the rest are thought to be involved in dark repair. The fact that so many genes participate in dark repair implies that these processes are under extensive genetic control in yeast. Studies with double mutants revealed that these genes can be classified into three epistatic groups (4,9) which are believed to represent three different repair pathways: 1) Excision repair, carried out by the genes in the *RAD1* group; 2) error prone repair, controlled by genes in the *RAD6* group; and 3) recombinational repair, controlled by the genes in the *RAD52* group. This last group of genes seems to be involved in several cellular processes as demonstrated by the fact that their mutants have complex phenotypes; not only are they defective in repair, but also they exhibit abnormalities in mitosis, meiosis, recombination and homothallic switching. However, very little is known about the function of these genes at the molecular level.

We have chosen to isolate these genes by cloning them into replicative plasmids, followed by *in vivo* and *in vitro* characterization. By such studies, we hope to determine if these genes are essential for the vegetative life of the yeast cell, how they are regulated and what are their products.

In this paper we describe the preliminary steps of these investigations, namely the isolation and characterization of plasmids that complement the mutations *rad50-1*, *rad51-1*, *rad54-3* and *rad55-3*.

MATERIALS AND METHODS

Strains: The yeast strains used for transformation are listed in Table 1; they were constructed by standard techniques of yeast genetics (21). Unless otherwise stated, the rest of the strains were obtained from the Yeast Genetic Stock Center (University of California, Berkeley).

Table 1. Transformable strains used

Strain	Genotype	Source
XS131-5A	*MATα rad50-1 leu2-112 leu2-3 ura3-52 trp1-289 his3-Δ1 lys1-1*	D. Schild
XS133-33	*MATα rad51-1 leu2-112 leu2-3 ura3-52 trp1-289 his3-Δ1*	D. Schild
XL4-22C	*MATα rad54-3 leu2-112 leu2-3 ura3-52 trp1-289*	this work
XL5-3C	*MATα rad55-3 leu2-112 leu2-3 ura3-52 trp1-289 his3-Δ1 his5*	this work

The *Escherichia coli* strains used were: HB101 (3), provided by A. J. Clark; JA300 (23), provided by J. Carbon, and DB6507 (a *pyrF* derivative of HB101), provided by D. Botstein.

Plasmids: YEp13 (5)is a derivative of the pBR322 plasmid of *E. coli* that also carries the *LEU2* gene of *S. cerevisiae* and a fragment of the yeast 2 plasmid with its origin of replication. pBR322 confers resistance to ampicillin and tetracycline to the host bacterial cell.

YRp7 (23) is also a pBR322 derivative, carrying the yeast origin of replication *ARS1* and the yeast *TRP1* gene; it was provided by D. Botstein.

YIp5 (20) consists of pBR322 and the *URA3* gene of yeast. This plasmid does not replicate autonomously in yeast but is able to transform it by integration; it was provided by D. Botstein.

Yeast pool 35 DNA was provided by K. Nasmyth; it was constructed in a similar way as that described by Nasmyth and Reed (17) but using the vector YEp13 instead of YRp7.

Media and Genetic Methods: Growth media and standard genetic procedures for yeast have been described elsewhere (21). Tests for auxotrophy were carried out in synthetic complete media lacking the particular requirement.

E. coli growth media have been described before (7). When necessary, ampicillin (60 μg/ml) and/or tetracycline (15 μg/ml) was added to the media.

Enzymes: Restriction enzymes and T4 DNA ligase were purchased from New England Biolabs and Bethesda Research Laboratories (Rockville, MD) and used as directed.

Plasmid extractions and purifications: Plasmid DNA from yeast was extracted by the method described by Nasmyth and Reed (17).

Two different methods were used to extract DNA from E. coli: for quick small-scale preps the method of Holmes and Quigley (12) was applied; when large quantities of plasmid were required, the procedure described by Birnboim and Doly (1) was used with minor modifications, scaled up to 1 liter cultures. Plasmid DNA was then purified by equilibrium density sedimentation in a CsCl-ethidium bromide gradient, as described by Davis et al. (7).

Transformation: Yeast transformation was carried out as described by Hinnen et al. (11) with minor modificatons.

E. coli strains were transforned using the procedure described in Davis et al. (7).

X-ray Treatment: The X-ray source and treatment used are described in Game and Mortimer (9), except that the dose rate used was 237.5 rad/sec and the standard dose was 57 Krad.

Electrophoresis: Electrophoresis was carried out in 1% agarose slab gels according to the method described by Davis et al. (7). Tris-borate buffer was used both in the gel and in the chambers.

RESULTS AND DISCUSSION

Isolation of plasmids

The pool 35 DNA was used to transform to Leu^+ the yeast strains listed in Table 1. The resulting transformants were tested for resistance to X-rays. In the case of rad50, rad51 and rad54, about 0.1 to 0.2% of the Leu^+ tested turned out to be Rad^+; in the case of rad55, the proportion was in the order of 1% (Calderon, Contopoulou and Mortimer, manuscript in preparation). If one considers that the genome of S. cerevisiae has enough DNA (13,500 Kb, according to Lauer et al. (13)) to code for around 10^4 genes, and that each of the plasmids of pool 35 carries on the average a 5-7 Kb insert (17), one would expect to find a particular gene represented at a frequency of about 0.05%. We find no simple explanation for the considerably high frequency obtained in the case of rad55.

In order to determine to what extent each plasmid restores the radiation resistance of the transformants to the wild type condition, haploid and homozygous diploid Rad^- strains carrying or lacking the plasmids were irradiated with increasing doses of X-rays. The resulting survival curves were compared with those of similar Rad^+ strains. In all cases studied, the presence of plasmids improved the resistance of the strains to X-rays, although the degree of improvement depended on the particular plasmid studied (Calderon, Contopoulou and Mortimer, manuscript in preparation). In no case did the degree of resistance exceed that of the wild type; we interpret this to mean that once the wild type level of repair has been reached, an excess of RAD gene product has no influence.

A phenotypic characteristic of some of the diploid Rad^- strains is poor sporulation and/or low viability of spores (22). In contrast to the restoration of X-ray resistance, the presence of the plasmids does not seem to improve these sporulation defects (Calderon, Contopoulou and Mortimer, manuscript in preparation). This result might be explained, at least partially, by the fact that the plasmids are lost very often during meiosis: only 20-60% of the resulting spores retain them.

Restriction Maps

Plasmid DNA was obtained from each yeast strain and used to transform *E. coli* strain HB101 to ampicillin resistance and leucine autotrophy. Plasmid DNA was extracted and purified from the transformed clones. The purified DNA was then digested with *EcoRI* and the resulting fragments were separated by agarose gel electrophoresis. Those plasmids having the smallest insert (one for each *RAD* gene) were chosen for subsequent studies. Detailed restriction maps of each are shown in Figure 1.

In order to demonstrate that the purified plasmids were able to retransform the yeast strains from which they were isolated, and to determine if there was any cross complementation, the plasmids illustrated in Figure 1 were used to transform each of the Rad$^-$ transformable strains (Table 1) to Leu$^+$. When tested for X-ray sensitivity, all were Rad$^+$ when the plasmid used carried the same putative *RAD* gene as that mutated in the recipient strains. Conversely, all Leu$^+$ transformants remained Rad$^-$ when the other plasmids were used.

Integration of Plasmids into the Genome

Complementation of a mutation by a plasmid may result from plasmids carrying the gene in question or another gene capable of "suppressing" either genetically or metabolically, the mutant phenotype. In the course of this investigation we have, in fact, isolated such a suppressor-bearing plasmid (to be described later) which complements the *rad50-1* mutation. It therefore became essential to prove the identity of the cloned *RAD* genes by means of homologous integration into the yeast genome. It is commonly accepted that integration at a certain locus demonstrates that the plasmid carries that particular gene.

In view of the restriction maps, two different approaches were chosen to pursue the integration:

Integration of the rad55-3 complementing plasmid

As shown in Figure 1, the plasmid [YEp13-*RAD55*-13C] has a unique *BamHI* site. According to the observation of Orr-Weaver *et al.* (18), cleavage of this plasmid at the *BamHI* site would increase the efficiency of its integration within a chromosomal DNA region homologous to the yeast DNA that was cut.

About 25 g of the plasmid [YEp13-*RAD55*-13C] were digested with *BamHI* and used to transform the strain XL5-3C (Table 1) to Leu$^+$. Out of 35 transformants studied, 3 (INT55-11, INT55-12 and INT55-30) failed to show loss of the Leu$^+$ Rad$^+$ characters indicating that the plasmid had been integrated. Strain INT55-11 was then crossed to strain XL39-5A (*MATa* *rad55-3* *leu2* *ura3*(and/or *4*) *met8* *trp1-289* *aro1D* *his3*(and/or *5*), constructed by us).

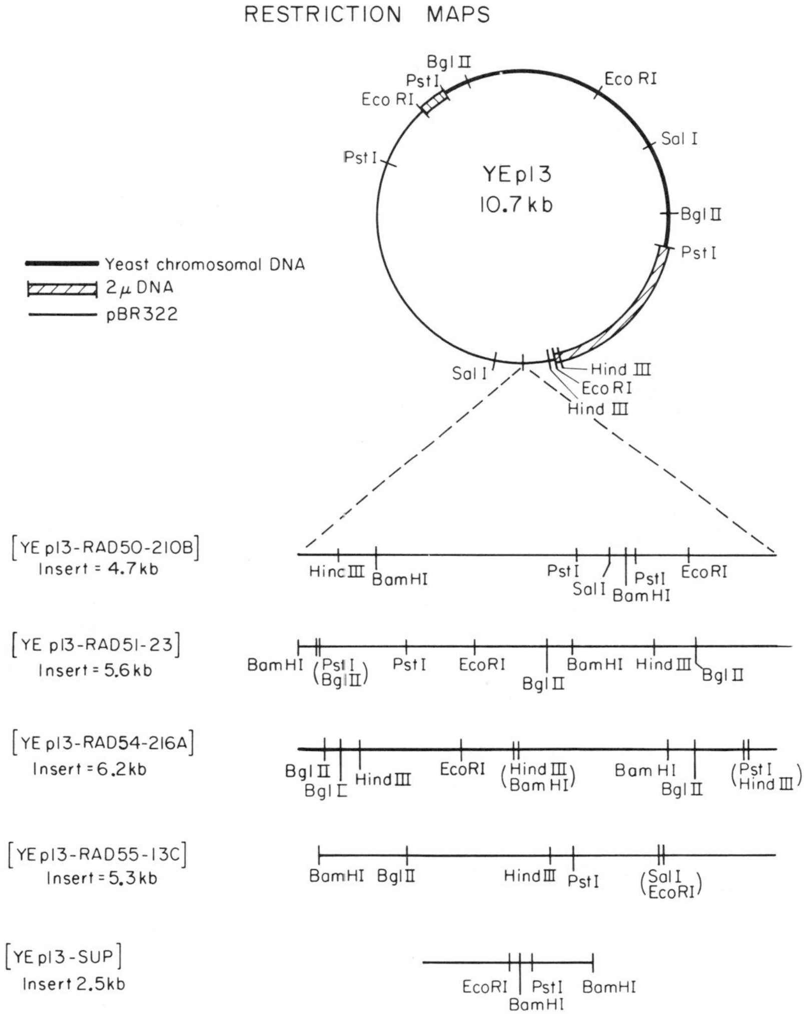

Figure 1. Restriction maps of the plasmids.

The resulting diploids are, therefore, homozygous for rad55-3 and leu2-112 leu2-3 but contain an integrated copy of the wild type allele of both of these genes. The genes RAD55 and ARO1 are located on the right arm of chromosome IV and are separated by a distance of 42.9 cM (16). About 60 asci derived from this cross were micromanipulated; only 52% of the spores were viable. Analysis of the asci containing 3 or 4 viable spores revealed that, as expected, both leu2 and rad55 markers segregated 2+ : 2- and all tetrads were parental ditypes; they mapped at a distance of about 40 cM from ARO1 (11 parental ditypes (PD): 1 non-parental ditype (NPD): 17 tetratypes (T)). These results indicate that the integration has taken place at the RAD55 locus and, consequently, that the RAD55 gene is present in plasmid [YEp13-RAD55-13C].

Integration of the rad50-1, rad51-1 and rad54-3 complementing plasmids

As shown in Figure 1, each of these plasmids has two BamHI sites that flank yeast DNA segments of 3 Kb, 3.4 Kb and 1.5 Kb, respectively. By subcloning these fragments into the single BamHI site of the integrative plasmid YIp5, then transforming the appropriate yeast strains with these plasmids and finally mapping the sites of integration in the yeast genome, one could demonstrate whether or not homology exists between the cloned fragments and the respective RAD genes. Before proceeding with the integration, we found it useful to determine if these BamHI fragments per se have Rad$^+$ function. This was carried out by subcloning them into the replicative plasmid YRp7; the BamHI fragments were excised and religated into the single BamHI site of YRp7. The ligation mixtures were used to transform the E. coli strain JA300 to Trp$^+$ Amp^R. Plasmid DNA was extracted from those transformants that were also Tet^S, using the "quick prep" method. This DNA was digested with restriction enzymes and the resulting fragments separated by agarose gel electrophoresis. The plasmids having the expected banding pattern for YRp7 plus the respective BamHI fragment, in the same orientation as in the original plasmids, were called [YRp7-RAD..-BB]; one each from RAD50, 51 and 54 was chosen for subsequent studies. Plasmid DNA was extracted, purified and used to transform the appropriate Rad$^-$ strains to Trp$^+$. In all three cases, the transformants tested were still Rad$^-$, indicating that the Rad$^+$ function is totally or partially outside the BamHI fragments.

The BamHI fragments were next subcloned into YIp5 in a way similar to that described for YRp7, except that the E. coli strain used for transformation was DB6507, and Ura$^+$ Amp^R transformants were selected. The plasmids obtained were called [YIp5-RAD..-BB] and were used to transform the yeast strain DBY947 (MAT ura3-52 ade2-101, provided by D. Botstein) to Ura$^+$. Integrants were called INT50-.., INT51-.. and INT54-.., respectively.

So far, 32 different INT50 strains have been analyzed. All were Ura$^+$ Rad$^+$. Crosses by XS132-1A (MATa ura3-52 rad50-1 his3 leu2 trp1-289, provided by D. Schild) revealed that, in all cases, the ura and the rad markers segregate 2+ : 2- but in an independent fashion, meaning that the integration has not taken place at the RAD50 locus. This result suggests that the cloned fragment lacks homology with the RAD50 gene.

Two different hypotheses could explain these findings: 1) The [YEp13-RAD50-210B] plasmid carries a gene that is able to complement rad50 mutations only when present in a high copy number, and 2) the insert of the rad50-1 complementing plasmid is made up of at least two Sau3A fragments which ligated together in the process of constructing the yeast pool but are normally separated in the genome; the BamHI fragment would carry a part that is neither homologous, nor continguous to the RAD50 gene. Presently we are working on these hypothesis.

When the plasmid used for transformation was [YIp5-RAD51-BB], seven integrants were obtained and all were Ura^+ Rad^+. One, INT51-B, was crossed to strain XS134-4B (MATa ura3-52 rad51-1 leu2 his3 trpl, provided by D. Schild). The analysis of 18 tetrads derived from this cross revealed that URA3 and RAD51 were closely linked: all of the tetrads were parental ditypes. The integrant INT51-B was also crossed to strain 3971-5B (MATa gal10 SUC mal trpl ura3 ura4 met8 ade5,7 leu2 lysl arolD ilvl canl). The genes RAD51 and ILV1 are located on chromosome V, about 11 cM apart (16). The results of the analysis of 15 tetrads showed that now URA3 and ILV1 were also linked (9 PD: 0 NPD : 6 T). We conclude that the integration has taken place at the RAD51 locus and that the plasmid that complements rad51-1 carries the RAD51 gene.

In the case of the INT54 transformants, two of them (INT54-A and INT54-D) were found to be Ura^+ Rad^+, while the other two (INT54-B and INT54-C) were Ura^+ Rad^-. Taking into account that the 1.5 Kb fragment carried by [YIp5-RAD54-BB] has no Rad^+ activity (as previously described), a possible explanation for the occurrence of Rad^- transformants would be that the BamHI fragment is totally internal to the RAD54 gene; its integration at the RAD54 locus would then give rise to two incomplete RAD54 genes, separated by the YIp5 sequences. This hypothesis implies that in INT54-B and INT54-C the integration occurred at the RAD54 locus. This in fact was shown to be the case: when INT54-C was crossed to strain X4004-3A (MATa ura3 lys5 trpl met2), a distance of 4.5 cM between the LYS5 and the URA3 loci was estimated (30 PD : 0 NPD : 3 T). According to J. Game (unpublished results), LYS5 and RAD54 are closely linked. In addition, this cross yielded only parental ditypes for the rad and ura markers.

When INT54-A, one of the Ura^+ Rad^+ transformants, was crossed to X4004-3A, the lys5 and ura3 markers segregated independently (3 PD : 1 NPD : 11 T), indicating that the integration has not occurred at the RAD54 locus. This might be explained by a gene conversion event between the URA3 allele on the plasmid and the chromosomal ura3 allele, or by integration of the plasmid at ura3. The latter explanation is unlikely because the mutation ura3-52 is thought to be a deletion and rarely undergoes homologous recombination (20).

The fact that the RAD54 gene can be completely disrupted by the plasmid and the cell still survives strongly suggests that it is a non-essential gene. In order to investigate how such a disruption affects the response of the cells to X-rays, INT54-B cells were irradiated with different doses of X-rays and the survival determined. The experiment was carried out at 23°C and 37°C as the rad54-3 mutation used throughout these studies is thermosensitive (9).

As can be seen from Figure 2 (top), the same survival curve was obtained at either temperature. Furthermore, the survival curves closely paralleled that obtained with cells carrying the rad54-3 mutation, when incubated at 37°C after irradiation, except for the resistant "tail" exhibited at high doses. Such "tail" effects are characteristic of mixed populations consisting by, in large, of sensitive cells with a small proportion of resistant ones. The most likely origin of the resistant subpopulation is by excision of the integrated plasmid, yielding back the original Rad$^+$ Ura$^-$ cells. This was tested by checking the Ura phenotype of the survivors. As can be seen in Figure 2 (bottom), the proportion of Ura$^-$ colonies increased with the dose of X-ray, until they made up almost 90% of the surviving population. Further confirmation for the occurrence of an excision event was obtained by showing that, indeed, all the Ura$^-$ clones, when tested for X-ray sensitivity, were again Rad$^+$.

Moreover, this excision seems to be induced by X-rays, as shown by the rise of sectored colonies Ura$^+$/Ura$^-$ among the surviving ones (Figure 2, bottom). Sectored colonies probably result from excision events occurring after surviving cells have divided. If no induction takes place, one would expect the proportion of sectored colonies to diminish with increasing doses, since the subpopulation of cells from which they derive, are being killed preferentially. The fact that the proportion increases rather than diminishes, would indicate that the excision is being induced. At high doses almost all the sensitive cells would have been killed; this would explain the eventual decline of the proportion of sectored colonies.

Isolation of a Suppressor

While testing the Leu$^+$ transformants, obtained with the pool 35, for other markers present in the transformable strains (Table 1), we found that a relatively high proportion of them (112/10,700 clones tested) were also Trp$^+$. Furthermore, in the case of the rad50 strain, most of the Rad$^+$ (42/44) turned out to be also Trp$^+$. The genes RAD50 and TRP1 are located on chromosome IV, but they are approximately 160 cM apart (16). Thus, the probability of having both genes together on the same plasmid, particularly at the high frequencies at which they were obtained, was thought extremely unlikely. A more likely possibility was that we had cloned a nonsense suppressor. Since the transformants were still Lys$^-$ and the lys marker present in the transformable strain, lys1-1, was an ochre mutation, we surmised that we were not dealing with an ochre suppressor. In fact, both rad50-1 and trp1-289 were demonstrated by standard procedures to be amber mutations and the plasmid that complements them, [YEp13-SUP], was shown to suppress other amber mutations, i.e., aro1D-1, met8-1, ade5,7 and trp-1 (Calderon, Contopoulou and Mortimer, manuscript in preparation). A restriction map of this plasmid is shown in Figure 1.

The most likely origin of this suppressor is the yeast strain AB320 (MATa/MATα [HO trp5-2 ura3-1 (ura1-1 ?) met4-1 ade2-1 leu2-1 lys2-1 can1-100]), which was the source of the insert DNA in pool 35. However, when a haploid ho derivative of AB320 (AB320 was provied by B. Hall) was crossed to strain 3971-5B that carries an assortment of nonsense alleles, no amber suppressor could be detected among the segregants of this cross. The possibility that AB320 carries a weak suppressor that is only active in high copy number, as is the case with plasmid YEp13, cannot be ruled out.

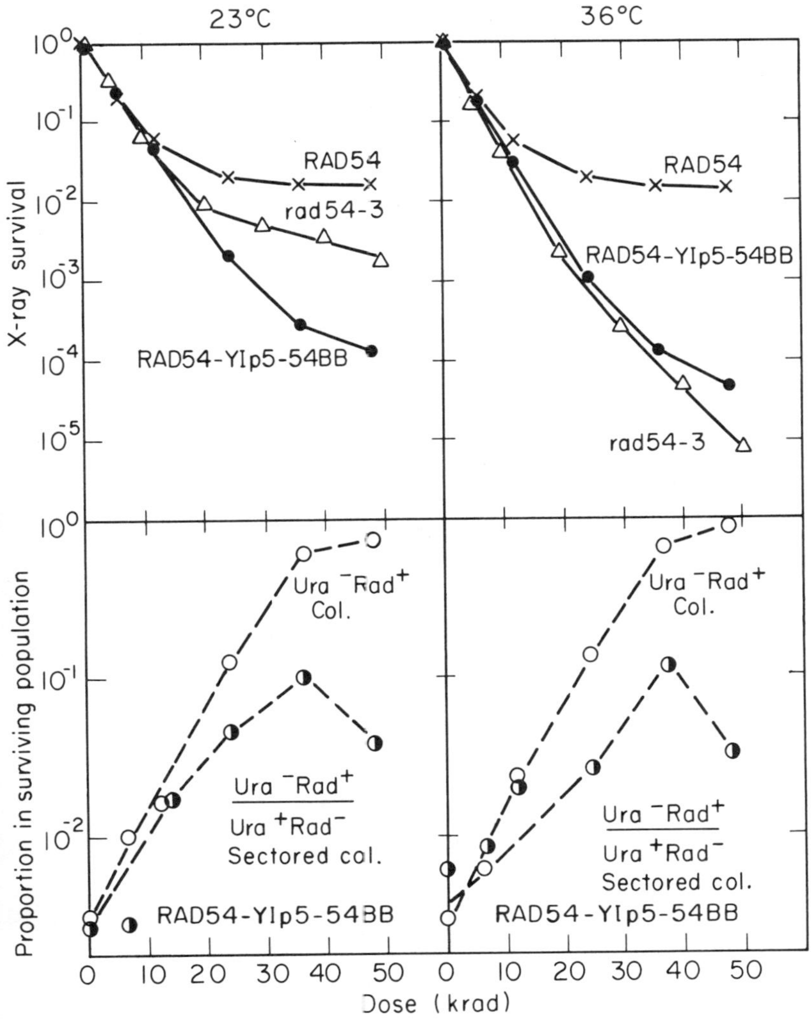

Figure 2. (top) Survival curves of strains carrying different alleles of RAD54; (bottom) proportion of Ura^- Rad^+ and Ura^- Rad^+/Ura^+ Rad^- sectored colonies arising among the INT54-B (RAD54-YIp5-54BB) cells that survived different doses of X-ray.

Another possible explanation would be that the plasmid carries the PSI factor which is known to enhance the expression of low level suppressors (6). One would then expect a low level supressor to be present in all transformable strains. This, however, seems not to be the case as demonstrated by crossing the strain XS131-5A (Table 1) to the strain MT193/3b (MATa SUQ5 ade2-1 his5-2 lys1-1 trp5-48 can1-100 ura3-1 leu1 [PSI], provided by B. Cox). Analysis of tetrads derived from this cross showed that the presence of PSI does not affect the normal segregation of the rad and the trp markers.

Two other possible hypotheses about the origin of the suppressor remain unproved:

1. Plasmid [YEp13-SUP] could carry a mitochondrial tRNA gene that behaves as a suppressor in cytoplasmically-directed protein synthesis; the high copy number of mitochondrial genomes would explain the high frequency at which this plasmid is found in the pool. This seems an unlikely explanation, however. The wobble rules for codon:anticodon recognition in mitochondrial translation would preclude a mitochondrial tRNA from functioning as an amber-specific suppressor (2).

2. The fragment that contains LEU2 in YEp13 is known to contain also a gene coding for a tRNA (8). This might well be SUP53, a gene closely linked to LEU2 (16), some of whose mutations are known to suppress amber mutations.

Presently, experiments to elucidate which of these hypotheses may be the correct one, are being carried out.

LITERATURE CITED

1. Birnboim, H C. and Doly, J. 1979. A rapid alkaline extraction procedure for screening recombinant plasmid DNA. Nucl. Acids Res. 7:1513-1523.

2. Bonitz, S. G., Berlani, R., Coruzzi, G., Li, M., Macino, G., Nobrega, F. G., Nobrega, M. P., Thalenfeld, B. E. and Tzagaloff, A. 1980. Codon recognition rules in yeast mitochondria. Proc. Natl. Acad. Sci. U.S.A. 77:3167-3170.

3. Boyer, H. W. and Roulland-Dussoix 1969. A complementation analysis of the restriction and modification of DNA in *Escherichia coli*. J. Mol. Biol. 41:459-472.

4. Brendel, M. and Haynes, R. H. 1973. Interactions among genes controlling sensitivity to radiation and alkylation in yeast. Molec. Gen. Genet. 125:197-216.

5. Broach, J. R., Strathern, J. N. and Hicks, J. B. 1979. Transformation in yeast: Development of a hybrid cloning vector and isolation of the *CAN1* gene. Gene 8:121-133.

6. Cox, B. S. 1965. , a cytoplasmic suppressor of super-supressors in yeast. Heredity 20:505-521.

7. Davis, R. W., Botstein, D. and Roth, J. R. 1980. Advanced Bacterial Genetics. Cold Spring Harbor Laboratory, Cold Spring Harbor, N.Y.

8. Dobson, M. J., Kingsman, S. M. and Kingsman, A. J. 1981. Sequence variation in the *LEU2* region of the *Saccharomyces cerevisiae* genome. Gene 16:133-139.

9. Game, J. C. and Mortimer, R. K. 1974. A genetic study of X-ray sensitive mutants in yeast. Mutat. Res. 24:281-292.

10. Haynes, R. H. and Kunz, B. A. 1981. DNA repair and mutagenesis in yeast, p. 371-414. In: J. N. Strathern, E. W. Jones and J. R. Broach (eds.), The Molecular Biology of the Yeast *Saccharomyces*. Cold Spring Harbor Laboratory, Cold Spring Harbor, N.Y.

11. Hinnen, A., Hicks, J. B. and Fink, G. R. 1978. Transformation of yeast. Proc. Natl. Acad. Sci. U.S.A. 75:1929-1933.

12. Holmes, D. S. and Quigley, M. 1981. A rapid boiling method for the preparation of bacterial plasmids. Anal. Biochem. 114:193-197.

13. Lauer, G. D., Roberts, T. M. and Klotz, L. C. 1977. Determination of the nuclear DNA content of *Saccharomyces cerevisiae* and implications for the organization of DNA in yeast chromosomes. J. Mol. Biol. 114:507-526.

14. Lawrence, C. W. 1982. Mutagenesis in *Saccharomyces cerevisiae*. Adv. Genet. 21:173-254.

15. Lemontt, J. F. 1980. Genetic and physiological factors affecting repair and mutagenesis in yeast, p. 85-120. In: F. J. deSerres, W. M. Generoso and M. D. Shelby (eds.), DNA Repair and Mutagenesis in Eukaryotes. Plenum Press, N.Y.

16. Mortimer, R. K. and Schild, D. 1980. Genetic map of *Saccharomyces cerevisiae*. Microbiol. Rev. 44:519-571.

17. Nasmyth, K. A. and Reed, S. I. 1980. Isolation of genes by complementation in yeast: Molecular cloning of a cell-cycle gene. Proc. Natl. Acad. Sci. U.S.A. 77:2119-2123.

18. Orr-Weaver, T. L., Szostak, J. W. and Rothstein, R. J. 1981. Yeast transformation: a model system for the study of recombination. Proc. Natl. Acad. Sci. U.S.A. 78:6354-6358.

19. Resnick, M. A. 1969. A photoreactivationless mutant of *Saccharomyces cerevisiae*. Photochem. Photobiol. 9:307-312.

20. Scherer, S. and Davis, R. W. 1979. Replacement of chromosome segments with altered DNA sequences constructed *in vitro*. Proc. Natl. Acad. Sci. U.S.A. 76:4951-4955.

21. Sherman, F., Fink, G. R. and Lawrence, C. W. 1977. Methods in Yeast Genetics. Cold Spring Harbor Laboratory, Cold Spring Harbor, N.Y.

22. Strike, T. 1978. Characterization of mutants sensitive to X-rays. Ph.D. thesis, University of California, Davis.

23. Tschumper, G. and Carbon, J. 1980. Sequence of a yeast DNA fragment containing a chromosomal replicator and the *TRP1* gene. Gene 10:157-166.

16. CLONING OF THE RAD52 GENE OF Saccharomyces cerevisiae

David Schild,[1] Boyana Konforti,[1] Carl Perez,[1] Warren Gish,[2] and Robert Mortimer[1]

Department of Biophysics and Medical Physics and Donner Laboratory[1]
and Department of Molecular Biology[2]
University of California
Berkeley, California 94720

Rec. Adv. Yeast Mol. Biol. (1982) 1:213-224

The RAD52 gene of Saccharomyces cerevisiae has previously been shown to be involved in both recombination and DNA repair. Here we report on the cloning of this gene. A plasmid containing a 5.9kb yeast insert in the YEp13 vector has been isolated and shown to complement the X-ray sensitive phenotype of the rad52-1 mutation. From this insert various fragments have been subcloned into the YRp7 vector. A spontaneous integration event of one of the subclones has been genetically mapped to the chromosomal location of RAD52, indicating that the structural gene has been cloned. A 1.97kb BamH1 fragment subcloned into YRp7 in one orientation complements the rad52-1 mutation, while the same fragment in the opposite orientation fails to complement. Various other subclones indicate that a BglII site, within the BamH1 fragment, is in the RAD52 gene. This BglII site has been deleted by S1-nuclease digestion and the resulting deletion inactivates the RAD52 gene. BAL31 deletions from one end of a 1.9kb Sal1-BamH1 fragment have been isolated; up to 0.9kb can be deleted without loss of RAD52 activity, demonstrating that the RAD52 gene is approximately 1kb or less in length.

INTRODUCTION

Several different DNA repair pathways have been identified in yeast, including a presumptive recombinational repair pathway (reviewed in ref. 7 and 12). The RAD52 gene is one of at least eight genes (the RAD52 epistasis group) already identified which are thought to be involved in recombinational repair (5,7). The rad52-1 mutation was originally identified as causing extreme sensitivity to X-rays, but only slight sensitivity to U.V. light (19). Strains carrying this mutation have also been shown to be partially or completely defective in meiosis and sporulation (5,6,18,27), meiotic recombination (6,18,27), mitotic gene conversion (6,18,25,27), double strand break repair (10,21), homothallic switching (13), and maintenance of chromosome stability, resulting in chromosome loss (15). In addition, vegetative cells with the rad52-1 mutation have been shown to have elevated rates of spontaneous mutation (30) and rad52/rad52 cells in meiosis have been shown to accumulate single

* This work was supported by N.I.H. grant GM07988 to D.S. and by the Office of Health and Environmental Research of the U.S. Department of Energy under contract DE-AC03-76SF00098.

strand breaks during premeiotic DNA synthesis (22). If one assumes that double strand breaks and other X-ray induced DNA lesions are repaired by a recombinational mechanism (20), it is possible to account for most of the phenotypes associated with the rad52-1 mutation by its defect in recombination.

Although many phenotypes of the rad52-1 mutation have been observed, no known enzymatic or structural activity of the RAD52 gene product is known. In order to study the RAD52 gene at the molecular level, we have cloned this gene. The cloning of E. coli DNA repair genes, such as RECA and LEXA, has already been very important in isolating and characterizing the primary products of these genes and in studying their transcriptional regulation (reviewed in ref. 23). The cloned RAD52 will be used to determine whether this gene is transcriptionally regulated and if so, what are the inducers of transcription. The cloned gene is currently being used to identify and isolate the RAD52 gene product and to determine whether this gene might code for an essential function.

MATERIALS AND METHODS

Strains. The transformable yeast strain XS95-6C (MATα rad52-1 ura3-52 leu2-3 leu2-112 trp1 and his3-1) was used in most of the experiments. It was constructed by first crossing g160/2d (MATα rad52-1 ade2-1 arg4 arg9 trp1 his5 leu1-1 ilv3 and leu2) from the Yeast Genetic Stock Center (Berkeley, CA.) to RH218 (MATa trp1-289) and then back crossing a resulting MATa rad52-1 trp1 derivative twice to DBY746 (MATα trp1-289 ura3-52 his3-1 leu2-3 leu2-112) (similar to SHY strains in ref. 2), kindly supplied by D. Botstein. The bacterial strains used were HB101 (CA600 leu$^-$ B1$^-$ thr$^-$ pro$^-$ lacz$^-$ Sm$^-$ recA$^-$, rB$^-$ mB$^-$ SUII) supplied by R. Davis, JA300 (thr$^-$ leuB6$^-$ thi$^-$ thyA$^-$ trpC117$^-$ hsrK$^-$ hsmK$^-$ strR) (29) supplied by John Carbon, and DB6507 (HB101 pyrF$^-$) supplied by D. Botstein.

Plasmids. A bank of near random (Sau3A) partial fragments cloned into the BamH1 site of the vector YEp13 was kindly supplied by K. Nasmyth (16). YRp7 and YIp5 were obtained from D. Botstein.

Yeast transformation. Yeast transformation was performed using a modified version of the Hinnen, Hicks and Fink method (9), as suggested by V. MacKay and R. Hitzeman. 100ml of YEPD culture, grown to 2×10^7 cells/ml, was collected and washed in water. The resulting pellet was resuspended in 5ml of S.E.D. (1M sorbitol, 25mM EDTA pH8.0 and 50mM dithiothreitol) and incubated 10 min. at 30°C. After a wash in 10ml 1M sorbitol, cells were resuspended in S.C.E. (1M sorbitol, 100mM Na citrate pH5 and 10mM EDTA) and 100 µl of zymolyase 60,000 (Kirin Brewery) at 1mg/ml was added. Cells were incubated at 30°C until 80-90% spheroplasting had occurred (1/2 to 1-1/2 hrs). Spheroplasting was monitored both microscopically and by loss of turbidity when spheroplasts were diluted into 10% SDS. Spheroplasts were washed 2 times in 1M sorbitol and once with CaS (1M sorbitol, 10mM $CaCl_2$ and 10mM Tris-HCl pH7.4). Spheroplasts were resuspended in 0.5ml CaS and divided into 100 µl aliquots. 5-20 µg plasmid DNA was added to cells and the mixture was incubated 15 min. at room temperature. 1ml of 20% polyethylene glycol (PEG-4,000, Sigma), 10mM $CaCl_2$ and 10mM Tris pH7.4 was added to each sample and incubated 15 min. at room temperature. After gently spinning the spheroplasts out of the PEG solution, they were resuspended in 150 μl of 1M sorbitol, 33.5% YEPD and 50mM $CaCl_2$. 50 μl aliquots were plated in yeast regeneration agar onto minimal dropout plates as described previously (9).

E. coli transformation and plasmid preparation. E. coli transformation was performed as described by R. Davis et al. (4). Small scale plasmid preparations were performed by the method of Holmes and Quigley (11) and large scale isolations plasmid were carried out using a protocol of Birnboim and Doly (1) scaled up to 1 liter by S. Conrad.

Restriction, ligation, S1 nuclease digestion and BAL31 digestion. All of the enzymes except BAL31 were obtained from Bethesda Research Laboratories and the B.R.L. suggested procedures were followed. BAL31 was obtained from New England Bio Labs and their procedure was followed.

X-ray source. A Machlett OEG-60 X-ray source operated at 50 kV and 25 mA was used. The exposure rate was 240 R/sec. The f factor for the exposure conditions used was .90 rad/R.

RESULTS AND DISCUSSION

Cloning of RAD52

A leu2 rad52-1 yeast strain (XS95-6C) was transformed with YEp13 (3) containing near random yeast DNA inserts (16) and Leu^+ transformants were selected. Approximately 2,200 Leu^+ transformant colonies were picked to master plates (50 per -leu plate) and grown up for 2 days at 30°C. They were then replica plated to YEPD plates and the replicas were treated with 50 krads of X-ray. Two transformants showed significant X-ray resistance and these were tested further. Survival curves on these transformants demonstrated that they had near wild type survival after X-ray treatment (data not shown). Cells derived from these transformants which had lost their plasmid became simultaneously leucine auxotrophs and radiation sensitive. Plasmid DNA was isolated from the two Rad^+ transformants by transforming E. coli with a crude yeast DNA preparation and selecting for ampicillin-resistant E. coli colonies. From these E. coli transformants, plasmid DNA was isolated and restriction analyses carried out. One plasmid contained a 5.9kb insert and the other one an insert slightly over 20kb. Preliminary restriction analysis of the two inserts was consistent with the smaller insert being part of the larger insert. Because of complications associated with obtaining large amounts of plasmid DNA and a detailed restriction map from the plasmid with the 20kb insert, we concentrated on the plasmid containing the smaller insert. A restriction map of the smaller plasmid is shown in Figure 1. Reintroduction of either of the 'RAD52' plasmids into XS95-6C (rad52-1 leu2) resulted in a Leu^+ Rad^+ phenotype for all of the transformant colonies tested (100 for each plasmid). These results are consistent with the plasmids containing either the RAD52 gene or a suppressor of the rad52-1 mutation. Although the rad52-1 mutation is not nonsense suppressible, it is possible that another gene on a high copy number plasmid might suppress the rad52-1 mutation or by-pass its defect. Experiments discussed below demonstrate that this is not the case and that at least the smaller plasmid contains the RAD52 gene and not a suppressor.

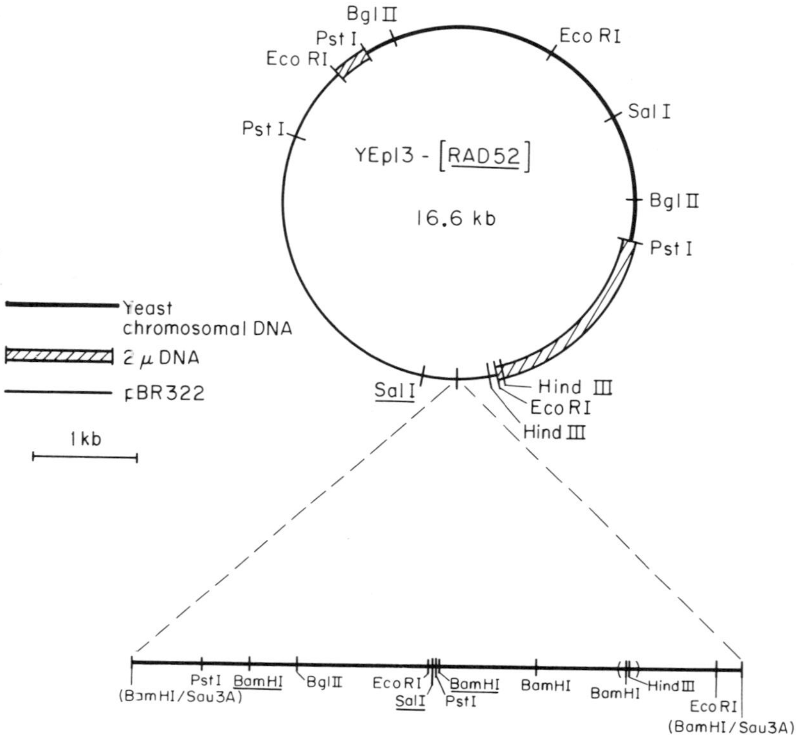

Figure 1. Restriction map of YEp13-[RAD52]. This plasmid consists of a 5.9kb yeast Sau3A fragment, containing the RAD52 gene, cloned into the BamHl site of the YEp13 vector (pBR322, the yeast LEU2 gene and the part of the yeast 2μ plasmid which includes the origin of replication) (3).

Subcloning of 'RAD52' and 'rad52' Fragments

From the original 5.9kb insert, we have subcloned several fragments into the YRp7 vector using standard techniques. A 3.3kb SalI fragment, consisting primarily of yeast DNA but with 275 base pairs from pBR322, and a 1.97kb yeast BamHl fragment were found to have 'RAD52' complementing activity (Figure 2). Unexpectedly, when this BamHl fragment was isolated in the opposite orientation in YRp7, it lacked activity. One trivial explanation for this difference was that the 1.97kb BamHl fragment in YRp7-C9 had been mutated or modified either by growth in E. coli or by the various biochemical manipulations during the subcloning procedures. We can rule this out because we have subcloned the 1.97kb fragment from YRp7-C9 back into YRp7 in the same orientation as YRp-C2-[RAD52] and this subclone now has RAD52 activity. The orientation difference we observe probably indicates that the activity of the BamHl fragment is dependent on part of the vector sequence, such as a promoter, or is being inhibited by part of the vector and that this dependence or inhibition is orientation specific. The BamHl-BglII subclones all lacked activity, indicating that the BglII site is probably either in the structural gene or between the gene and its regulatory region.

Integration Analysis of YRp7-A4 Sal-[RAD52]

In order to prove that we had cloned the RAD52 gene, rather than a suppressor of rad52-1, we examined integration events of one of the subclones. It has been demonstrated that integration of plasmids in yeast occurs via homologous recombination (8,17,25). Therefore, integration events at the chromosomal location of RAD52 would indicate that the plasmid contains the RAD52 region. Three independent spontaneous integration events were found in which the YRp7-A4-Sal-[RAD52] plasmid had integrated into the genome of XS95-6C. Two of these integration events were at the chromosomal location of TRP1 (data not shown) but one integrated at the chromosomal site of RAD52. The chromosomal location of this integrant was established by crossing it to a Rad^+ haploid of the opposite mating type and dissecting 10 asci from the resulting diploid (rad52[RAD52]/RAD52). All eight asci with four viable spore colonies yielded 4 Rad^+:0 rad^- segregants and the two asci with three viable spore colonies yielded 3 Rad^+:0 rad^- segregants. This segregation pattern establishes that the integration event occurred at or very close to the RAD52 chromosomal site on chromosome XIII. If the YRp7-A4-Sal-[RAD52] plasmid had not integrated at or very near the site of rad52-1 one would have expected some asci with 3 Rad^+:1 rad52-1 or 2 Rad^+: 2 rad52-1 segregations. The original integrant was also crossed to a rad52-1 lys7 trpl strain and asci from this cross were dissected. The Rad^+ Trp^+ phenotypes segregated together as expected for an integrated plasmid containing both TRP1 and RAD52. Both were also linked to lys7 (P:NP:T = 4:0:4) and to a centromere (FDS:SDS = 5:2) at distances consistent with the map position of RAD52 (14).

The 1.9kb BamHl fragment has also been subcloned into YIp5, the URA3 integrating vector (2,28), and preliminary results indicate that this plasmid also integrates at the chromosomal location of RAD52.

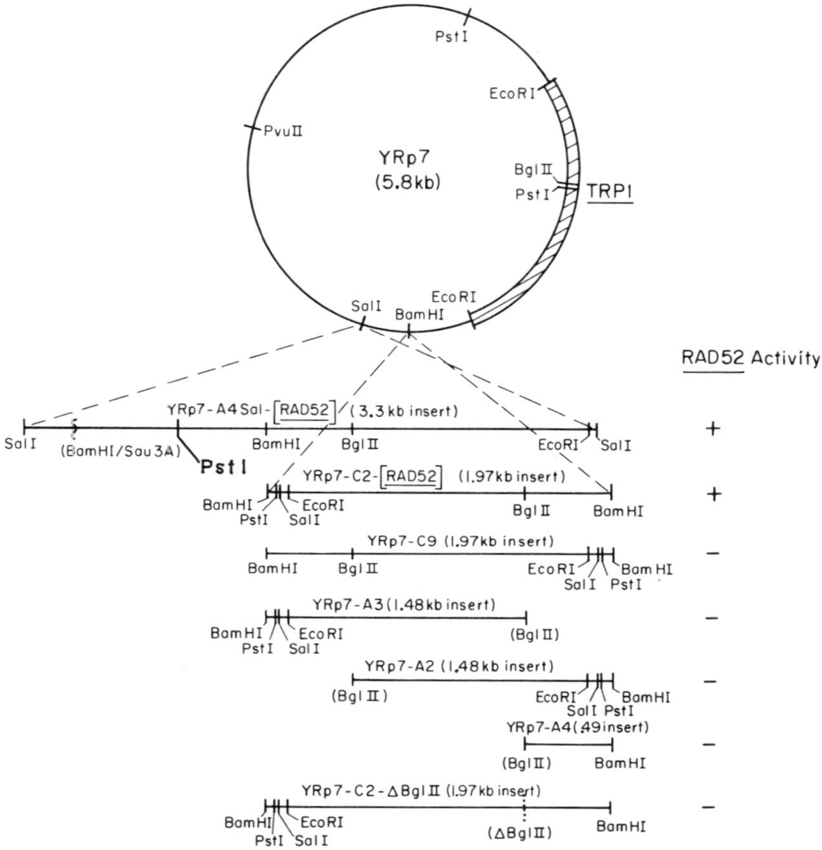

Figure 2. Subclones with and without RAD52 activity. Various segments of the original 5.9kb fragment subcloned into the SalI or BamHl site of the YRp7 vector (pBR322 with the yeast TRP1-ARS1 region). The inserts in the YRp7-C2-[RAD52] and YRp7-C9 plasmids are identical except for orientation.

BAL31 Deletions

In order to more precisely define where the RAD52 gene is located within the 1.97kb BamHl fragment, we have isolated deletions into the insert from one end. Since the subcloning experiments (Figure 2) and the BglII deletion experiment (see below) indicated that the RAD52 region included the BglII site, we isolated deletions into the insert from the side furthest away from the BglII site (the left side in YRp7-C2-[RAD52], see Figure 3a). Because the isolation of BAL31 deletions is simplified by starting from a restriction site that is unique in the plasmid, we first deleted the short SalI fragment of YRp7-C2-[RAD52], using standard restriction and religation procedures. The resultant plasmid (YRp7-C2-ΔSal-[RAD52]) contains both unique Sall and BamHl sites (Figure 3a). This plasmic still complements the rad52-1 mutation, which is expected since the previously subcloned 3.3kb Sall fragment, which lacks the short yeast BamHl to SalI region, has RAD52 activity (Figure 2). The YRp7-C2-ΔSal-[RAD52] plasmid was linearized by restriction with Sall and digested with BAL31 for various lengths of time from 15 min. to 2 hr. Most of the time points gave large deletions; all of the deletions we used were from the 15 and 30 min. time points. Since BAL31 causes deletions in both directions from the Sall site, we restricted with PvuII before ligations so that we could later determine the length of the deletion within the RAD52 region. PvuII leaves blunt ends, but since BAL31 ends are frequently not blunt we increased the number of blunt ends by filling in some of the overhangs using the Klenow fragment of DNA PolI, prior to blunt end ligation. Following ligation, we transformed E. coli and used restriction analysis (PstI and BamHl double digest) on mini plasmid preparations to screen for plasmids with different sized deletions up to about 1.5kb. Large scale plasmid preparations were done on 24 plasmids and these plasmids were reintroduced into the rad52-1 yeast strain XS95-6C in order to score for RAD52 activity. The deletions which were reintroduced into yeast are diagrammed in Figure 3b. The size of the deletions are approximations since agarose gel electrophoresis does not give exact lengths, but each deletion diagrammed did appear to be slightly different in length. Some deletions appeared to be of the same size and these duplications are not listed in Figure 3b although the RAD52 activity results for these apparent repeats were consistent. Some of the sizes of the deletions might be overestimates if BAL31 had actually digested much further in the SalI to PvuII direction and had deleted past the PvuII site. This seems unlikely for most of the deletions since the distance from SalI to PvuII is ∿1.4kb and most of our deletions were 1kb or shorter. The results for two deletions (Gl and 02) are inconsistent with the results from the other deletions; although they appeared to be short deletions they lacked activity. These cases might represent longer deletions which appear shorter because of the addition of random short DNA fragments during ligation. The rest of the deletions are consistent with deletions of about 900 bases retaining RAD52 activity, while longer deletions inactivate the gene. Therefore, the RAD52 gene is at most 1kb in length.

Deletion of BglII Site

A deletion-frameshift mutation in RAD52 has been isolated to determine whether RAD52 codes for an essential function; only two rad52 alleles exist (5) and neither is known to be nonsense suppressible. If both alleles are

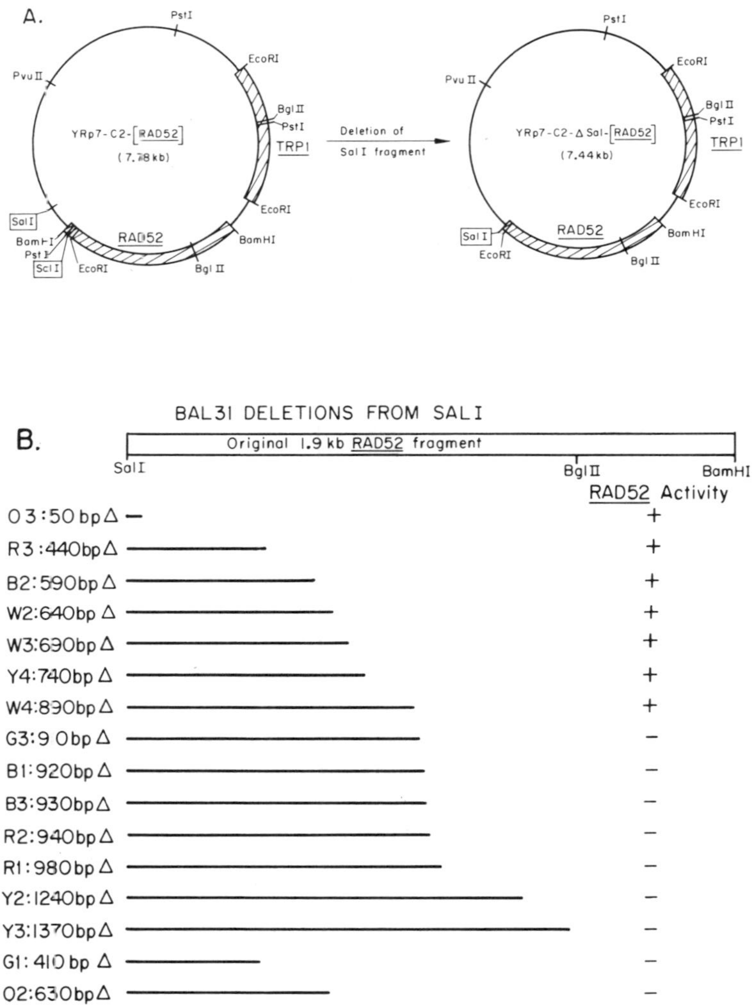

Figure 3. BAL31 deletions. A. Removal of short SalI fragment from YRp7-C2-[RAD52] before BAL31 deletions isolated. B. BAL 31 deletions from SalI site into RAD52 region. Line represents length of deleted segment.

leaky, then it is possible that RAD52 might encode an essential function. The rad52-2 allele is clearly leaky, but it is not known if rad52-1 is also leaky (5,27). The rad54-3 mutation is more X-ray sensitive than rad52-1 (J. Game, personal communication) and since rad52 and rad54 are in the same epistasis group, it might indicate that rad52-1 is a slightly leaky allele. It is known that diploids homozygous for rad52-1 have extremely low (~10-20%) plating efficiency and also spontaneously undergo chromosome loss (15) indicating a probable role for this gene during normal cell division.

Since the previously discussed subcloning experiments indicated that the BglII site in the 1.97kb BamHl fragment was probably in the RAD52 gene, we decided to isolate a "null" allele of rad52 by constructing a four base pair deletion of the BglII site using SI nuclease. The 1.97kb BamHl fragment was first subcloned into pBR322 so that BglII site in the BamHl fragment was the only BglII site in the plasmid. This plasmid was restricted with BglII, treated with Sl single-strand exonuclease, blunt-end ligated, recut with BglII (to decrease transformation ability of plasmids still containing a BglII site), and transformed into HB101. Mini plasmid preparations on the transformants showed that the BglII site was missing from plasmids in most of the transformants. Following large scale preparations of two plasmids with independently derived deletions of BglII, we subcloned these deletions into both YRp7 and YIp5 in the proper orientation for RAD52 activity. When the deletions in YRp7 were introduced into yeast they failed to complement rad52-1 (Figure 2). This demonstrates that the BglII site is either in the structural gene or in a region essential for transcription or translation. We have recently integrated these deletions in YIp5 into a wild type strain of yeast and are currently examining excision events in both haploids and diploids in order to determine whether RAD52 codes for an essential function. If these experiments show that RAD52 codes for a nonessential function, our BglII deletion allele will still be valuable as a nonleaky allele.

NOTE ADDED IN PROOF

Kenji Adzuma (personal communication), in the laboratory of Dr. H. Ogawa (Osaka University, Osaka, Japan), has independently isolated a 1.98kb BamHl fragment which complements the rad52-1 mutation. This fragment appears to be identical to our 1.97kb BamHl fragment, since both share a common restriction map.

ACKNOWLEDGEMENTS

We especially wish to thank Isabel Calderon, Rebecca Contopoulou and Michael Botchan for useful discussions during this research, Noel Fong and Tommy McKey for excellent technical assistance, and the Molecular Biology 101 laboratory class for help in the BAL31 deletion experiments.

LITERATURE CITED

1. Birnboim, H. C. and Doly, J. 1979. A rapid alkaline extraction procedure for screening recombinant plasmid DNA. Nucl. Acids Res. 7: 1513-1523.

2. Botstein, D., Falco, S. C., Stewart, S. E., Bohnnan, M., Scherer, S., Stinchcomb, D. T., Struhl, K. and Davis, R., 1980. Sterile host yeasts (SHY): A eucaryotic system of biological containment for recombinant DNA experiments. Gene 8: 17-24.

3. Broach, J. R., Strathern, J. N. and Hicks, J. B., 1979. Transformation in yeast: Development of a hybrid cloning vector and isolation of the CAN1 gene. Gene 8: 121-133.

4. Davis, R. W., Botstein, D. and Roth, J. R., 1980. Advanced Bacterial Genetics. Cold Spring Harbor Laboratory, Cold Spring Harbor, N.Y.

5. Game, J. C. and Mortimer, R. K., 1974. A genetic study of X-ray sensitive mutations in yeast. Mutation Res. 24: 281-292.

6. Game, J. C., Zamb, T. J., Braun, R. J., Resnick, M. and Roth, R. M., 1980. The role of radiation (rad) genes in meiotic recombination in yeast. Genetics 94: 51-68.

7. Haynes, R. H. and Kunz, B. A., 1981. DNA repair and mutagenesis in yeast, p. 371-414. In: J. N. Strathern, E. W. Jones and J. R. Broach (eds.), The Molecular Biology of the Yeast Saccharomyces: Life Cycle and Inheritance, Cold Spring Harbor Laboratory, Cold Spring Harbor, N.Y.

8. Hicks, J. B., Hinnen, A. and Fink, G. R., 1978. Properties of yeast transformation. Cold Spring Harbor Symp. Quant. Biol. 43, 1305-1313.

9. Hinnen, A., Hicks, J. B. and Fink, G. R., 1978. Transformation of yeast. Proc. Natl. Acad. Sci. USA 75: 1929-1933.

10. Ho, K.S.Y., 1975. Induction of DNA double-strand breaks by X-rays in a radiosensitive strains of the yeast Saccharomyces cerevisiae. Mutation Res. 30: 327-334.

11. Holmes, D. S. and Quigley, M., 1981. A rapid boiling method for the preparation of bacterial plasmids. Anal. Biochem. 114: 193-197.

12. Lemmott, J. F., 1980. Genetic and physiological factors affecting repair and mutagenesis in yeast, p. 85-120. In: W. M Generoso, M. D. Shelby and F. J. DeSerres (ed.), DNA Repair and Mutagenesis in Eucaryotes. Plenum Press, New York.

13. Malone, R. E. and Esposito, R. E., 1980. The RAD52 gene is required for homothallic interconversion of mating types and spontaneous mitotic recombination in yeast. Proc. Natl. Acad. Sci. USA 77: 503-507.

14. Mortimer, R. K. and Schild, D., 1980. Genetic map of Saccharomyces cerevisiae. Microbiol. Rev. 44: 510-571.

15. Mortimer, R. K., Contopoulou, R. and Schild, D., 1981. Mitotic chromosome loss in a radiation sensitive strain of the yeast *Saccharomyces cerevisiae*. Proc. Natl. Acad. Sci. (USA) 78:5778-5782.

16. Nasmyth, K. A. and Tatchell, K., 1980. The structure of transposable yeast mating type loci. Cell *19*: 753-764.

17. Orr-Weaver, T. L., Szostak, J. W., Rothstein, R. J., 1981. Yeast transformation: A model system for the study of recombination. Proc. Natl. Acad. Sci. (USA) *78*: 6354-6358.

18. Prakash, S., Prakash, L., Burke, W. and Montelone, B., 1980. Effects of the *rad52* gene on recombination in *Saccharomyces cerevisiae*. Genetics *94*: 31-50.

19. Resnick, M. A., 1969. Genetic control of radiation sensitivity in *Saccharomyces cerevisiae*. Genetics *62*: 519-531.

20. Resnick, M. A., 1976. The repair of double-strand breaks in DNA: A model involving recombination. J. Theor. Biol. *59*: 97-106.

21. Resnick, M. A. and Martin, P., 1976. The repair of double-stranded breaks in the nuclear DNA of *Saccharomyces cerevisiae* and its genetic control. Molec. Gen. Genet. *143*: 119-129.

22. Resnick, M. A., Kasimos, J. M., Game, J. C., Braun, R. J., and Roth, R. M., 1981. Changes in DNA during meiosis in a repair-deficient mutant (*rad52*) of yeast. Science *212*: 543-545.

23. Rupp, W. D., 1982. Genetic engineering and DNA repair, p. 205-215. In: C. Helene, M. Charlier, Th. Montenay-Garestier and G. Laustriat (eds.), *Trends in Photobiology*, Plenum Press, N.Y.

24. Rykowski, M. C., Wallis, J. W., Choe, J. and Grunstein, M., 1981. Histone H2B subtypes are dispensable during the yeast cell cycle. Cell *25*: 477-487.

25. Saeki, T., Machida, I. and Nakai, S., 1981. Genetic control of diploid recovery after -irradiation in the yeast *Saccharomyces cerevisiae*. Mut. Research *73*: 251-265.

26. Scherer, S. and Davis, R. W., 1979. Replacement of chromosome segments with altered DNA sequences constructed *in vitro*. Proc. Natl. Acad. Sci. USA *76*: 4951-4955.

27. Strike, T. L., 1978. Characterization of mutants of yeast sensitive to X-rays. Ph.D. Thesis, University of California, Davis.

28. Struhl, K., Stinchcomb, D. T., Scherer, S. and Davis, R. W., 1979. High-frequency transformation of yeast: Autonomous replication of hybrid DNA molecules. Proc. Natl. Acad. Sci. USA *76*: 1035-1039.

29. Tschumper, G. and Carbon, J., 1980. Sequences of a yeast DNA fragment containing a chromosomal replicator and the TRP1 gene. Gene 10: 157-166.

30. Von Borstel, R. C., Cain, K. T. and Steinberg, C. M., 1971. Inheritance of spontaneous mutability in yeast. Genetics 69: 17-27.

17. CLONING OF A DNA REPAIR GENE IN YEAST*

L. Prakash, R. Polakowska, and B. Slitzky

Department of Radiation Biology and Biophysics
University of Rochester School of Medicine and Dentistry
Rochester, New York 14642

Rec. Adv. Yeast Mol. Biol. (1982) 1:225-241

A fragment of DNA which restores resistance to UV, gamma rays, and methyl methanesulfonate in both rad6-1 and rad6-3 mutants has been isolated on a recombinant plasmid, YEp13. Recombinant plasmids containing such DNA segments were obtained by transforming a leu2-3 leu2-112 rad6-1 strain to LEU+ and screening for UV resistance among the LEU+ transformants. Three classes of recombinant plasmids, based on restriction with BamHI, were obtained. The cloned DNA segment complementing rad6 was transferred to an integrating plasmid containing the yeast URA3 gene, and then used to determine the site of recombination of the cloned DNA. Preliminary genetic experiments suggest that the cloned segment integrates at the rad6 locus. Subcloning of this segment has yielded a 1.9 kb fragment which still functions in complementation of rad6. When this fragment is used as a probe for hybridization to total yeast RNA, two transcripts are observed.

INTRODUCTION

Procaryotic and eucaryotic organisms possess many different mechanisms which enable them to repair damage induced in their DNA by a wide variety of physical and chemical agents. Modification of the DNA damage, however, depends on the genetic constitution of the organism as well as the post-exposure conditions. In *Escherichia coli*, repair of ultraviolet light (UV) induced damage is governed by genes in two epistasis groups - one controlled by the uvrA+, uvrB+, uvrC+ system involved in excision of pyrimidine dimers, and the other controlled by the recA+-lexA+ system (10,51), involved in postreplication repair. The sensitivity of double mutants consisting of uvrA and recA is such that one pyrimidine dimer is sufficient to kill the cell, indicating that no other pathways play a major role in repair of UV damage in *E. coli* (14).

Many cellular repair processes, such as excision repair and postreplication repair, also occur in eucaryotes. However, relatively few DNA repair mutants are known in higher eucaryotes. In the yeast

*Support was obtained from the National Institutes of Health grant GM 19261 and from contract No. DE-AC02-76EV03490 with The U.S. Department of Energy at the University of Rochester Department of Radiation Biology and Biophysics and has been assigned Report No. UR-3490-2184.

Saccharomyces cerevisiae, which has a well-characterized genetic system, a large number of radiation sensitive (rad) mutants representing over 30 distinct genetic loci have been isolated and many have been characterized extensively in terms of their effect on UV induced mutations (19,21,39), chemically induced mutations (31,33), and their repair defects. Comparisons of sensitivities to killing of various double mutant combinations indicate that three epistasis groups exist for repairing UV induced DNA damage (6, see 11 for review). One group consists of nine genes, RAD1, RAD2, RAD3, RAD4, RAD7, RAD10, RAD14, RAD16, and MMS19, involved in excision of UV-induced pyrimidine dimers (32,36,42,43,48,49). Like the excision defective uvrA and uvrB mutants of E. coli, the yeast excision defective mutants show enhanced frequencies of mutations following UV irradiation. Experiments with double mutants consisting of the rad mutants defective in excision coupled with cdc9, which is temperature sensitive for growth and lacks detectable DNA ligase activity in vitro (18), suggest that rad1, rad2, rad3, rad4, and rad10 mutants are defective in an initial incision step required for pyrimidine dimer removal and rad14 is defective in a step subsequent to incision (50). In addition, it has been shown that cell-free extracts from rad1, rad2, rad3, rad4, and rad10 strains are capable of removing dimers from preincised UV irradiated DNA (44). However, it has not been possible to demonstrate pyrimidine dimer incising activity in cell-free extracts of yeast.

The RAD52 epistasis group consists of eight genes (RAD50 to RAD57) which confer sensitivity primarily to ionizing radiation (8), while UV sensitivity is enhanced mainly in the presence of a defect in excision (6). Mutability induced by UV or gamma rays in mutants of genes of the RAD52 group is similar to that observed in RAD+ strains (19,25). Many of the mutants in this group, particularly rad52, affect recombination and DNA strand break repair. The rad52-1 mutant is defective in spontaneous, UV and ionizing radiation induced homologous mitotic recombination (23,38,40), mitotic gene conversion (17), meiotic recombination between homologous chromosomes (9,38), UV induced sister chromatid recombination (37), and in mating type interconversion (23) which occurs by transposition of DNA copies from silent a and α loci to the mating type locus. The rad52-1 mutation also blocks chromosomal integration of gapped-linear or linear molecules but not of circular molecules (30). In addition, both the rad51-1 and rad52-1 mutants are defective in the repair of DNA double strand breaks (16,41). The rad52-1 mutants are also not as proficient as wild type in postreplication repair of UV damaged DNA (34).

Mutants of genes in the third epistasis group affect sensitivity to both UV and ionizing radiation and consist of rad6, rad8, rad9, rad18, rev1, rev2, rev3, and mms3. Many of these mutants show reduced UV mutability for at least some, if not all, loci tested (19,22,24). In addition, some of them have pleiotropic effects on spontaneous mitotic gene conversion, spontaneous mutation, and DNA repair (see 11 for review). The rad18-3 mutant increases rates of spontaneous mitotic recombination (1). Both alleles of the rad6 locus, rad6-1 and rad6-3, enhance spontaneous and UV induced mitotic heteroallelic recombination in diploids and are proficient in spontaneous and UV induced unequal sister chromatid recombination in haploids (26). Both rad6 mutants show no induced mutations with UV, ionizing radiation (19,25), ethyl methanesulfonate, N-methyl-N'-nitro-nitrosoguanidine, and a wide variety of other chemical mutagens (31). The rad6-1 and rad18-2 mutants show a great inhibition of postreplication repair of UV damaged DNA while the rev3-1 mutation does not affect it (34).

Our aim has been to study the genes and gene products involved in DNA repair and mutagenesis in order to elucidate the molecular mechanisms of these important cellular responses. We have begun this study by cloning one of the genes, RAD6, which probably plays an important role in the regulation of these two cellular processes and report here our preliminary results.

MATERIALS AND METHODS

Strains: The yeast strains LP2530-2A: MATa leu2-3 leu2-112 his5-2 trp1-1 rad6-1 and LP2652-9C: MATa leu2-3 leu2-112 trp5-c ura3-52 rad6-3 were constructed by standard techniques of yeast genetics and used for transformation. Strain AB320 HO met4-1 ade2-1 lys2-1 trp5-2 leu2-1 can1-100 ura3-1 and/or ura1 was obtained from J. Hopper and S150-2B MATa leu2-3 leu2-112 ura3-52 trp1-289 his3Δ1 from S. Baim. Escherichia coli strain HB101 was obtained from H. Eberle.

Plasmids used for cloning: YEp13 (2), a 10.7 kb hybrid of pBR322 containing a yeast DNA fragment with the 2 μ circle origin of replication, enabling it to replicate autonomously in yeast, the yeast LEU2 gene, which complements E. coli leuB, the col E1 origin of replication, and the genes conferring ampicillin (amp^R) and tetracycline resistance (tet^R). A single BamHI site, used as the cloning site, is present in the tet gene; insertion in the BamHI site inactivates the tet gene.

YIp5 (45) a 5.5 kb plasmid, also derived from pBR322, replicates autonomously in E. coli but not yeast, contains the amp and tet genes, and the yeast URA3 gene.

pAB108, a 6.1 kb derivative of pBR322, contains the amp gene and replicates autonomously in both E. coli and yeast because of the yeast chromosomal replicator ARS2, and contains the URA3 gene for selection in yeast. pAB108 contains unique EcoRI, HindIII, BamHI sites, was constructed by and obtained from S. Baim. pBEU49 is a recAo281 multicopy plasmid obtained from J. Clark and has a BamHI fragment containing the recA gene of E. coli.

DNA purifications: Plasmid DNA from yeast was purified by a modified method of Hirt (13). Yeast DNA was purified by a mini-prep method obtained from G.S. Roeder (unpublished results). Growth of E. coli containing plasmids was carried out as described (29) but with nucleosides omitted. Plasmid DNA was then purified by the cleared lysate method of Clewell and Helinski (5) but with 0.1 percent Triton X-100 instead of Brij 58. For screening of large number of E. coli containing plasmids, the procedure of Ish-Horowicz and Burke (15) for DNA extraction from plasmid or cosmid cultures was used.

Pool DNA for transformation: The pool of yeast DNA sequences used for transformation, and designated pool 35, was constructed by K. Nasmyth and S. Reed and obtained from J. Hopper. Genomic DNA was purified from strain AB320, digested partially with Sau3A, and fragments of average size of 10 kb purified by sucrose gradients. These fragments were ligated into the BamHI site of YEp13 and used to transform *E. coli* to amp^R. Transformation of yeast was carried out as previously described (12).

Scoring of UV, X-ray and methyl methanesulfonate sensitivity: Methods were as described by Prakash and Prakash (35).

Gel electrophoresis, transfer to nitrocellulose filters, preparation of probes and hybridization: Restriction endonucleases were purchased from New England BioLabs and used in 0.09 M Tris-HCl, pH 7.4 - 0.01 M $MgCl_2$ buffer. Electrophoresis in agarose slab gels (Model 800, purchased from Aquebogue Machine Shop, Aquebogue L.I., N.Y.) to separate restriction fragments was carried out in 89 mM Tris-HCl, 23 mM phosphoric acid, 2.5 mM EDTA buffer, pH 8.3. Agarose gels also contained 0.5 μg/ml ethidium bromide. The restriction fragments were photographed under short-wave ultraviolet light with Polaroid type 57 film. DNA fragments to be used as probes were isolated by electroelution from agarose gels as described by Zaret and Sherman (52). Nick translation of fragments for radioactive labeling was carried out by using the nick translation kit purchased from Amersham. The specific activities of various probes ranged from 1.4×10^7 to 6.1×10^7 cpm/μg DNA and total of about 2×10^6 cpm was used for each Southern blot. The method of Southern (46) was used to transfer restriction fragments to nitrocellulose paper and hybridizations were carried out as described by Zaret and Sherman (52).

RESULTS AND DISCUSSION

Isolation of *RAD6*-containing plasmids: We cloned the *RAD6* gene by screening for UV resistant colonies in a population of transformed *rad6-1* mutants cells. This method of complementation has been used successfully to clone various other yeast genes, such as *cdc10*, *cdc28*, and *rad52* (4,28 and Schild and Mortimer, personal communication). However, before making use of this procedure, several precautions had to be taken because of the nature of the only existing *rad6* alleles. Both *rad6-1*, isolated by Cox and Parry (7), and *rad6-3*, isolated in our laboratory (35), are nonsense alleles suppressed by both amber and ochre suppressors (20,47). Therefore, we included an amber allele, *trp1-1*, and an ochre allele *his5-2*, in the *rad6-1* strain to be transformed for isolation of the *RAD6* gene. Any putative *RAD6*-containing clones could then be tested for the presence of a suppressor activity. The restoration of the *RAD+* phenotype to a *rad6* mutant could arise from either the presence of the *RAD6* gene on the cloned DNA fragment, or the presence of any other gene which could suppress the *rad6* phenotype. Any strains which were *RAD+* as a result of cloning of a translational suppressor could be eliminated by testing *RAD+* clones for ability to grow without histidine or tryptophan. A second precaution taken was to determine that strain AB320, from which pool 35 DNA had been constructed, did not contain any genes that might suppress either *rad6-1* or *rad6-3*. This was done by crossing AB320 to *rad6-1* and *rad6-3* suitably marked haploids and analyzing the segregation pattern for the rad6 phenotype in the resulting progeny. In both crosses, rad6-1 and rad6-3 segregated 2:2, indicating that strain AB320 did not contain any suppressors of either rad6-1 or rad6-3.

Strain LP2530-2A, MATa leu2-3 leu2-112 his5-2 trp1-1 rad6-1, was transformed to LEU+ with pool 35 DNA. LEU+ transformants were then transferred to media lacking leucine and tested for their response to UV irradiation as well as for growth in the absence of histidine or tryptophan. Nine independent LEU+ clones were obtained which were UV resistant and retained dependence on histidine and tryptophan for growth. However, pool 35 contains a suppressor present in about 1 percent of LEU+ transformants which suppresses trp1-1 poorly, does not act on the amber allele ade5-7, or the ochre alleles lys1-1, ilv1-1, and can1-100, but suppresses trp1-289, aro1B, and met8-1 very well (R. Mortimer, personal communication). In order to eliminate the possibility that the putative RAD6 clones contained this suppressor which was also suppressing rad6-1 and rad6-3, we transformed strain S150-2B to LEU+ with each plasmid and tested 30 LEU+ transformants from each experiment. None of the LEU+ transformants were TRP+, indicating that the putative RAD6 clones did not contain suppressors of trp1-289.

Plasmid DNA obtained from each of the nine clones was used to transform E. coli strain HB101 to ampicillin resistance. Plasmid DNA from each of the nine transformed strains was then purified by by amplification with chloramphenicol and centrifugation through CsCl. The purified DNA was used to transform LP2530-2A (rad6-1) and LP2652-9C (rad6-3) to LEU+. All LEU+ transformants were UV resistant and still required tryptophan and histidine for growth. In addition, resistance to the lethal effects of X-rays and methyl methanesulfonate was obtained and sporulation ability was restored in diploids obtained by crossing rad6-1 haploids to transformed rad6-1 haploids of opposite mating type. These results indicate that the nine plasmids, designated pTB12 through pTB20, all contain a DNA fragment which restores RAD6 function to rad6-1 and rad6-3 strains.

Restriction maps of putative RAD6-containing plasmids: Digestion with BamHI was used to determine whether any similarities existed among the nine yeast DNA inserts. Three groups were obtained, based on the types of fragments obtained following digestion with BamHI: (a) one insert fragment in pTB13, pTB14, pTB15, pTB16, and pTB20 (b) two insert fragments in pTB17 (c) and no insert fragment in pTB13, pTB18, and pTB19. The extent of homology among the cloned yeast DNA fragments was determined by mapping several restriction sites in pTB16, one of five plasmids in class (a), and pTB17, the only plasmid in class (b). The results are presented in Figure 1. The 5.2 kb BamHI fragment of pTB17 is homologous to the 5.2 kb BamHI fragment of pTB16 and both fragments contain the RAD6 function since transfer of each fragment in both orientations into the BamHI site of YEp13 resulted in plasmids capable of transforming rad6-1 and rad6-3 strains to RAD+ with high efficiency. Restriction maps for members of class c have not yet been completed, however, hybridization results indicate that this class contains a small deletion within the 4.2 kb HindIII fragment present in both pTB16 and pTB17 (Figure 2). For this experiment, the 5.2 kb BamHI fragment from pTB17 was used as a probe to hybridize to HindIII or BamHI digested

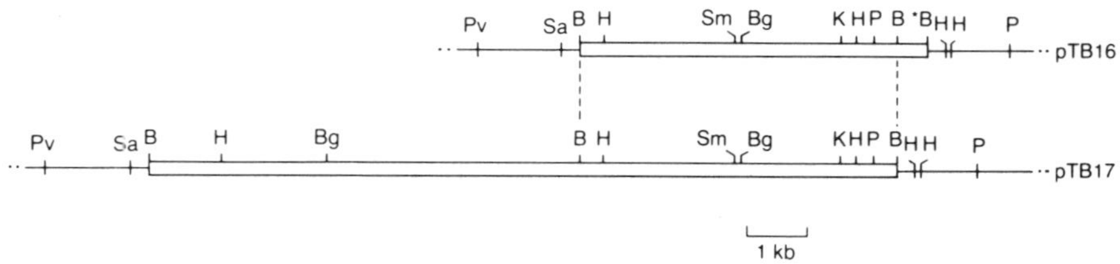

Figure 1. Restriction maps of recombinant plasmids containing the RAD6 gene.

Insert of yeast DNA segments, given by open bars, is in the BamHI site of the tet gene of YEp13. The thin line represents YEp13 DNA. The extent of overlap and orientation was determined by analysis of restriction sites and Southern gels. Symbols for restriction enzymes are as follows: B, BamHI; *B, BamHI/Sau3A junction; Bg, BglII; H, HindIII; K, KpnI; P, PstI; Pv, PvuII; Sa, SalI. The 5.2 kb BamHI fragment of pTB16 is homologous to the 5.2 kb BamHI fragment of pTB17, which contains an additional 6.6 kb BamHI fragment, as shown. In pTB16, the insertion of the Sau3A fragment occurred such that the left end, near the SalI site, generated a BamHI site, while the other end, near the HindIII site (on the right hand in the figure), generated a site not cut by BamHI and is designated *B. However, an additional BamHI site occurs about 0.5 kb to the left of *B and BamHI digestion of pTB16 yields an insert fragment of about 5.2 kb.

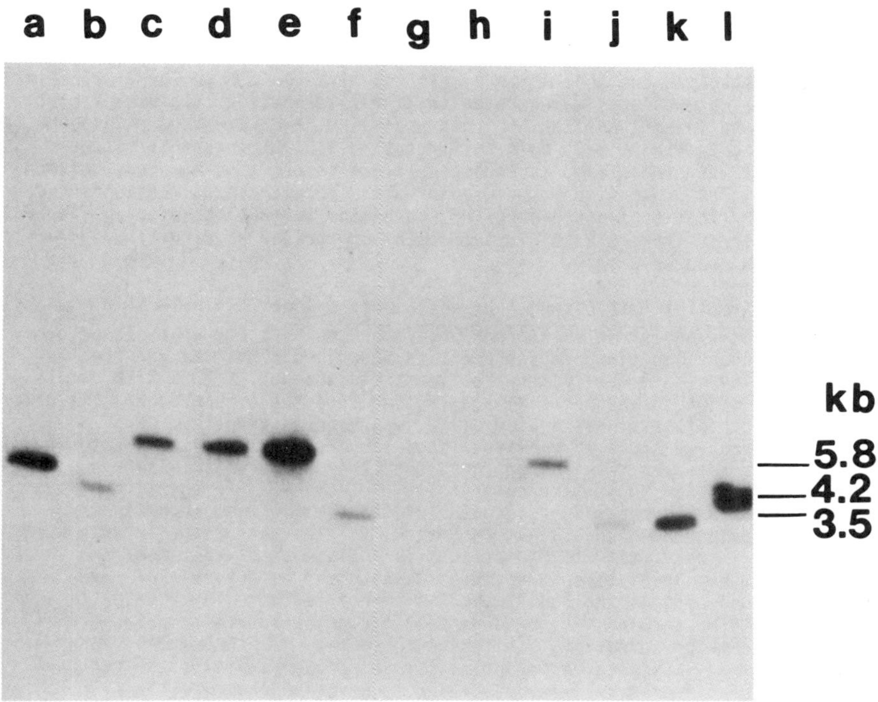

Figure 2. Structural homologies among RAD6 plasmids.

The 5.2 kb BamHI fragment of pTB17 was purified by electroelution and labeled by nick translation. Digested plasmid DNA samples were separated by agarose gel electrophoresis, transferred to nitrocellulose, and hybridized with the ^{32}P labeled probe. Lane a: plasmid DNA from pTB60, which contains the 4.2 kb HindIII fragment of pTB16 cloned into pAB108 (see below), digested with HindIII; lanes b, f, j, and k, plasmid DNA from pTB13, pTB13, pTB18 and pTB19, respectively, digested with HindIII; lanes c, d, e, and i: plasmid DNA from pTB17, pTB16, pTB16 and pTB14, respectively, digested with BamHI. Lane d contains about 1/4 the DNA present in lane e. Lanes g and h contain HindIII digested YEp13 and BamHI digested pBEU49 DNA, respectively. Lane 1 contains the purified HindIII fragment obtained from pTB60. The left hand side of the gel is somewhat skewed upward, giving those fragments an apparently larger size than they actually have, as observed in many other gels.

plasmids from all three classes. Lanes b, f, j, and k contain plasmids of class c, and the probe hybridizes to only one fragment of HindIII digested DNA in each case, and the size of that fragment is 3.5 kb. Lanes c, d, e, and i, on the other hand, contain BamHI digested plasmids of class a, and in these cases, the fragment which hybridizes with the 5.2 kb BamHI probe is about 5.2 kb. Lane g, which contains HindIII digested YEp13 (the vector without any insert) and lane h, which contains BamHI digested pBEU49 DNA, a plasmid containing the _E. coli recA+_ gene on a BamHI fragment, do not hybridize with the probe, as expected, since there is no homology between the yeast DNA probe with either vector alone or with vector containing _E. coli_ DNA. Identical patterns of hybridization to that observed in Figure 2 are obtained if the 5.2 kb BamHI fragment or the 4.2 kb HindIII fragment of pTB16 are used as probes.

Confirmation that cloned fragments contain RAD6 DNA: Since complementation is not sufficient proof for the identity of a cloned gene, that _RAD6_ was contained in the plasmids described above was verified by mapping the cloned fragment to the yeast chromosome . The 5.2 kb BamHI fragment of pTB16 was purified and ligated into the single BamHI site of YIp5. This plasmid, designated pTB33, was used to transform LP2652-9C (_rad6-3 ura3-52 trp5-c_) to _URA+_. Since YIp5 does not replicate autonomously but transforms yeast by integration, and since the _ura3-52_ allele does not recombine and is thought to be a small deletion (45), integration of the _URA3_ gene should occur in the region of the yeast genome which shows homology with the cloned yeast DNA fragment. The _URA+_ transformants were tested for sensitivity to UV irradiation: six were UV resistant and three UV sensitive; the latter three might have arisen as a result of gene conversion. One of the _rad6-3 URA3+ trp5-c_ integrants was crossed to a _RAD+ ura3-52 TRP5+_ strain. The _URA3_ gene is located on chromosome V, whereas the _TRP5_ gene is on chromosome VII, about 20 cM away from _RAD6_ (27). Analysis of 20 tetrads obtained after sporulation of diploids yielded 14 parental ditype and 6 tetratype tetrads, giving 15 percent recombination. The linkage of _ura3_ to _trp5_ indicates that _ura3_ had integrated at the _rad6_ locus and that the _RAD6_ gene was present on the plasmid pTB33.

Integration at the _RAD6_ locus would generate a fragment of about 10.5 kb following digestion of total yeast genomic DNA with BglII, as indicated in Figure 3, whereas integration elsewhere in the genome would yield a BglII fragment whose size cannot be predicted since the relation of the flanking BglII sites to the BglII site in the integrated segment is not known. Further evidence for the identity of the cloned fragment as the _RAD6_ gene was obtained by digesting genomic DNA obtained from two _RAD+_ integrants and three _rad6_ integants with BglII and hybridizing to a pBR322 probe (Figure 4). As predicted for integration at the _RAD6_ locus, the size of the BglII fragment hybridizing to the probe is about 10.5 kb. Also as expected, vector sequences are present in the genomic DNA of all integrants (lanes a to e) whereas no hybridization was observed with genomic DNA prepared from two non-transformed strains (lanes f and g).

Subcloning and preliminary characterization of the RAD6 gene: In order to identify the smallest unit having _RAD6_ function, it was necessary to obtain smaller fragments of insert DNA. The HindIII fragment present within the two BamHI sites of pTB16 was cloned into the single HindIII site of

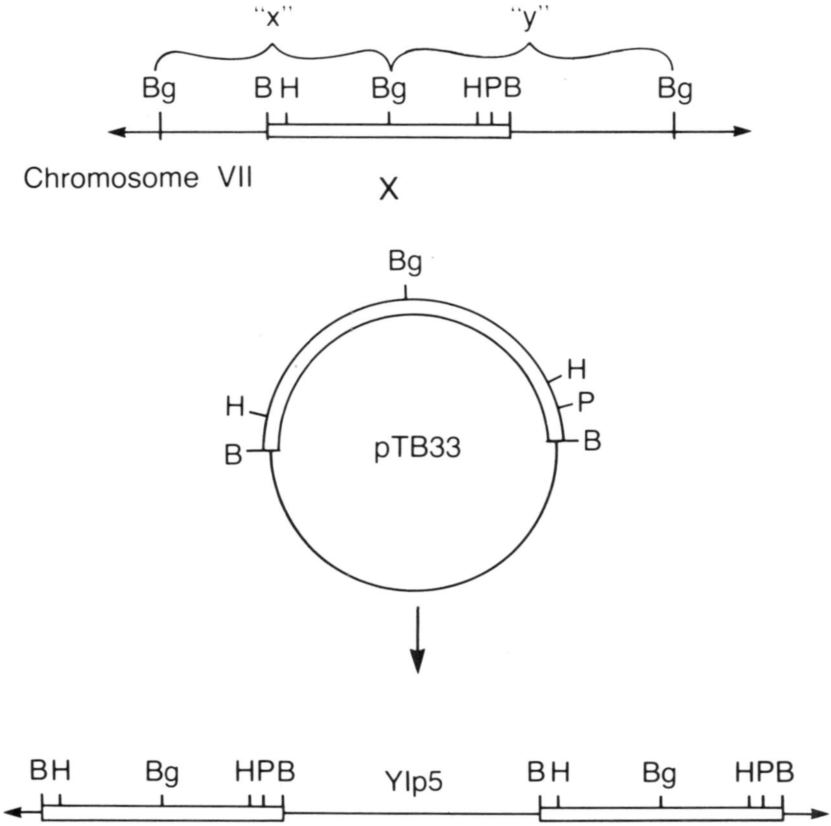

Figure 3. Integration of a DNA segment on a plasmid by recombination with its region of homology in the genome of yeast.

Plasmid pTB33 was constructed by insertion of the BamHI fragment of pTB16 into the BamHI site of YIp5. The structure of the resulting plasmid has not yet been confirmed but is assumed to be as given above. The open bar refers to the cloned segment on pTB33 and its corresponding region of homology on chromosome VII, the location of the RAD6 gene. "X" and "Y" refer to the distance between the BglII site in the cloned segment and the next BglII site on chromosome VII, going leftward and rightward, respectively. The thin line respresents chromosomal DNA. The lower linear chromosome represents the integration of the cloned DNA segment into chromosome VII with two copies flanking the vector sequence of YIp5. The predicted size of the fragment generated by BglII digestion of chromosomal DNA from integrants is about 10.5 kb. Symbols for restriction enzymes are as given in Figure 1.

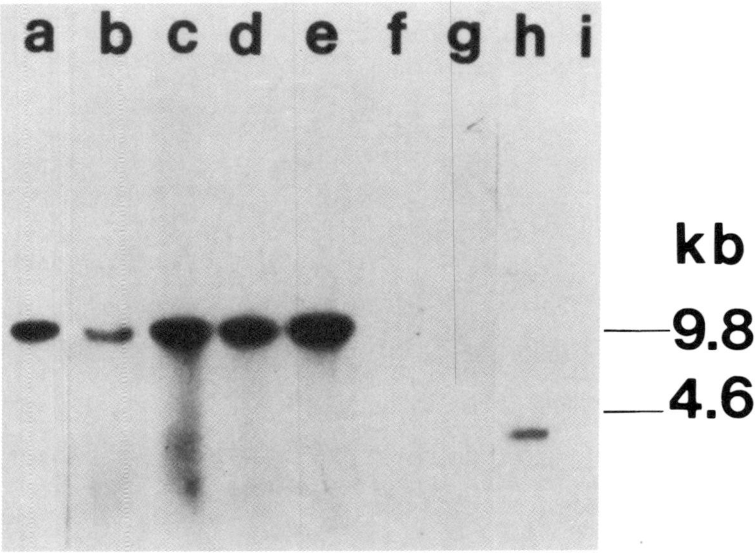

Figure 4. Detection of pBR322 sequences in RAD+ URA+ integrants.

Chromosomal DNA from two UV resistant integrants, BES-2 and BES-5, lanes a and b, respectively; three UV sensitive integrants, BES-7, BES-8, BES-9, lanes c, d, and e, respectively, and two RAD+ non-transformed strains, AB320, lane f, and #264, lane g, were digested with BglII and fragments separated by electrophoresis on agarose gels. The DNA fragments were transferred to nitrocellulose and hybridized to ^{32}P-labeled pBR322 DNA. Lane h contains pBR322 DNA and lane i contains lambda DNA digested with HindIII.

pAB108, generating the plasmid pTB60 (Figure 5). pTB60 DNA was used to transform LP2652-9C to URA+ and all URA+ transformants obtained were also found to be UV resistant, indicating that the RAD6 gene was contained in the HindIII fragment. This 4.2 kb HindIII fragment, when hybridized to HindIII digested genomic yeast DNA, hybridizes to only one band of about 4.2 kb, indicating that no repeated sequences corresponding to this fragment are present in the yeast genome (Figure 6). Further subcloning of the RAD6 gene was achieved by deleting the rightward BglII/HindIII fragment of pTB60 by digestion with both BglII and BamHI (Figure 5). The resulting products were ligated and the mix used to transform HB101 to ampicillin resistance. Ampicillin resistant colonies were screened for the presence of the appropriate deletion. The plasmid generated in this way, pTB64, was also shown to contain RAD6 function as determined by the ability to restore UV resistance to the rad6-3 strain, LP2652-9C, indicating that this leftward HindIII/BglII fragment contains the RAD6 gene. Experiments are in progess to determine whether further subcloning of the RAD6 gene is possible while still retaining function.

As a prerequisite for identification of the RAD6 protein product by *in vitro* translation, we will identify the RAD6 transcript. In preliminary experiments, the 4.2 kb HindIII fragment of pTB60 was purified and used as a probe to hybridize to total yeast RNA. Three transcripts of about 1.7, 1.1, and 0.8 kb, were observed. The HindIII/BglII fragment present in pTB64 was isolated from HindIII/BglII double digestion of pTB60 by electroelution, since in the creation of pTB64, the BglII/BamHI ligation generates a sequence not recognized by either restriction enzyme. When this 1.9 kb HindIII/BglII fragment is used as a probe for hybridization to total yeast RNA, the largest transcript disappears and the two smaller ones remain. The possibility remains that further subcloning of the 1.9 kb HindIII/BglII fragment will yield an even smaller DNA segment which has RAD6 function and that when this smaller fragment is used as a probe for hybridization to yeast RNA, one of the two transcripts will not be seen. However, it is also possible that the RAD6 gene specifies two mRNAs, as has been found for the SUC2 gene, which specifies invertase (3); one transcript encodes the secreted, glycosylated form of invertase while the other encodes the intracellular form of invertase. In the case of the RAD6 gene, one transcript may be inducible, in response to DNA damaging agents, and the other may be constitutive.

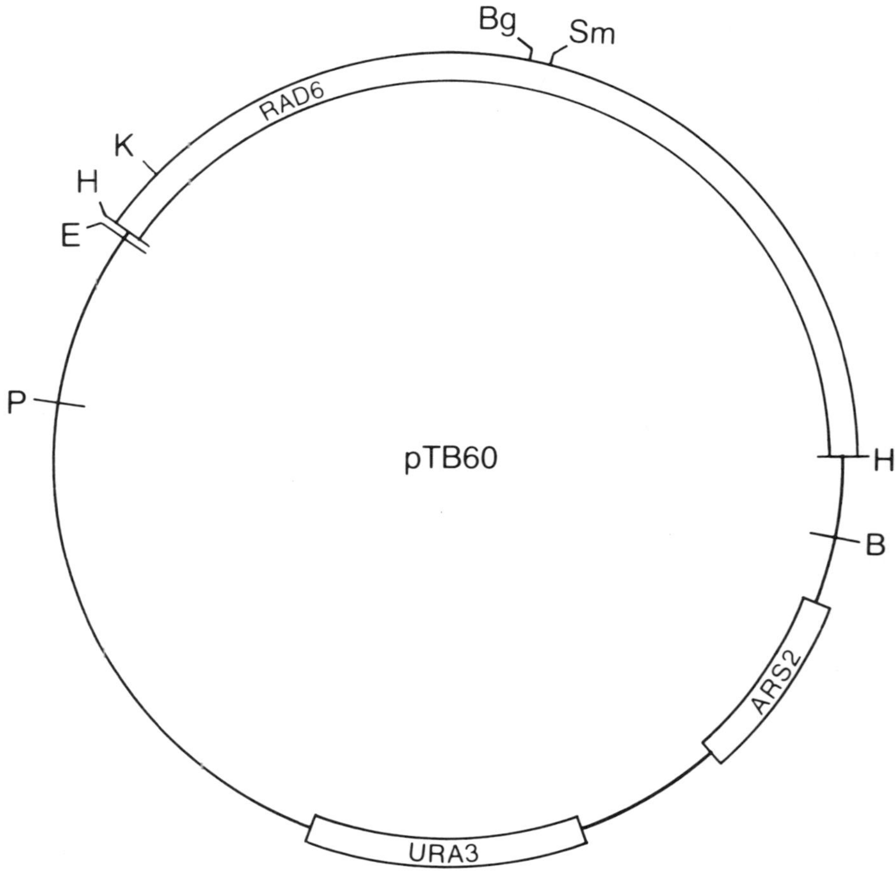

Figure 5. Restriction map of pTB60.

The HindIII fragment from pTB16 was cloned into the single HindIII site of pAB108, a 6.1 kb derivative of pBR322, and designated here by the thin line and the two open bars, one with the URA3 gene and one with the ARS2 sequence of yeast DNA. The open bar marked RAD6 represents the HindIII fragment of pTB16. Symbols for restriction enzymes are as given in Figure 1.

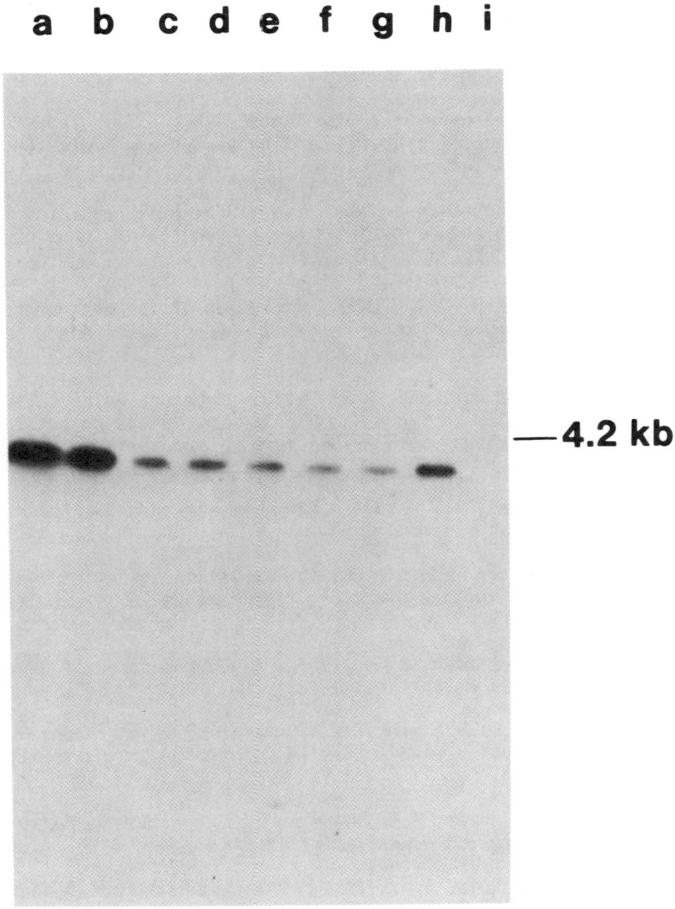

Figure 6. Autoradiogram of Southern blot of total yeast DNA hybridized with the 4.2 kb HindIII fragment of pTB60 labeled with ^{32}P.

Total DNA purified from strains BES-2, a RAD+ integrant, lane a; BES-7, a rad integrant, lane b; RAD+ non-transformed strains: AB320, lane c, A364A, lane d, #264, lane e, B-635, lane g, LP1777-9A, lane h, and a rad6-3 non-transformed strain MD-65, lane f, all digested with HindIII. Lane i contains a HindIII digest of lambda DNA. DNAs were separated by agarose gel electrophoresis, transferred to nitrocellulose, and hybridized to ^{32}P-labelled probe.

LITERATURE CITED

1. Boram, W.R. and Roman, H. 1976. Recombination in _Saccharomyces cerevisiae_: a DNA repair mutation associated with elevated mitotic gene conversion. Proc. Natl. Acad. Sci. U.S.A. 73:2828-2832.

2. Broach, J.R., Strathern, J.N. and Hicks, J.B. 1979. Transformation in yeast: development of a hybrid cloning vector and isolation of the _CAN1_ gene. Gene 8:121-133.

3. Carlson, M. and Botstein. 1982. Two differently regulated mRNAs with different 5' ends encode secreted and intracellular forms of yeast invertase. Cell 28:145-154.

4. Clarke, L. and Carbon, J. 1980. Isolation of the centromere-linked _CDC10_ gene by complementation in yeast. Proc. Natl. Acad. Sci. U.S.A. 77:2173-2177.

5. Clewell, D.B. and Helinski, D.R. 1969. Supercoiled circular DNA-protein complex in _Escherichia coli_: purification and induced conversion to an open circular DNA form. Proc. Natl. Acad. Sci. U.S.A. 62:1159-1166.

6. Cox, B.S. and Game, J.C. 1974. Repair systems in Saccharomyes. Mutat. Res. 26:257-264.

7. Cox, B.S. and Parry, J.M. 1968. The isolation, genetics and survival characteristic of ultraviolet-sensitive mutants in yeast. Mutat. Res. 6: 37-55.

8. Game, J.C. and Mortimer, R.K. 1974. A genetic study of X-ray sensitive mutants in yeast. Mutat. Res. 24:281-292.

9. Game, J.C., Zamb, T.J., Braun, R.J., Resnick, M. and Roth, R.M. 1980. The role of radiation (_rad_) genes in meiotic recombination in yeast. Genetics 94:51-68.

10. Hanawalt, P.C., Cooper, P.K., Ganesan, A.K. and Smith, C.A. 1979. DNA repair in bacteria and mammalian cells. Ann. Rev. Biochem. 48:783-836.

11. Haynes, R.H. and Kunz, B.A. 1981. DNA repair and mutagenesis in yeast, p. 371-414. In: J.N. Strathern, E.W. Jones and J.R. Broach (eds.), The Molecular Biology of the Yeast Saccharomyces. Cold Spring Harbor Laboratory, Cold Spring Harbor, N.Y.

12. Hinnen, A.H., Hicks, J.B. and Fink, G.R. 1978. Transformation in yeast. Proc. Natl. Acad. Sci. U.S.A. 75:1919-1933.

13. Hirt, B. 1967. Selective extraction of polyoma DNA from infected mouse cell cultures. J. Mol. Biol. 26:365-369.

14. Howard-Flanders, P. 1968. DNA repair. Ann. Rev. Biochem. 37:175-200.

15. Ish-Horowicz, D. and Burke, J.F. 1981. Rapid and efficient cosmid cloning. Nucl. Acids Res. 9:2989-2998.

16. Jachymczyk, W.J., von Borstel, R.C., Mowat, M.R.A. and Hastings, P.J. 1981. Repair of interstrand cross-links in DNA of *Saccharomyces cerevisiae* requires two systems for DNA repair: the *RAD3* system and the *RAD51* system. Mol. Gen. Genet. 182:196-205.

17. Jackson, J.A. and Fink, G.R. 1981. Gene conversion between duplicated genetic elements in yeast. Nature (London) 292:306-311.

18. Johnston, L.H. and Nasmyth, K.A. 1978. *Saccharomyces cerevisiae* cell cycle mutant *cdc9* is defective in DNA ligase. Nature (London) 274:891-893.

19. Lawrence, C.W. and Christensen, R.B. 1976. UV mutagenesis in radiation-sensitive strains of yeast. Genetics 82:207-232.

20. Lawrence, C.W. and Christensen, R.B. 1979. Metabolic suppressor of trimethoprim and ultraviolet light sensitivities of *Saccharomyces cerevisiae rad6* mutants. J. Bacteriol. 139:866-876.

21. Lawrence, C.W., Stewart, J.W., Sherman, F. and Christensen, R. 1974. Specificity and frequency of ultraviolet-induced reversion of an iso-1-cytochrome *c* ochre mutant in radiation sensitive strains of yeast. J. Mol. Biol. 85:137-162.

22. Lemontt, J.F. 1971. Mutants of yeast defective in mutation induced by ultraviolet light. Genetics 68:21-33.

23. Malone, R.E. and Esposito, R.E. 1980. The RAD52 gene is required for homothallic interconversion of mating types and spontaneous mitotic recombination in yeast. Proc. Natl. Acad. Sci. U.S.A. 77:503-507.

24. Martin, P. Prakash, L. and Prakash, S. 1981. a/α-specific effect of the *mms3* mutation on ultraviolet mutagenesis in *Saccharomyces cerevisiae*. J. Bacteriol. 146:684-691.

25. McKee, R.H. and Lawrence, C.W. 1979. Genetic analysis of gamma-ray mutagenesis in yeast. I. Reversion in radiation-sensitive strains. Genetics 93:361-373.

26. Montelone, B.A., Prakash, S. and Prakash, L. 1981. Recombination and mutagenesis in *rad6* mutants of *Saccharomyces cerevisiae*: evidence for multiple functions of the *RAD6* gene. Mol. Gen. Genet. 184:410-415.

27. Mortimer, R.K. and Schild, D. 1980. Genetic map of *Saccharomyces cerevisiae*. Microbiol. Rev. 44:519-571.

28. Nasmyth, K.A. and Reed, S.I. 1980. Isolation of genes by complementation in yeast: molecular cloning of a cell-cycle gene. Proc. Natl. Acad. Sci. U.S.A. 77:2119-2123.

29. Norgard, M.V., Emigholz, K. and Monahan, J.J. 1979. Increased amplification of pBR322 plasmid deoxyribonucleic acid in Escherichia coli K-12 strains RR1 and X1776 grown in the presence of high concentrations of nucleoside. J. Bacteriol. 138: 270-272.

30. Orr-Weaver, R.L., Szostak, J.W. and Rothstein, R.J. 1981. Yeast transformation: a model system for the study of recombination. Proc. Nat. Acad. Sci. U.S.A. 78:6354-6358.

31. Prakash, L. 1974. Lack of chemically induced mutation in repair-deficient mutants of yeast. Genetics 78:1101-1118.

32. Prakash, L. 1975. Repair of pyrimidine dimers in nuclear and mitochondrial DNA of yeast irradiated with low doses of ultraviolet light. J. Mol. Biol. 98:781-795.

33. Prakash, L. 1976. Effect of genes controlling radiation sensitivity on chemically-induced mutations in Saccharomyces cerevisiae. Genetics 83:295-301.

34. Prakash, L. 1981. Characterization of postreplication repair in Saccharomyces cerevisiae and effects of rad6, rad18, rev3 and rad52 mutations. Mol. Gen. Genetics 184:471-478.

35. Prakash, L. and Prakash, S. 1977. Isolation and characterization of MMS sensitive mutants of Saccharomyces cerevisiae. Genetics 86:33-55.

36. Prakash, L. and Prakash, S. 1979. Three additional genes involved in pyrimidine dimer removal in Saccharomyces cerevisiae: RAD7, RAD14 and MMS19. Mol. Gen. Genet. 176: 351-359.

37. Prakash, L. and Taillon-Miller, R. 1981. Effects of the rad52 gene on sister chromatid recombination in Saccharomyces cerevisiae. Curr. Genet. 3:247-250.

38. Prakash, S., Prakash, L., Burke, W. and Montelone, B. 1980. Effects of the RAD52 gene on recombination in Saccharomyces cerevisiae. Genetics 94:31-50.

39. Resnick, M.A. 1969. Induction of mutations in Saccharomyces cerevisiae by ultraviolet light. Mutat. Res. 7:315-332.

40. Resnick, M.A. 1975. The repair of double strand breaks in chromosomal DNA of yeast. p. 549-556. In: P.C. Hanawalt and R.B. Setlow (eds.), Molecular Mechanisms for Repair in DNA. Plenum Press, New York.

41. Resnick, M.A. and Martin, P. 1976. The repair of double-strand breaks in the nuclear DNA of Saccharomyces cerevisiae and its genetic control. Molec. Gen. Genet. 143:119-129.

42. Resnick, M.A. and Setlow, J.K. 1972. Repair of pyrimidine dimer damage induced in yeast by ultraviolet light. J. Bacteriol. 109:979-986.

43. Reynolds, R.J. 1978. Removal of pyrimidine dimers form *Saccharomyces cerevisiae* nuclear DNA under non-growth conditions as detected by a sensitive enzymatic assay. Mutat. Res. 50:43-56.

44. Reynolds, R.J., Love, J.D. and Friedberg, E.C. 1980. Molecular mechanisms of pyrmidine dimer excision in *Saccharomyces cerevisiae*: excision of dimers in cell extracts. J. Bacteriol. 147:705-708.

45. Scherer, S. and Davis, R.W. 1979. Replacement of chromosome segments with altered DNA sequences constructed *in vitro*. Proc. Natl. Acad. Sci. U.S.A. 76:4951-4955.

46. Southern, R.M. 1975. Detection of specific sequences among DNA fragments separated by gel electrophoresis. J. Mol. Biol. 98:503-517.

47. Tuite, M.F. and Cox, B.S. 1981. *RAD6+* gene of *Saccharomyces cerevisiae* codes for two mutationally separable deoxyribonucleic acid repair functions. Molec. Cell Biol. 1:153-157.

48. Unrau, P., Wheatcroft, R. and Cox, B.S. 1971. The excision of pyrimidine dimers from the DNA of ultraviolet irradiated yeast. Mol. Gen. Genet. 113:359-362.

49. Waters, R. and Moustacchi, E. 1974. The disappearance of ultraviolet-induced pyrimidine dimers from the nuclear DNA of exponential and stationary phase cells of *Saccharomyces cerevisiae* following various post-irradiation treatments. Biochim. Biophys. Acta 312:626-632.

50. Wilcox, D.R. and Prakash, L. 1981. Incision and postincision steps of pyrimidine dimer removal in excision-defective mutants of *Saccharomyces cerevisiae*. J. Bacteriol. 148:618-623.

51. Witkin, E. 1976. Ultraviolet mutagenesis and inducible DNA repair in *Escherichia coli*. Bacteriol. Rev. 40:869-907.

52. Zaret, K.S. and Sherman, F. 1982. DNA sequence required for efficient transcription termination in yeast. Cell 28:563-573.

18. MOLECULAR MECHANISMS OF DNA RECOMBINATION: TESTING MITOTIC AND MEIOTIC MODELS*

Michael S. Esposito and Carlo V. Bruschi

Biology and Medicine Division
Lawrence Berkeley Laboratory
University of California
Berkeley, California 94720

Rec. Adv. Yeast Mol. Biol. (1982) 1:242-253

A hyperhaploid n + 1 strain of Saccharomyces cerevisiae (LBL1) disomic for chromosome VII was employed to isolate hyper-rec and hypo-rec mutations affecting spontaneous mitotic gene conversion and intergenic recombination. The genotype of LBL1 permits simultaneous and independent identification of rec mutations that enhance or diminish gene conversion and those that enhance or diminish intergenic recombination. Five phenotypic groups of rec mutants were isolated following ultraviolet light mutagenesis. Rec mutations that simultaneously abolish or enhance both classes of recombinational events were detected. These results demonstrate that gene conversion and intergenic recombination are under joint genetic control in mitotic cells. Conversion-specific and intergenic recombination-specific rec mutants were also recovered. Their properties indicate that conversion and intergenic recombination are separable phenomena dependent upon discrete REC genes. The rec mutants isolated in LBL1 provide a method to test molecular models of mitotic and meiotic recombination.

INTRODUCTION

Interest in the mechanisms of mitotic and meiotic recombination in Saccharomyces cerevisiae has been heightened by recent demonstrations that genetic recombination plays a significant role in the regulation of gene expression in yeast and other eukaryotes (4,10). The mechanisms and pathways of yeast mitotic and meiotic recombination are exceptionally amenable to study. Molecular studies of DNA recombination and detailed genetic analyses suggest the existence of a variety of recombinational intermediates and modes of resolution (6). While several rec mutants have been isolated in yeast the extant collection does not appear to include variants defective in each step of recombination (1,6). We constructed an n + 1 hyperhaploid strain disomic for chromosome VII (5) for the isolation of a broad spectrum of rec mutations. The genotype of LBL1 permits detection of rec mutations affecting the initiation of recombination, establishment of Holliday structures or related intermediates, isomerization of Holliday structures and resolution of Holliday structures by endonucleolytic scission (11) or DNA replication (3). In this report we discuss the phenotypes of recently isolated rec mutants and analytic methods for identification of their putative defects. The latter methods were developed to characterize the recombinational phenotype of the spo11-1 mutant, as described in the accompanying article by Bruschi and Esposito.

---

* This work was supported by the Director, Office of Energy Research, Office of Biology and Medicine of the U.S. Department of Energy under Contract No. DE-AC03-76SF00098, NIH Grant GM 29002 and NIHES Grant ES 02756.

MATERIALS AND METHODS

Yeast strains. The genotype of LBL1, the n + 1 hyperhaploid disomic for chromosome VII, is given in Figure 1.

Media and techniques. The recipes for media have been reported previously (8). Strain LBL1 was maintained on slants of leucineless tryptophanless omission medium or frozen in 50% (v/v) glycerol at -70°C. Before each experiment single colony isolates of LBL1 were obtained by streaking on leucineless trytophanless omission medium or synthetic complete medium followed by incubation at 24°C. The genotypes of putative *ade5*, *ade6*, and *ade5 ade6* segregants of LBL1 were verified by complementation analysis employing *ade5* and *ade6* test strains. Segregation of *cly8*, a heat sensitive lethal mutation, was monitored by failure of growth on glucose nutrient medium at 36°C.

Determination of mitotic rates of restitution of haploidy, intergenic recombination and gene conversion. The rate of occurrence of each type of mitotic event was determined by the method of the median of Lea and Coulson (12) as previously reported (5,9,13). The LBL1 strain was plated on synthetic complete medium and incubated at 24°C. Twenty-five colonies having a diameter of 2.5 mm were employed in rate determinations. Colonies of the same size were used because such colonies have undergone approximately the same number of cell divisions. Each colony was resuspended in sterile distilled water and plated on synthetic complete medium, leucineless tryptophanless medium, cycloheximide medium and leucineless tryptophanless cycloheximide medium.

Induction of Rec mutations by ultraviolet light mutagenesis. The LBL1 strain was plated on leucineless tryptophanless medium and exposed to a 65% killing dose of ultraviolet light. Surviving red colonies of 2.5 mm diameter were resuspended and plated on leucineless tryptophanless cycloheximide medium to detect *rec* mutations affecting spontaneous gene conversion and intergenic recombination. Red cycloheximide resistant segregants of LBL1 detected on leucineless trptophanless cycloheximide medium result from gene conversion of *CYH2*^{s} to *cyh2*^{r} while white resistant colonies result from intergenic recombination in the *CYH2* - *TRP5* interval (5).

RESULTS

Chromosomal Recombination and Segregation in Strain LBL1

Strain LBL1 is a n + 1 hyperhaploid disomic for chromosome VII which we employed for the isolation of *hyper-rec* and *hypo-rec* mutations affecting mitotic gene conversion and/or intergenic recombination. The genotype of LBL1 is shown in Figure 1. The disomic chromosome is heterozygous at six loci distributed on both sides of the centromere. The strain forms red colonies on chemically defined and complex glucose nutrient media due to the presence of the *ade2-1* mutation that confers adenine auxotrophy and accumulation of a cell-limited red pigment (15). The *trans* arrangement of markers at *LEU1* and *TRP5* is employed for maintenance of the disomic strain. Haploid segregants resulting from breakdown of the disome are unable to grow on media lacking leucine and tryptophan. The $cyh2^r$ mutation, a recessive marker, permits simultaneous monitoring of gene conversion, intergenic recombination and restitution of haploidy.

Gene conversion, intergenic recombination and restitution of haploidy (due to chromosomal loss of nondisjunction) result in rare cycloheximide resistant segregants that can be detected selectively as summarized in Table 1 and previously reported (5). The rates of occurence of each class of mitotic event resulting in cycloheximide resistant segregants were determined by the method of the median of Lea and Coulson (12) as described in the previous section. Twenty-five colonies of the LBL1 strain grown at 24°C were plated on diagnostic media to determine the number and types of cycloheximide resistant segregants that had arisen during growth. The results of this analysis are summarized in Table 2. The data demonstrate that the chromosome VII disome is highly stable and that gene conversion of $CYH2^s$ to $cyh2^r$ and intergenic recombination in the interval CYH2 - TRP5 can be detected separately from events resulting in restitution of haploidy by plating cells on leucineless tryptophanless cycloheximide medium. Two classes of cycloheximide resistant colonies (Classes B and C of Table 1) arise on this medium. Red colonies are of the Class B type and are due to gene conversion of $CYH2^s$ to $cyh2^r$ with maintenance of disomy. White colonies are of the Class C type and result from intergenic recombination in the interval CYH2 - TRP5. They are white due to homozygosity for the ade5 mutation which blocks accumulation of the red pigment precursor.

Ultraviolet light induced rec mutations affecting spontaneous mitotic gene conversion, as measured by the appearance of Class B segregants, and/or intergenic recombination, as measured by the appearance of Class C segregants, were obtained as described below.

Isolation of Recombination Mutants

The LBL1 strain was plated on leucineless tryptophanless medium and mutagenized by exposure to ultraviolet light as described in MATERIALS AND METHODS. After growth at 24°C, surviving red colonies were resuspended and plated on leucineless tryptophanless cycloheximide medium to assay the frequencies of Class B and Class C segregants within each colony. A total of 2,467 survivors of UV mutagenesis were examined. Fifty-four mutants that exhibit a reproducible hypo-rec or hyper-rec phenotype were obtained. The phenotypes of the mutants are summarized in Table 3. Five phenotypic groups were observed. Mutants in Group I exhibit reduced levels of both spontaneous mitotic gene conversion and intergenic recombination. This mutant class demonstrates that gene conversion and intergenic recombination are under coordinate genetic control. Group II mutants retain the capacity for gene conversion but exhibit a reduced level of intergenic recombination. The existence of this class indicates there are REC gene products required for detection of intergenic recombination that are not required for gene conversion. Mutants of Group III display normal levels of spontaneous gene conversion and enhanced intergenic recombination. The existence of Group III, like that of Group II, demonstrates that intergenic recombination is affected by gene products that are not involved in gene conversion.

Table 1. Chromosomal loss, nondisjunction, gene conversion and intergenic recombination of chromosome VII markers.

Event	Genotypes of cycloheximide resistant segregants		Phenotypic class[a]
Chromosomal loss or nondisjunction	Haploid	ade5 cyh2^r trp5 LEU1 ade6 cly8	A
Gene conversion of CYH2^s to cyh2^r	Disome	ADE5 cyh2^r TRP5 leu1 ADE6 CLY8 ade5 cyh2^r trp5 LEU1 ade6 cly8	B
Recombination in the interval CYH2 - TRP5	Disome	ade5 cyh2^r TRP5 leu1 ADE6 CLY8 ade5 cyh2^r trp5 LEU1 ade6 cly8	C
Recombination in the interval TRP5 - LEU1	Disome	ade5 cyh2^r trp5 leu1 ADE6 CLY8 ade5 cyh2^r trp5 LEU1 ade6 cly8	D
Recombination in the interval LEU1 - centromere	Disome	ade5 cyh2^r trp5 LEU1 ADE6 CLY8 ade5 cyh2^r trp5 LEU1 ade6 cly8	D
Gene conversion and chromosomal loss or nondisjunction	Haploid Haploid	ade5 cyh2^r trp5 LEU1 ade6 cly8 ADE5 cyh2^r TRP5 leu1 ADE6 CLY8	A E
Recombination and chromosomal loss or nondisjunction	Haploid Haploid Haploid Haploid	ade5 cyh2^r trp5 LEU1 ade6 cly8 ade5 cyh2^r trp5 LEU1 ADE6 CLY8 ade5 cyh2^r TRP5 leu1 ADE6 CLY8 ade5 cyh2^r trp5 leu1 ADE6 CLY8	A D F G

[a] Class A is diagnostic of chromosomal loss or nondisjunction. Class B is diagnostic of gene conversion of CYH2^s to cyh2^r with maintenance of disomy. Class C is diagnostic of recombination with maintenance of disomy. Class D is diagnostic or recombination. Class E is diagnostic of gene conversion and resititution of haploidy. Classes F and G are diagnostic of recombination and restitution of haploidy.

Table 2. Rates of spontaneous mitotic restitution of haploidy, gene conversion and intergenic recombination of chromosome VII markers.

Event detected	Phenotypic class	Mitotic rates: Cyh medium	Mitotic rates: Leu^-Trp^-Cyh medium
Chromosomal loss or nondisjunction	A	1.37×10^{-5}	
Gene conversion of $\underline{CYH2^s}$ to $\underline{cyh2^r}$	B	1.45×10^{-6}	2.03×10^{-6}
Recombination in the interval <u>CYH2</u> - <u>TRP5</u>	C	3.66×10^{-6}	3.30×10^{-6}
Recombination in the interval <u>TRP5</u> - <u>ADE6</u>	D	4.47×10^{-6}	
Total of all events[a]		$2.44 \pm 0.24 \times 10^{-5}$	$5.33 \pm 0.64 \times 10^{-6}$

[a]A total of 25 colonies of LBL1 grown on synthetic complete medium at 24°C were plated on both cycloheximide synthetic complete medium and leucineless tryptophanless cycloheximide medium. The total rates per cell division of the ensemble of events that result in Cyh^r and $Leu^+Trp^+Cyh^r$ resistant segregants were determined by the median method of Lea and Coulson (12). A total of 1033 colonies recovered on Cyh medium and 314 colonies recovered on Leu^-Trp^-Cyh medium were further characterized to determine the rates for each phenotypic class. Classes A,B,C, and D accounted for 98% of the events detected on Cyh medium. Classes B and C accounted for 98% of the events detected on Leu^-Trp^-Cyh medium (5).

Table 3. Mitotic recombination mutants isolated following ultraviolet light mutagenesis of LBL1.

Phenotypic Groups	No. Mutants Obtained	Average Spontaneous Mitotic Recombination Frequencies[a]	
		Cyh^r Conv./10^6 cells	Cyh^r Inter. Rec./10^6 cells
I. Conv.$^-$ Recip. Exch.$^-$	13	1.5	0.8
II. Conv.$^+$ Recip. Exch.$^-$	24	22.8	0.8
III. Conv.$^+$ Recip. Exch.$^{+++}$	13	15.3	121.9
IV. Conv.$^-$ Recip. Exch.$^+$	1	0.8	19.7
V. Conv.$^{+++}$ Recip. Exch.$^{+++}$	3	94.7	1006.4
Control LBL1 (25 colonies)		19.4	29.1

[a]The average values for each phenotypic group of mutants represent the results of four independent determinations per mutant. For each determination a 2.5 mm colony grown on synthetic complete medium was resuspended and plated on leucineless tryptophanless cycloheximide medium to determine the frequency of red $Leu^+Trp^+Cyh^r$ convertants and white $Leu^+Trp^+Cyh^r$ intergenic recombinants. Phenotypic symbols are as follows: - = hypo-rec; + = wildtype; and +++ = hyper-rec.

Groups IV and V contain the fewest number of mutants. Group IV, containing one mutant, fails to exhibit conversion but does exhibit intergenic recombination. The phenotype of Group IV demonstrates that gene conversion involves processes that are not required for intergenic recombination. Group V mutants exhibit enhanced levels of both gene conversion and intergenic recombination. Like Group I, Group V demonstrates the existence of REC gene products that coordinately control the levels of gene conversion and intergenic recombination in mitotic cells. More detailed information regarding the functions of the REC genes comprising the five phenotypic classes which we have identified can be obtained by examining their meiotic phenotypes, as discussed below.

DISCUSSION

Several molecular models of recombination have been proposed to explain the properties of mitotic and meiotic gene conversion and reciprocal recombination (2,11,14,16). Most share three elements in common: 1) Gene conversion reflects the establishment of heteroduplex DNA, which may persist or undergo mismatch repair; 2) Reciprocal recombination results from the formation, isomerization, and endonucleolytic cleavage of Holliday structures; and 3) Chromosomal recombination is initiated postreplicationally in G2 of mitosis and meiosis. Recent studies, however, have provided genetic evidence that most, if not all, mitotic recombination between homologous chromosomes of diploid cells is initiated prereplicationally (3,7,18) and that postreplicational mitotic sister chromatid exchange also occurs (17). We proposed a molecular model of prereplicational mitotic recombination to explain the properties of mitotic recombination between homologs (3,9). According to the model prereplicational recombination is initiated by heteroduplex DNA formation involving unreplicated homologs. The formation of heteroduplex DNA results in gene conversion of heterozygous markers and can be detected both with and without mismatch repair (6). Gene conversional events, for example, result in Class B red cycloheximide resistant segregants of the LBL1 disome (Table 1). Mitotic reciprocal intergenic recombination, according to our model, requires the formation of a Holliday structure, does not require isomerization, and can only be detected when the Holliday structure is not endonucleolytically cleaved and persists through chromosomal DNA replication. Class C segregants of LBL1 presumably arise in this manner by formation of a Holliday structure in the CYH2 - TRP5 interval.

The critical functions required for detection of prereplicational mitotic gene conversion and intergenic recombination are summarized in Table 4. The collection of hyper-rec and hypo-rec mutants isolated in the LBL1 strain (Table 3) includes all of the phenotypic groups anticipated following loss of functions affecting prereplicational mitotic gene conversion and reciprocal recombination, i.e., Groups I, II, and III of Table 3. Group IV of Table 3, a rare mutational class, may result in preferential mismatch repair of $CYH2^s$:$cyh2^r$ heteroduplexes to the $CYH2^s$:$CYH2^s$ homoduplex state, thereby diminishing the frequency of Class B red cycloheximide resistant comvertants. Group V mutants of Table 3, which enhance both gene conversion and intergenic recombination are likely to result from increased overall initiation of mitotic recombination. We have previously described a mitosis-specific hyper-rec mutation of this type, rem1-1 (8,9).

Future studies of Group I, II and III *rec* mutants will focus on a comparison of their mitotic and meiotic phenotypes. Group I is expected to include mutants that are *hypo-rec* with respect to both meiotic gene conversion and reciprocal recombination owing to loss of a function required for initiation of heteroduplex DNA formation in both mitosis and meiosis (Table 5). Group II mutants are expected to be of two types. Those which diminish Holliday structure formation in both meiosis and mitosis are expected to affect mitotic and meiotic recombination in the same manner (Table 5). Those which are defective in mitotic reciprocal recombination due to failure to resolve mitotic Holliday structures by replication are not expected to affect meiotic reciprocal recombination (Table 5). Group III mutants that exhibit a *hyper-rec* phenotype with respect to reciprocal recombination due to loss of a function required for both mitotic and meiotic endonucleolytic scission of Holliday structures are expected to exhibit a *hypo-rec* phenotype with respect to meiotic reciprocal recombination. Evidence that the *spoll-1* mutation may belong to this class of *rec* mutations is discussed in this volume by Bruschi and Esposito.

The isomerization of Holliday structures may play no important role in prereplicational recombination (Tables 4 and 5). Consequently *rec* mutants isolated in LBL1 are not expected to include variants defective in isomerization of Holliday structures. Putative isomerase defective mutants can, however, be detected as variants that fail to exhibit reciprocal sister chromatid exchange and reciprocal meiotic intergenic recombination. Selective systems to isolate such mutants are presently available.

Table 4. Prereplicational versus postreplicational recombination.

Functions involved in recombination	Prereplicational recombination	Postreplicational recombination
1. Formation of heteroduplex DNA	required	required
2. Establishment of Holliday structures	required	required
3. Isomerization of Holliday structures	not required	required
4. Endonucleolytic scission of Holliday structures	not required and antagonistic	required
5. Replicational resolution of Holliday structures	required	impossible

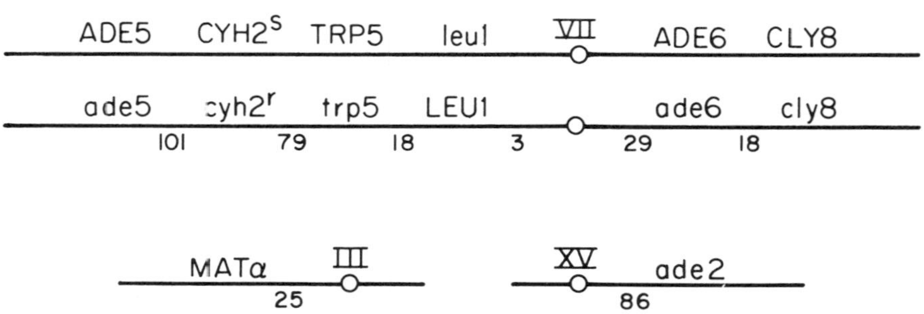

Figure 1. Genotype of the chromosome VII disomic strain LBL1. The map distances shown are the standard meiotic values.

Table 5. Predicted phenotypes of _rec_ mutations affecting prereplicational and/or postreplicational recombination[a].

Absent function	Prereplicational recombination		Postreplicational recombination	
	Conversion	Recip. Exch.	Conversion	Recip. Exch.
1. Formation of heteroduplex DNA	-	-	-	-
2. Establishment of Holliday structures	+	-	+	-
3. Isomerization of Holliday structures	+	+	+	-
4. Endonucleolytic scission of Holliday structures	+	+++	+	-
5. Replicational resolution of Holliday structures	+	-	+	+

[a]Phenotypic symbols are as follows: - = hypo-rec; + = wildtype; and +++ = hyper-rec

LITERATURE CITED

1. Baker, B. S., Carpenter, A.T.C., Esposito, M. S., Esposito, R. E., and Sandler, L. 1976. The genetic control of meiosis. Ann. Rev. Gent. 10: 53-134.

2. Catcheside, D. G. 1977. The genetics of recombination. University Park Press, Baltimore.

3. Esposito, M. S. 1978. Evidence that spontaneous mitotic recombination occurs at the two-strand stage. Proc. Natl. Acad. Sci. USA 75:436-4440.

4. Esposito, M. S. and Esposito, R. E. 1977. Gene conversion, paramutation and controlling elements: a treasure of exceptions. pp. 59-92. In: Goldstein, L. and Prescott, D. M. (eds.) Cell Biology, A Comprehensive Treatise, Vol. 1, Academic Press, New York.

5. Esposito, M. S., Maleas, D. T., Bjornstad, K. A. and Bruschi, C. V. 1982. Simultaneous detection of changes in chromosome number, gene conversion and intergenic recombination during mitosis of *Saccharomyces cerevisiae*: spontaneous and ultraviolet light induced events. Current Genetics 5: *in press*.

6. Esposito, M. S. and Wagstaff, J. E. 1981. Mechanisms of mitotic recombination. pp. 341-370, In: Strathern, J. N., Jones, E. W., and Broach, J. R. (eds.), Molecular Biology of the Yeast Saccharomyces: Life Cycle and Inheritance. Cold Spring Harbor Laboratory, Cold Spring Harbor, N.Y.

7. Fabre, F. 1978. Induced intragenic recombination in yeast can occur during the *G1* mitotic phase. Nature 272:795-798.

8. Golin, J. E. and Esposito, M. S. 1977. Evidence for joint genic control of spontaneous mutation and genetic recombination during mitosis in *Saccharomyces cerevisiae*. Mol. Gen. Genet. 150:127-135.

9. Golin, J. E. and Esposito, M. S. 1981. Mitotic recombination: mismatch correction and replicational resolution of Holliday structures formed at the two-strand stage in *Saccharomyces*. Mol. Gen. Genet. 183:252-263.

10. Herskowitz, I. and Oshima, Y. 1981. Control of cell type in *Saccharomyces cerevisiae*: mating type and mating-type interconversion. pp. 181-209. In: Strathern, J. N., Jones, E. W. and Broach, J. R. (eds.), Molecular Biology and the Yeast Saccharomyces: Life Cycle and Inheritance. Cold Spring Harbor Laboratory, Cold Spring Harbor, N.Y.

11. Holliday, R. 1964. A mechanism for gene conversion in fungi. Genet. Res. 5:282-304.

12. Lea, D. E. and Coulson, C. A. 1948. The distribution of numbers of mutants in bacteria populations. J. Genetics 49:264-284.

13. Malone, R. E., Golin, J. E., and Esposito, M. S. 1980. Mitotic *versus* meiotic recombination in *Saccharomyces cerevisiae*. Current Genetics 1:241-248.

14. Meselson, M. S. and Radding, C. M. 1975. A general model for genetic recombination. Proc. Natl. Acad. Sci. USA 72:358-361.

15. Roman, H. 1956. Studies of gene mutation in *Saccharomyces*. Cold Spring Harbor Symp. Quant. Biol. XXI:175-183.

16. Stahl, F. W. 1979. Genetic recombination. Thinking about it in phage and fungi. Freeman, San Francisco.

17. Szostak, J. W. and Wu, R. 1980. Unequal crossing over in the ribosomal DNA of *Saccharomyces cerevisiae*. Nature 284:426.

18. Wildenberg, J. 1970. The relation of mitotic recombination to DNA replication in yeast pedigrees. Genetics 66:291-304.

19. RECOMBINATION PROCESSES IN A SPORULATION-DEFECTIVE MUTANT OF S. cerevisiae: ROLE OF HOLLIDAY STRUCTURE RESOLUTION*

Carlo V. Bruschi and Michael S. Esposito

Biology and Medicine Division
Lawrence Berkeley Laboratory
University of California, Berkeley, California 94720

Rec. Adv. Yeast Mol. Biol. (1982) 1:254-268

A mutant of the yeast Saccharomyces cerevisiae, defective in the production of ascospores, is used for comparative analysis of the formation and resolution of Holliday structures in meiotic and mitotic recombination. Diploid hybrids homozygous for the mutation spo11-1 exhibit not only a sporulation-deficient phenotype, but also a very low level of ascosporal viability, due to nondisjunction and chromosomal loss. Ascosporal survival is ca. 0.02% while in SPO11/SPO11 diploids ascosporal survival typically exceeds 95%. The genetic characterization of spo11-1/spo11-1 hybrids has provided evidence supporting the hypothesis that failure to cut Holliday structures, during both mitosis and meiosis results in a hyper-rec phenotype in mitosis and hypo-rec phenotype in meiosis. Diploid spo11-1/spo11-1 cells are hyper-rec with respect to spontaneous mitotic intergenic recombination and hypo-rec with respect to spontaneous meiotic intergenic recombination. The spontaneous level of mitotic intergenic recombination at 24°C, a semi-permissive temperature for the sporulation process in spo11-1/spo11-1 hybrids, is ca. 5 fold higher in spo11-1/spo11-1 than in wild-type SPO11/SPO11 hybrids, while no significant difference exists at the restrictive temperature of 34°C. No activity on spontaneous intragenic heteroallelic recombination (gene-conversion), which does not depend upon resolution of Holliday structures, is detected at both temperatures. Meiotic intergenic recombination in the mutant, in contrast, is ca. 2 fold lower than that of wild type hybrids.

The phenotype of spo11-1 homozygous diploids precisely fulfills the expected phenotype predicted according to the model of prereplicational mitotic recombination versus postreplicational meiotic recombination for cells defective in the endonucleolytic scission of Holliday structures in both mitosis and meiosis. Preliminary studies of mitotic and meiotic recombination of the 2-micron DNA plasmid in spo11-1/spo11-1 and SPO11/SPO11 cells have been initiated to characterize the recombinational defect of spo11-1 cells at the DNA molecular level.

---

* This work was supported by the Director, Office of Energy Research, Office of Biology and Medicine of the U.S. Department of Energy under Contract No. DE-AC03-76SF00098. The authors' research is supported by the National Institutes of Health, Grant No. GM 29002. We thank Dr. S. Klapholz for providing yeast strains employed in this study.

INTRODUCTION

Different mechanisms of genetic recombination in meiosis and mitosis and the existence of common *REC* gene functions can be genetically studied in yeast by the isolation of mutants affected in one or both processes with respect to the formation and resolution of Holliday structures (7). In meiosis, in which genetic recombination appears to occur after DNA synthesis and chromosomal duplication, in the G2 phase, the cleavage of Holliday structures formed in the G2 phase is the hypothesized basis for the generation of meiotic intergenic recombinants (11,15). Failure to resolve Holliday structures that engage nonsister chromatids would be expected to reduce the level of detectable meiotic intergenic recombination, to cause extensive meiotic nondisjunction and to result in drastic ascospore lethality owing to the consequent aneuploidy.

The effect of failure to resolve Holliday structures in mitosis is expected to differ from that observed in meiosis. We have recently presented genetic data indicating that spontaneous mitotic recombination between homologous chromosomes occurs at the two-strand stage, i.e., in the G1 phase, before chromosomal duplication, in both wild type hybrids and hybrids homozygous for the mitotic *hyper-rec* mutation *rem1-1* (6,10). A specific prereplicational model of mitotic recombination at the two-strand stage was advanced by Esposito (6), who proposed that intergenic recombination is initiated prereplicationally by formation of Holliday structures. Some of these structures, that are not resolved by endonucleolytic scission, are instead resolved by chromosomal DNA replication (Figure 1). According to this model, failure to cut Holliday structures by enzymatic activity before DNA replication would allow more of them to be resolved by DNA synthesis and would increase the level of mitotic recombination. These resolution events change the linkage relationships of heterozygous markers bracketing the site of the Holliday structure and are detected by standard methods which rely on the appearance of homozygous sectors. Holliday structures resolved by endonucleolytic cleavage, in contrast, do not produce sectors or mitotic intergenic recombinants and thus escape detection.

A critical analysis of the involvement of resolution of Holliday structures in meiotic and mitotic recombination processes can thus be carried out with mutants that show high levels of lethality of the meiotic products due to nondisjunction and consequent chromosomal loss leading to aneuploidy. Such mutants may include variants defective in the resolution of Holliday structures formed during both mitosis and meiosis. They would be expected to exhibit a *hypo-rec* phenotype with respect to spontaneous intergenic recombination and a *hyper-rec* phenotype with respect to spontaneous mitotic intergenic recombination. Furthermore, the level of gene-conversion measured as intragenic heteroallelic recombination should be normal, since conversion depends upon formation of heteroduplex DNA rather than the mode of Holliday structure resolution. The extent of mitotic *hyper-rec* activity of such mutants would depend upon the efficiency with which Holliday structures are normally cleaved in the wild type.

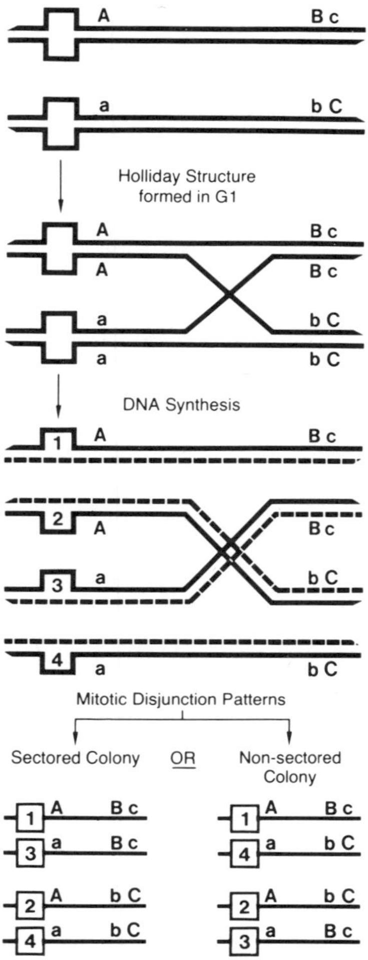

Figure 1. The prereplicational model of mitotic recombination proposed by Esposito (6). Prior to chromosomal duplication, a Holliday structure (11) is formed by exchange of single strands of DNA of the same chemical polarity. When the Holliday structure is not cleaved prior to duplication of the chromosomes the resulting chromatids consist of a pair with markers in the non-recombinant configuration, *i.e.* 1 and 4 and a pair of recombinant chromatids, *i.e.* 2 and 3. When chromatids 1 and 3 separate from 2 and 4 a sectored colony results, exhibiting segregation of the recessive markers *b* and *c*. The arrows indicate the 3' termini of the deoxyribonucleotide chains. The broken lines indicate DNA synthesized during the S phase of mitosis.

The data described below demonstrate that diploids homozygous for the mutation spo11-1 exhibit the genetic characteristics expected for cells defective in both mitotic and meiotic endonucleolytic scission of Holliday structures.

MATERIALS AND METHODS

Strains. Mitotic and meiotic recombination was studied in the MATa/MATα heterothallic spo11-1/spo11-1 hybrid, CBX1, and the MATa/MATα heterothallic SPO11/SPO11 congenic hybrid, CBX2, having the genotypes shown in Table 6.

Media and genetic techniques. Genetic procedures and the compositions of the glucose nutrient (YPD), synthetic growth media, and sporulation media have been reported (8,14).

Plating of parallel vegetative cultures. The diploids used in recombination experiments were plated on YPD medium and incubated at 24°C or 34°C to obtain single colonies. 100 ml liquid YPD cultures supplemented with 60 mg/liter of adenine were inoculated with one colony of approximately 3 mm in diameter (ca. 1.5 - 3.0 x 10^7 cells). The cultures were grown at 24°C or 34°C, respectively, harvested during the exponential phase of growth, concentrated by centrifugation, washed once in 10 ml of sterile distilled water and plated on the appropriate synthetic media. The plates were incubated for 7 days prior to counting.

Caclulation of recombination rates. Mitotic recombination rates were calculated by the methods of Drake (1) and Lea and Coulson (13). The pertinent recombination rates for each culture were obtained from the equation $r=0.4343 f/(\log N-\log N_0)$, (where r=rate, f=frequency of recombinants, N=final total cell number, and N_0=initial cell number). The median rate was then determined.

RESULTS

Isolation and Properties of spo11-1

The spo11-1 mutation was isolated following ultraviolet mutagenesis of ascospores derived from a homothallic diploid strain of *Saccharomyces cerevisiae* and subsequently introduced into heterothallic stocks (2,3). Diploids homozygous for the spo11-1 mutation exhibit reduced sporulation in comparison to wild type diploids at both 24°C and 34°C. At 24°C spo11-1/spo11-1 diploids exhibit ca. 25% asci in comparison to wild type diploids which yield ca. 70% asci at this temperature. At 34°C spo11-1/spo11-1 diploids yield ca. 10% asci in comparison to congenic wild type diploids that exhibit ca. 60% sporulation.

We have characterized the effect of the spo11-1 mutation on mitotic and meiotic recombination at both 24°C and 34°C in order to determine the temperature dependence of the spo11-1 defect. To perform the analysis it was necessary to determine the effect of the spo11-1 mutation on ascosporal viability at 24°C. In a preliminary experiment we dissected 81 four-spored asci and 1 three-spored ascus obtained following sporulation of the _spo11-1/spo11-1_ hybrid at 24°C. None of the ascospores gave rise to viable colonies at 24°C. This result indicated the _spo11-1_ mutation results in drastically reduced ascospore viability.

In order to further assess the effect of the _spo11-1_ mutation on survival of cells exposed to sporulation medium, a sporulated suspension of the _spo11-1/spo11-1_ hybrid was diluted and plated on solid YPD medium. A total of 1026 plated cells plus asci were observed microscopically over a 20 hr period during incubation at 24°C on YPD plates. The results of this microscopic analysis are summarized in Table 1 below.

Table 1. Viability of _spo11-1/spo11-1_ after sporulation at 24°C.

Micro-Colonies	Dead Cells	Dead Asci	Total Observed
41	728	257	1026
(4.0%)	(71.0%)	(25.0%)	

Only 41/1026 surviving microcolonies were observed. This observation demonstrates that the _spo11-1_ mutation results in reduced viability of unsporulated as well as sporulated cells. To obtain a more precise estimate of ascosporal viability of _spo11-1/spo11-1_ segregants, a total of 855 microcolonies were characterized by light microscopic observation in order to distinguish between apparently diploid cell colonies and haploid cell colonies. Haploid cells can be recognized by their smaller size and haploid growth habit. A total of 4/855 presumptive haploid colonies were observed. Subsequent genetic analysis of putative haploids has supported the conclusion that they are indeed haploid or hyperhaploid. These observations and the knowledge that the sporulated suspension consisted of 25% asci allow us to calculate that ascospore survival of the _spo11-1/spo11-1_ haploid is ca. 0.02%.

Effect of the _spo11-1_ Mutation on Meiotic Recombination

We employed the CBX1 _spo11-1/spo11-1_ hybrid to determine the effect of the _spo11-1_ mutation on meiotic intergenic recombination for markers on chromosome VII following sporulation at 24°C. Owing to the fact that the _spo11-1_ mutation drastically reduces ascospore viability, it was not possible to observe its effect upon meiotic recombination by standard tetrad analysis. We performed a modified random spore analysis taking advantage of the fact that the diploids are can1/CAN1 heterozygotes sensitive to canavanine. Meiotic _can1_ segregants resistant to canavanine can be randomly selected out of sporulated populations by replica plating on arginineless plates containing 60 ug/ml of canavanine. 40 independent colonies of the _spo11-1/spo11-1_ hybrid were sporulated and replicated on canavanine medium. Single canavanine-resistant colonies were

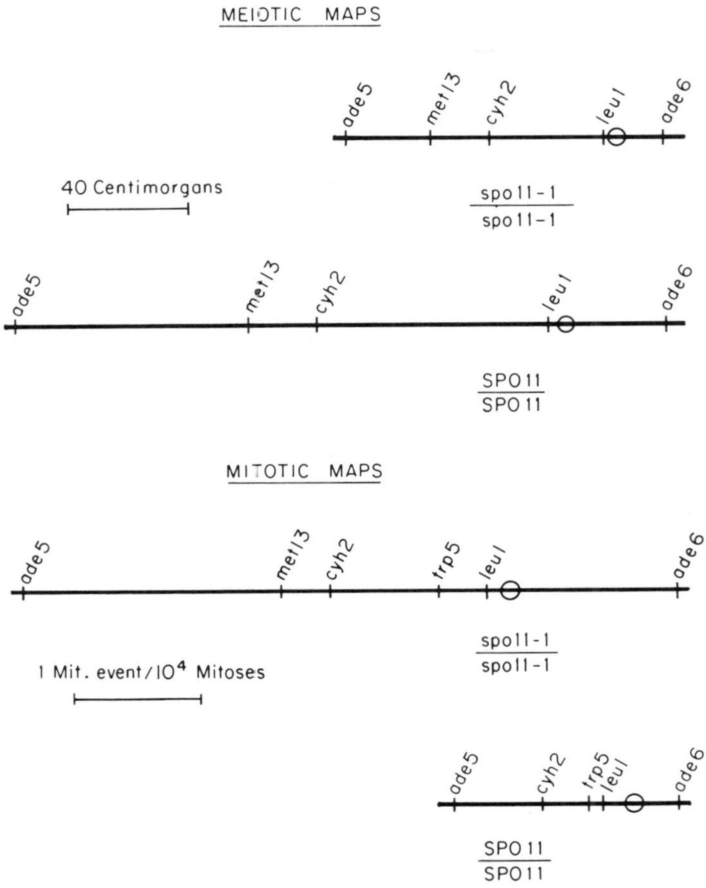

Figure 2. Meiotic and mitotic maps of chromosome VII intervals of the wild type SPO11/SPO11 and the spo11-1/spo11-1 mutant hybrids. Both maps of the wild type were determined in previous studies from our laboratory (8,9,10,14).

then picked up randomly, grown on YPD master plates, replicated to the complete and drop-out media and mated to 4 different adel tester strains. Two of these were a and α mating type tester strains and the two others were chromosome VII testers. Mating between the tester strains and spoll-1 meiotic segregants resulted in adenine prototrophic colonies that could be scored following replica plating to adenine-deficient medium. We analyzed 1160 colonies that mated with one of the two mating type testers. The sample included both monosomic chromosome VII segregants and disomics for chromosome VII. The analysis provided an estimate of the percent meiotic recombination among surviving ascospores of the spoll-1/spoll-1 hybrid for the intervals between the chromosome VII markers ade5-met13-cyh2-leul-ade6 and the chromosome V markers canl-hom3. The results are shown in Table 2.

Table 2. Single strand analysis of meiotic segregants of the spoll-1/spoll-1 hybrid (CBX1)[a]

Interval	+ +		+ −		− +		− −		Total	Corrected % Rec.
ade5-met13	R_1	223	P_2	139	P_1	367	R_2	94	823	28
met13-cyh2	R_1	117	P_2	429	P_1	162	R_2	61	769	19
cyh2-leul	R_1	80	P_2	199	P_1	370	R_2	120	769	38
leul-ade6	P_1	406	R_1	86	R_2	68	P_2	261	821	20
canl-hom3	P_1	1083	R_1	77					1160	9

a P_1 and P_2 are the parental categories; R_1 and R_2 are the recombinant ones. The percentages of recombination have been corrected for the presence of disomics for chromosome VII that increase the real number of chromosomes involved in the analysis. P_2 and R_2 for the interval between canl and hom3 on chromosome V are not given because canavanine sensitive strands are not rescued.

On the basis of the values of meiotic recombination reported in Table 2, we have constructed the meiotic map of the spoll-1/spoll-1 hybrid which is compared with that of the SPO11/SPO11 wild type in Figure 2. As one can see there is an overall reduction of ca. 50% of the map length between the ade5 and ade6 markers: the total length of the wild type map is 216 cM while in the mutant it is only 105 cM. The reduction of recombination is not homogeneous in the various intervals measured on chromosome VII. Recombination in the leul-ade6 and cyh2-leul intervals is 50% of the wild type level and 64% of the wild type level between ade5 and met13. The greatest reduction was observed in the interval between the can1 and hom3 markers on chromosome V in which the spo11-1 value is of the order of 14 fold lower than that of the wild type. The interval met13-cyh2 on chromosome VII exhibited no apparent reduction in spo11-1/spo11-1.

The gene to centromere distance for leu1 was also estimated but by a different approach, since it cannot be obtained from random spore analysis. We isolated a total of 104 microcolonies following sporulation and growth of the spo11-1/spo11-1 hybrid on YPD medium. In this sample, 98 microcolonies consisted of diploid cells i.e., they were non-maters, were able to sporulate poorly, and exhibited heterozygosity for one or more markers originally present in heterozygous or heteroallelic condition. Among these 98, two exhibited reciprocal recombination in the leu1 to centromere map interval, providing an estimate of the meiotic recombination frequency for this interval. Previous studies have demonstrated that map values obtained among diploids following extensive exposure to sporulation conditions approximate true meiotic map distances (4,5,16).

Effect of the spo11-1 Mutation on Mitotic Recombination

The spontaneous rates of mitotic intergenic recombination for markers on chromosome VII were measured by plating aliquots from five parallel cultures of the spo11-1/spo11-1 CBX1 hybrid and the SPO11/SPO11 CBX2 congenic wild type on synthetic complete medium and selective media.

The rate of heteroallelic (intragenic) recombination at trp5, located on chromosome VII, at 24°C was determined from the total number of prototrophs in each of the five parallel cultures as described in MATERIALS AND METHODS. The rate of heteroallelic recombination leading to prototrophy is the same in the spo11-1/spo11-1 hybrid and the wild type. The spo11-1 mutation does not confer a hyper-rec phenotype with respect to heteroallelic intragenic recombination events leading to prototrophy (Table 3). The spo11-1 mutation does, however, exhibit an enhancement of the rate with which heteroallelic recombination at trp5 is associated with intergenic recombination in the interval trp5-cyh2 leading to the production of red-white and white Trp^+ colonies. The spo11-1 mutation also confers hyper-rec activity in the case of intergenic recombination events unselected with respect to association with gene conversion, as discussed below.

The rates of mitotic events resulting in the production of canavanine resistant segregants at 24°C and of cycloheximide resistant segregants at 24°C and 34°C are reported in Tables 4 and 5, respectively. No substantial differences between spo11-1/spo11-1 and SPO11/SPO11 were observed. Since canavanine resistant and cycloheximide resistant segregants arise by both conversion and intergenic recombination in unknown proportions we proceeded to measure intergenic recombination on chromosome VII.

Intergenic recombination on chromosome VII was monitored by the production of white and red-white sectored colonies on synthetic complete medium at the two temperature of 24°C and 34°C. Since the hybrids are homozygous for the ade2-1 mutation, mitotic events that result in homozygosity for either ade5 or

Table 3. Spontaneous tryptophan prototrophic segregants recovered from cultures following growth of the spo11-1/spo11-1 and SPO11/SPO11 hybrids at 24°C.

Diploid	Culture Number	Prototrophic segregants, No/10^6 Red	White	Sectored	Total	Cells/culture x 10^{-9}
spo11-1 / spo11-1	1	16.9	65.2	0.0	82.2	1.56
	2	18.9	3.4	1.0	23.2	1.03
	3	6.7	2.2	0.0	8.9	1.19
	4	18.9	3.3	0.0	22.1	1.22
	5	22.5	111.0	0.0	133.5	1.18
	6	28.1	2.2	0.1	30.4	3.19
	7	16.3	1.1	0.2	17.5	5.95
	8	18.6	3.0	0.1	21.7	4.98
	9	21.2	4.9	0.0	26.0	3.39
	10	14.4	2.1	0.5	17.0	4.28
	Rate	1.00 x 10^{-6}	1.70 x 10^{-7}		1.27 x 10^{-6}	
SPO11 / SPO11	1	20.3	1.4	0.0	21.7	3.42
	2	16.2	4.2	0.0	28.3	2.51
	3	19.2	2.5	0.0	21.7	3.69
	4	111.9	1.5	0.2	113.6	7.83
	5	233.1	2.2	0.4	235.7	3.02
	Rate	0.93 x 10^{-6}	1.11 x 10^{-7}		0.99 x 10^{-6}	

Table 4. Spontaneous canavanine resistant segregants recovered from cultures following growth of spo11-1/spo11-1 and SPO11/SPO11 hybrids at 24°C.

Diploid	Culture Number	Resistant segregants, No/10^5 Red	White	Sectored	Total	Cells/culture x 10^{-9}
spo11-1 / spo11-1	1	35.8	0.3	1.5	37.6	1.56
	2	31.0	2.3	0.0	33.3	1.03
	3	39.3	0.0	0.0	39.3	1.19
	4	142.0	2.3	0.3	144.6	1.22
	5	54.2	8.8	0.0	63.1	1.18
	6	23.2	0.0	0.0	23.2	3.19
	7	42.2	0.1	0.0	42.3	5.95
	8	25.6	0.0	0.0	25.6	4.98
	9	26.0	0.3	0.1	26.5	3.39
	10	33.7	0.0	0.3	34.0	4.28
	Rate	1.90 x 10^{-5}	1.50 x 10^{-7}		1.97 x 10^{-5}	
SPO11 / SPO11	1	17.0	0.6	0.0	17.6	3.42
	2	38.2	14.9	0.0	53.1	2.51
	3	38.0	0.4	0.0	38.4	3.69
	4	29.4	0.2	0.0	29.6	7.83
	5	151.8	349.7	9.5	511.0	3.02
	Rate	1.76 x 10^{-5}	3.00 x 10^{-7}		2.99 x 10^{-5}	

Table 5. Spontaneous cycloheximide resistant segregants recovered from colonies of the spo11-1/spo11-1 and SPO11/SPO11 hybrids grown at 24°C and 34°C.

Diploid	Colony Number	Colonies grown at 24°C: Resistant segregants No. 10^5	Colonies grown at 24°C: Cells/colony $x10^{-7}$	Colonies grown at 34°C: Resistant segregants No. 10^5	Colonies grown at 34°C: Cells/colony $x10^{-7}$
spo11-1/spo11-1	1	12.3	2.6	24.1	3.9
	2	18.6	2.2	18.0	1.8
	3	13.5	2.5	18.2	2.4
	4	22.0	2.2	18.0	2.1
	5	10.9	2.6	47.5	1.7
	Rate	0.8×10^{-5}		1.1×10^{-5}	
SPO11/SPO11	1	6.2	1.2	7.0	1.8
	2	7.0	1.4	19.4	0.9
	3	29.6	1.6	19.8	1.9
	4	23.9	0.8	10.1	2.3
	5	10.7	1.6	9.5	2.8
	Rate	0.7×10^{-5}		0.6×10^{-5}	

ade6 result in the production of white and red-white sectored colonies. Red-white sectored colonies represent events that occurred in the cell divisions following plating. Their frequency thus provides a direct estimate of the rate of mitotic events resulting in ade5/ade5 and ade6/ade6 segregants (Table 6). Comparing the rates of the mutant and the wild type at 24°C we found a five fold higher rate for the mutant that is statistically significant at the 99% confidence level. At 34°C, however, there is only a small

Table 6. Mitotic recombination rates of spo11-1/spo11-1 and SPO11/SPO11 hybrids at 24°C and 34°C

Diploids	Temperature	Red-white sectored colonies on synthetic complete	Cyh^r	Trp$^+$	Can^r
spo11-1 / spo11-1 (CBX1)	24°C	5.2 ± 1.2x10^{-4}(19/36821)	0.8x10^{-5}	1.3x10^{-6}	2.0x10^{-5}
	34°C	5.3 ± 1.4x10^{-4}(14/26668)	1.1x10^{-5}		
SPO11 / SPO11 (CBX2)	24°C	1.4 ± 0.6x10^{-4}(05/34808)	0.7x10^{-5}	0.9x10^{-6}	2.9x10^{-5}
	34°C	4.3 ± 1.3x10^{-4}(11/25288)	0.6x10^{-5}		

					VII			V		XV	
ADE5	met13-c	cyh2^r	trp5-c	leu1-c	0	ade6	hom3	0	CAN1^s	0	ade2
ade5	MET13	CYH2^s	trp5-d	LEU1		ADE6	HOM3		can1^r		ade2

difference between the two rates of recombination that is in the range of the standard deviations and is not significant. It thus seems that the wild type gene product is temperature sensitive and that the spo11-1 mutation reduces the range in which the enzymatic activity is optimal. At 34°C, the SPO11/SPO11 hybrid is a phenocopy of the spo11-1/spo11-1 mutant diploid.

The genotypes of 41 red-white sectored spo11-1/spo11-1 colonies were determined to locate the exchanges resulting in homozygosity for the distal markers ade5 and ade6. Five subclones from the red and white portions of the sectored colonies were sporulated at 24°C and backcrossed to haploid chromosome VII testers. Their genotypes were inferred from the phenotypes of the back-cross hybrids. The results of this analysis are summarized in Figure 2, which illustrates the mitotic hyper-rec activity of spo11-1/spo11-1 with respect to intergenic events.

In order to demonstrate the involvement of the resolution of Holliday structures by DNA synthesis at the molecular level during mitotic recombination, we have begun a structural study of the yeast 2-micron DNA plasmid in the spo11-1/spo11-1 hybrid in which we expect the formation of dimers and multimers as a product of DNA replication through uncleaved Holliday structures. Preliminary results obtained by electron-microscopy of covalently closed DNA molecules in spo11-1/spo11-1 diploids did show the frequent presence of dimers and tetramers. Size determinations by gel-electrophoresis are currently in progress.

DISCUSSION

The data presented above provide strong evidence in favor of the conclusion that the product of the SPO11 gene is directly involved in endonucleolytic cleavage of Holliday structures in both mitosis and meiosis. This fact identifies a common step in the mitotic and meiotic pathways of recombination. The mitotic hyper-rec and meiotic hypo-rec phenotype with respect to intergenic recombination of the spo11-1/spo11-1 hybrid lends support to the view that recombination between homologs of mitotic cells occurs prereplicationally while meiotic recombination between homologs occurs postreplicationally. Given these results, genetically marked haploid and diploid spo11-1 strains can now be employed to test a corollary of the G1 model of mitotic recombination, viz, that sister chromatid exchange, like meiotic homologous chromosomal recombination, is dependent upon endonucleolytic cleavage of Holliday structures. Several yeast systems amenable to this type of analysis have already been developed in other laboratories (cf. 7 for review).

The spo11-1 mutation also provides a basic genetic approach to determine whether the overall sequence of molecular events postulated to occur by molecular models of genetic recombination is correct (6). The data discussed above suggest that the SPO11 gene encodes a late function in the recombination sequence. Previous studies have uncovered both hyper-rec and hypo-rec mutations that may affect early stages of recombination, e.g., initiation of single-strand transfer (7). The phenotypes of doubly mutant strains incorporating spo11-1 and those previously isolated rec mutations may challenge models of genetic recombination. Reciprocal temperature shift studies (12) employing ts-rec mutants may also provide information with respect to the order of function in recombination of REC gene products.

LITERATURE CITED

1. Drake, J. 1970. The molecular basis of mutation. (Holden Day, San Francisco).

2. Esposito, M. S. and Esposito, R. E. 1969. The genetic control of sporulation in Saccharomyces. I. The isolation of temperature-sensitive sporulation-deficient mutants. Genetics 61:79-89.

3. Esposito, R. E., Frink, N., Bernstein, P., and Esposito, M. S. 1972. The genetic control of sporulation in Saccharomyces. II. Dominance and complementation of mutants of meiosis and spore formation. Mol. Gen. Genet. 114:214-248.

4. Esposito, R. E. and Esposito, M. S. 1974. Genetic recombination and commitment to meiosis in Saccharomyces cerevisiae. Proc. Natl. Sci. USA 71:3172-3176.

5. Esposito, R. E., Plotkin, D. and Esposito, M. S. 1974. The relationship between genetic recombination and commitment to chromosomal segregation at meiosis. pp. 277-285, In: R. Grell (ed.), Molecular Mechanisms of Recombination. Plenum, New York.

6. Esposito, M.S. 1978. Evidence that spontaneous mitotic recombination occurs at the two-strand stage. Proc. Natl. Acad. Sci. USA 75:4436-4440.

7. Esposito, M.S. and Wagstaff, J. E. 1981. Mechanisms of mitotic recombination. pp. 341-370, In: Strathern, J. N., Jones, E. W., and Broach, J. R. (eds.), Molecular Biology of the Yeast Saccharomyces: Life Cycle and Inheritance. Cold Spring Harbor Laboratory, Cold Spring Harbor, New York.

8. Golin, J. E. and Esposito, M. S. 1977. Evidence for joint genic control of spontaneous mutation and genetic recombination during mitosis in Saccharomyces. Mol. Gen. Genet. 150:127-135.

9. Golin, J. E. 1979. The properties of spontaneous mitotic recombination in Saccharomyces cerevisiae. Doctoral thesis, University of Chicago.

10. Golin, J. E. and Esposito, M. S. 1981. Mitotic recombination: mismatch correction and replicational resolution of Holliday structures formed at the two strand stage in Saccharomyces. Mol. Gen. Genet. 183:252-263.

11. Holliday, R. 1964. A mechanism for gene conversion in fungi. Genet. Res. 5:282-304.

12. Jarvik, J. and Botstein, D. 1973. A genetic method for determining the order of events in a biological pathway. Proc. Natl. Acad. Sci. USA 70:2046-2050.

13. Lea, D. E. and Coulson, C. A. 1948. The distribution of numbers of mutants in bacteria populations. J. Genet. 49:264-284.

14. Malone, R. E., Golin, J. E. and Esposito, M. S. 1980. Mitotic *versus* meiotic recombination in *Saccharomyces cerevisiae*. Current Genet. 1:241-248.

15. Meselson, M. S. and Radding, C. M. 1975. A general model for genetic recombination. Proc. Natl. Acad. Sci. USA 72:358-361.

16. Plotkin, D. 1978. Commitment to meiotic recombination: a temporal analysis. Doctoral thesis, University of Chicago.

20. THE MOLECULAR GENETICS OF NON-TANDEM DUPLICATIONS AT ADE8 IN YEAST*

S. Fogel, T. Choi, D. Kilgore, K. Lusnak and M. Williamson

Department of Genetics
University of California, Berkeley, California 94720

Rec. Adv. Yeast Mol. Biol. (1982) 1:269-288

The ADE8 locus of chromosome IV has been cloned. Various subclones, principally those with fragment lengths of 2.5 and 4.0 kb, confer ADE8 function on recipient strains carrying the unique allele ade8-18. A visual screen for detecting integrations of the autonomously replicating vector is described along with diagnostic genetic tests that identify the genomic integration sites. Most integrants generate non-tandem duplications at the ade8-18 site, though some also occur at trp1. None were found at ura3. The frequency of integration via homologous recombination into a genomic site is proportional to the physical length of the corresponding DNA segment, carried in the YRp17 vector. Similarly, overall plasmid excision rates are proportional to the total length of the integrated segment and the distribution of events for a plasmid with a given DNA insert is determined by the position of the mutant site within the genetic fine structure map. Meiotic gene conversion, intrachromosomal conversion, and postmeiotic segregation were studied in several hybrids containing two, three or four ADE8 sequences within conventional chromosomes or non-tandem duplications that are either isosequential or heterosequential.

****

Gene conversion is typically defined as a non-reciprocal transfer of genetic information from one parental DNA segment to the corresponding DNA segment of the homologous chromosome (10-17,21,25). In yeast, it is generally agreed that conversion events can occur both in meiosis and mitosis. The meiotic process involves specific synaptic pairings at the zygotene or pachytene stage of prophase I, along with the formation of heteroduplex DNA. These heteroduplexes may isomerize, as suggested by the Aviemore model (24,28) to yield associated exchanges that recombine the outside markers. Alternatively, isomerization might not occur and the somewhat more probable parental marker array is maintained. Thus, because gene conversion and genetic recombination are mechanistically linked, conversion has been viewed as a signal of the basic recombination event (15,16,17).

* Supported by an NIH research grant (GM17317) to S.F. and by the Berkeley Agricultural Experiment Station of the University of California.

More recently, the definition and significance of gene conversion has been enlarged and extended--particularly by perspectives and insights stemming from the experimental potentialities inherent in recombinant DNA technology coupled to yeast transformation (2). Several major studies impinge on current molecular recombination schemes and these are based on a combined genetical-recombinant DNA approach. The importance of these works is summarized here.

Introduced into sites on non-homologous chromosomes, by means of recombinant DNA procedures, various his3 alleles were shown to undergo heteroallelic recombination and restoration of prototrophy (31). Clearly, homologous sequences could experience informational transfer or gene conversion regardless of their position in the genome. Of course, the frequency of such heterologous events is several orders of magnitude lower than controls involving heteroalleles occupying corresponding sites on homologous chromosomes. Similar findings were reported by others (1).

Related to this extension of gene conversion's definition are the reports concerning sister chromatid exchange (23,34). Such studies were facilitated by the coupling of critical genetic analyses with recombinant DNA procedures that permitted differential marking of highly iterated sequences. To account for the events observed, as well as the more recently described instance of gene amplification at the cupl locus[1] in yeast, it is assumed that misalignment between homologous iterated segments is followed by heteroduplex DNA formation, isomerization and an appropriate scission (4). Taken collectively, these events lead to a reciprocal exchange that generates quantitatively dissimilar products i.e., unequal crossing over. This event sequence may occur in mitotic cells and involve sister chromatids. Alternatively, it could occur in meiotic cells where homologous chromosomes might be engaged interactively during synapsis. With regard to DNA replication, conversion is viewed as prereplicative in mitosis (7,8,9,18) and as postreplicative in meiosis (2-4).

A final approach entails the analysis of non-tandem duplications. These originate as a consequence of integrational events, i.e. homologous recombination events which lead to the incorporation of a plasmid bearing a cloned DNA sequence into a site within the corresponding genomic region. Several studies (19,20,22,23) may be taken as representative. Such genomic reorganizations allow for the possibility of intrachromosomal gene conversion with and without associated crossing over when the interactive sequences are separated by a physical distance equal to the number of base pairs contained in the cloning vehicle.

The present study aims to clone, analyze and sequence the structural and regulatory sequences of the ARG4 and ADE8 loci. Of special interest are the alleles arg4-16, ade8-14, ade8-18 and others for which we have accumulated an extensive, highly detailed background of biochemical and recombinational data (5,6,8,15,16,17). Our program seeks to address, at the DNA sequence level, the generalizations based on genetic analysis that relate to parity, fidelity, polarity, frequencies of associated outside marker exchange, the position of associated exchanges, the conversional behavior of insertions,/deletions and perhaps most importantly the marker effects of specific mutants (5,17). Marker effects may reflect DNA structural- organizational features relating to

1. Fogel and Welch, in press, PNAS, September 1982.

heteroduplex DNA correction mechanisms that sense, detect and repair mismatched base pairs as in single base substitution or frameshift mutants, or extensive non-homologies as are found at the MAT locus, or non-pairings as in deletions or insertions several kilobases in extent (16).

The production of non-tandem duplications and some of their salient features have been reported (19,20,22,23). The present approach seeks to exploit the resolving power of recombinant DNA-transformation systems with a view towards advancing our understanding of recombination from the self-contained black box level of genetic fine structure analysis to the physical-molecular level. Current state of the art techniques make it practical to isolate and analyze specific genes or gene fragments (2). Moreover, isolated genes, carried in plasmid vectors with known but varied replicative and integrative behavior, may be subjected to *in vitro* or *in vivo* mutagenesis and then returned to the host cell by DNA transformation for studying the effects of such alterations on gene organization, expression and regulation.

Cloning the ADE8 locus

A recipient conventional diploid strain of the following genotype was synthesized after several rounds of mating, sporulation, ascus dissection, scoring ascosporal colonies and selection. Except for MAT, each of the following markers was homozygous ade2-1 trp1, ura3 ade8-18 arg4-16, cup1. Comparable strains carrying the wild type allele at ade2 were also prepared. When these studies were initiated by the senior author, in the laboratory of R.W. Davis at Stanford University, it was presumed that transformation studies in diploids immediately amenable to meiotic characterization would confer several advantages over transformation in haploids. Accordingly, these strains were monitored and selected for abundant sporulation with a high proportion of 4-spored tetrads, high ascosporal survival equal to or better than 95%, and ease of spheroplasting and transformability. From a vantage point of hindsight, it is apparent that congenic haploids appropriately marked to facilitate subsequent analytical studies would also be useful.

A random library of DNA fragments was prepared from the DNA isolated from a standard yeast strain--S288C. Conventional restriction enzyme digestion and sticky end ligation were involved. Yeast DNA was partially digested with the 6 base pair recognition enzyme BamHI and the fragments were ligated into the unique BamHI site of the YRp17 vector. This plasmid was constructed and kindly made available in purified form by W.M. McDonnell and R.W. Davis. The plasmid YRp17 shown in Fig. 1 is derived from YRp12. It contains a unique BamHI site located within the tetracycline resistance gene. This site is also present in YRp7, a TRP1/pBR322 vector (33). YRp17 also harbors the yeast URA3 locus, ARS1--an autonomous replication sequence, and an adjacent, unique EcoRl site (32). Pools containing about 25,000 independent plasmids, i.e. each plasmid bearing a different inserted DNA fragment, were obtained by transforming *E. coli* to ampicillin resistance and tetracycline sensitivity. The pool's informational content was assessed by transforming the above described recipient yeast strain and selecting for complementation with respect to the TRP1 and URA3 functions. Among the latter, colonies displaying complementation relative to ade8-18 could be readily identified in the recipient ADE2 strain, since their replica-transfer prints would grow confluently on synthetic complete media deficient for adenine. Single yeast colonies containing presumptive ADE8 fragments were purified, grown, lysed and used to infect *E. coli*

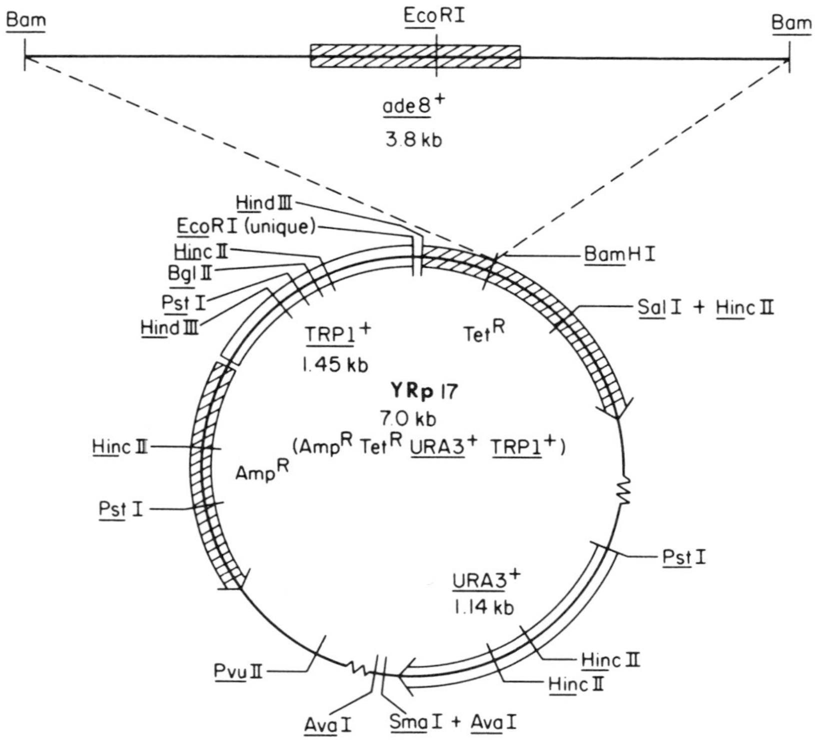

Fig. 1. The YRp17 cloning vehicle (constructed by W.M.McDonnell and R.W. Davis, pers. comm.) is a 7.0 kb plasmid carrying pBR322 sequences and the wild type sequences for the yeast genes TRP1 and URA3 in addition to an autonomous replication sequence, ars. A 4.0 kb BamH1 insert carrying ADE8 function is shown.

cultures for the isolation of purified plasmid preparations. A YRpl7 plasmid containing a 4kb wild type DNA fragment inserted into the unique BamHI site comprises the initial hybrid DNA species on which the present study is based.

The 4kb fragment was subjected to restriction enzyme digestion, electrophoretic analysis, and subcloning. The restriction map is given as Fig. 2. Several different subclones were isolated and two will be discussed here. The first is about 2.5 kb in length. It was generated by cleaving the 4kb BamHI fragment with the enzyme EcoRl--a procedure that yields two fragments; i.e., 2500bp and 1500bp. When these were subcloned and tested, only the larger fragment provided ADE8 function. Like the parental 4.0 kb segment, it integrates preferentially into the chromosome IV region marked by ade8-18. A smaller, secondary subclone of 1750 bp (from SalI to BamHI shown in Fig. 2) fails to provide ADE8 function in URA3, TRPl transformants. But, the subclones of the 2500bp fragment recombine intragenically within the ade8-18 sequence, at an extremely low frequency, to yield functional ADE8 prototrophs. Thus, the coding region and probably the regulatory sequence of ADE8 are available to us within a combined physical length corresponding to about 2500 bp. The ade8-18 site falls within the 1750 bp fragment. In contrast, the mutant, ade8-10, is located within a cluster adjacent to a terminus of the ADE8 genetic fine structure map (5). Recombinational integration resulting in prototrophy occurs between this ochre nonsense mutant and the smaller 750 bp segment. Accordingly, the 750 bp fragment includes ade8-10 but not ade8-18, since it recombines only with the former heteroallele, while the larger 1750 fragment must subtend and include the non- revertible, high PMS ade8-18 site situated in the approximate center of the fine structure map; see Fig. 3. A DNA sequence analysis of the 1750 bp fragment containing the mutant ade8-18 compared to wild type will almost certainly provide a molecular basis that accounts for the unusual attributes of this particular allele, i.e., its high frequency of postmeiotic segregation when heterozygous, its marker effects in recombination, and its nonrevertible character (6).

Ultimately, the ade8-18 sequence may be compared to ade8-14, an adjacent, effectively inseparable ochre mutant known to revert and which nonetheless exhibits equivalent PMS patterns when heterozygous. However, ade8-14 does not exhibit the pronounced marker effect of stimulating recombination rates with alleles to the left or right as is the case with ade8-18 (5). Clearly, DNA sequence data can be expected to illuminate our insight concerning recombination at the intragenic level. Currently, we can orient the cloned fragment relative to Esposito's fine structure map shown in Fig. 3. The ade8-10 site falls between the restriction sites EcoRl and SalI and the ade8-18 site is localized on the 1750 bp segment bounded by SalI and BamHI. Determined also is the orientation of the 2.5kb subclone. The ade8-10 allele located at the low end of the conversion polarity gradient is near the EcoRI site and the ade8-18 allele distinguished by high conversion and high PMS rates (17) falls on the opposite side of the SalI site. The availability of two KpnI sites and single sites for HpaI, XhoI, and BglII provide additional subcloning opportunities.

A similar combined genetic-molecular approach is projected for the specific mutant allele arg4-16. This mutant also displays an exceptionally high meiotic conversion rate and a high proportion of these events, 30%, are postmeiotic segregations (35). The isolation and cloning of the ARG4 locus was definitively established earlier by others (3), and we have obtained an

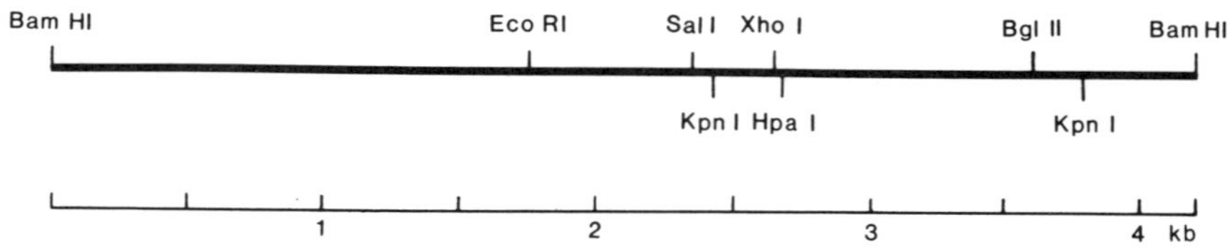

Fig. 2. Restriction map of the 4 kb fragment.

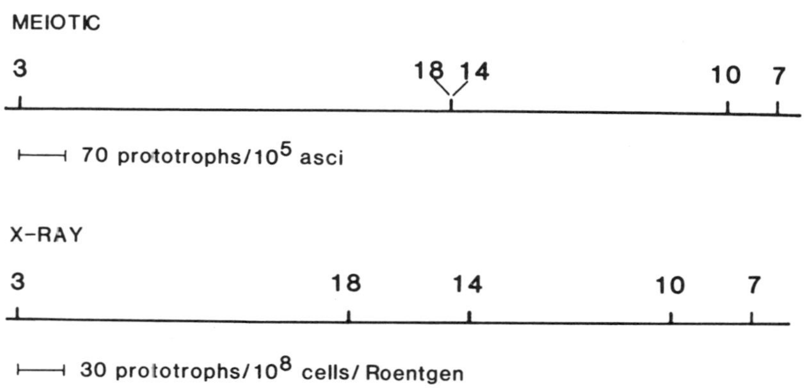

Fig. 3. Genetic fine structure map of Adenine8, after Esposito (1967).

independent clone. Experiments with the *ARG4* clone will be discussed elsewhere.

ADE8 integration--a visual screen

When the YRp17 plasmids carrying either the 2.5 or 4.0 kb cloned fragments are employed to transform spheroplasts of a normal *MAT*a/*MAT*α diploid strain homozygous for the markers *ade2*, *ade8-18*, *ura3* and *trp1*, selection for positive *URA3* and *TRP1* function is achieved by plating the protoplast-plasmid mixture in soft agar on normal solidified media deficient for tryptophane and uracil. Such transformants display behavior expected of cells containing an autonomously replicating plasmid. When the selective pressure is relaxed, by growing transformed cells on a rich medium such as YEPD, a large proportion of the daughter cells, and their subsequent progeny, lack the plasmid and consequently display a *trp1 ura3* phenotype. These observations are fully consistent with those made earlier (32,33).

Transformed *ade2*, *ade8-18* cells, i.e., containing the autonomously replicating plasmid described above, appear white rather than red on YEPD and other media. Since *ADE8* function is undoubtedly provided by the plasmid, as indicated by transformation of *ade8-18 ADE2* strains to prototrophs, it is appropriate to inquire concerning the underlying causes. Factors contributing to this altered phenotypic expression are as follows:

1. Cells that have lost the plasmid, or failed to acquire it, are *ade2 ade8-18* genotypically. Such cells are devoid of red pigment since the metabolic block imposed by *ade8-18* occurs prior to the *ade2* block in the *de novo* biosynthetic pathway.

2. White cells typically outgrow red cells.

3. Transformed cells may excrete uracil and tryptophane and these nutrilites are utilized for growth by *ade2*, *ade8-18* white cells.

Transformed cells are, at best, faint pink on trp^-ura^- media and moderately pink on *trp ura ade* media. Such white or pink patches often display numerous clearly defined sectors which are dark red by comparison to the adjacent pink/white background and exhibit a pigmentation level characteristic of *ade2 ADE8* cells. Subsequent tetrad analysis indicates that each red sector represents a single, independent integration event. Accordingly, the white-red or red-white shift, here designated as the *Roman effect*, in honor of Prof. H. Roman who first exploited the red-white system in yeast genetics (29,30), serves as a visual screen for the detection and ready isolation of homologous recombination events in which a plasmid is incorporated into the recipient's genome. Furthermore, the red-white Roman effect forms the basis for quantifying intrachromosomal gene conversion rates as well as plasmid excision rates.

The system's clarity and inherent objectivity make it unusually adaptable as a novel means for identifying and characterizing specialized DNA sequence of yeast, or other organisms, that modulate, regulate or control general as well as site specific recombination. Even greater resolving power may be introduced by coupling our systems to other recombinant DNA constructions that are highly selective. In this manner, only excision products would survive.

Among these, the proportions of red and white colonies could be determined with high speed counting devices. Accordingly, integration and excision might be amenable to a highly detailed molecular and genetic characterization.

Thirty two (32) independent red sectors were isolated and sporulated. Ascus dissection and tetrad analyses were performed. Twenty-eight red-sectored clones were heterozygous for a single integration event--ADE8 cosegregated with URA3, TRP1 in the 4+:4 ratio expected for simple Mendelian behavior. Three integrants were apparently homozygous. These might represent two simultaneous integration events, mitotic co-conversions, or mitotic recombination between the centromere and the site of integration. Finally, a single sector was simply heterozygous for ADE8/ade8-18. Here, we suppose that an intrachromosomal excision event resulted in the loss of the segment containing URA3, TRP1 ade8-18. Alternatively, it might more likely represent an informational transfer from the plasmid to the chromosomal site. Transfers in the opposite direction may occur also. This isolate, Bam5, serves as our standard conventional heterozygote. It displays a basic conversion frequency of 10% among 621 fully analyzed tetrads. Of the 63 aberrant segregations, 40 or 63.5% displayed postmeiotic segregations, detected as sectored ascosporal colonies by means of our standard plate-dissection-replica-plating procedure.

Site of integration

Red, URA3, TRP1, ADE8 segregants from each of twenty-one integrants involving the cloned 4kb fragment, were crossed to a single tester strain displayed in Table 1. Single zygotes were isolated, grown, sporulated, and subjected to dissections and tetrad analysis. Table 1 presents the diagnostic screen for identifying the site of integration. Of the twenty-one integrant hybrids analyzed in this manner, 18 displayed segregations consistent with integration at or near the ade8-18 site on chromosome IV. In addition, three integrations occurred close to the centromere at or near trp1. To a first approximation, it appears as though the frequency of integration at a given site is proportional to the physical length of the homologous segment carried in the plasmid. Thus, the TRP1 segment and the Bam H1 fragment containing ADE8 within the YRp17 vector are respectively 1.43 and 4.0 kb; or a ratio of about 3. Also, it may be noted that when the ADE8 segment is reduced to about 2.5 kb, a significantly higher proportion of the integration events (about 50%) occur at trp1 compared to ade8-18. These findings suggest that the probability of localized effective mitotic pairing, as might be required for integration events involving mitotic recombination between homologous sequences, increases as a simple function of the physical length of the interactive sequences. Finally, as regards homologous sequence integration events, the three apparent homozygotes, about 10%, suggest that integration itself may be recombinogenic. Experiments specifically designed to test this notion in a critical manner are warranted.

Rates of Mitotic Intrachromosomal Excision

Events leading to the excision of YRp17 plasmid sequences containing ADE8 inserts may be studied quantitatively with respect to the actual recombination rate at which the integrated plasmid sequence is removed from the genome. Mitotic excision rates are readily estimated from protocols based on the method of the median (7,9). In our studies, eleven independent haploid colonies, each containing about 1.5×10^7 cells, were assayed via dilution-

TABLE 1. Diagnostic segregation and genetic screen for identifying site of integration

Integrant	trpl	TRP4	ade8-18 URA3 TRPl ADE8	RNA3	ura3	ade2
X						
Tester	TRPl	trp4	ade8-14	rna3	URA3	ade2

Site of Integration	Tetrad segregation on synthetic complete medium deficient for: tryptophane	adenine	uracil
ura3	$2^+:2^-$ $1^+:3^-$ $0^+:4^-$	18 18 + + or 14 14 + + or 18 14 + +	$4^+:0^-$
trpl	$2^+:2^-$	14 14 + + & 18 18 + + or 18 14 + +	$2^+:2^-$ $3^+:1^-$ $4^+:0^-$
ade8	$2^+:2^-$ $1^+:3^-$ $0^+:4^-$	14 14 + +	$2^+:2^-$ $3^+:1^-$ $4^+:0^-$

plating on non-selective media and subsequent replica-transfers to diagnostic media. Each colony arose from a single cell isolated by micromanipulation. Incubations were at 30° for three days. Three quantities are ascertained: a) Numbers of cells in each of the eleven parental colonies, b) Proportions of cells within each colony that yield trp ura colonies and c) Proportion of the auxotrophic colonies that are red (ADE8) or white (ade8-18). White colonies are classified as ade8-18 only on the basis of appropriate complementation tests. Such estimates allow us to compute a recombination rate for excision in terms of a probability per cell division.

Excision rate experiments were conducted wth non-tandem duplication haploid strains derived from the meiotic analyses of integrants containing ADE8 fragment of 4.0, 2.5 and 1.7 kb in length. The integrations occurred at ade8-18, ade8-10 or trp1.

Taken collectively, the data obtained are consistent with the following simple model: We assume that the mitotic excision rate is determined primarily by the sequence homology length between the recipient and inserted sequences in a given non-tandem duplication. Exchanges leading to excision are assumed to occur randomly along the homologous segment. Accordingly, with a particular integration array e.g., ADE8-URA3-TRP1-ade8-x or ade8-x-URA3-TRP1-ADE8), the proportion of ade8-x-ura3-trp1 to ADE8-ura3-trp1 excision products would be determined by the position of the ade8-x mutant site within the region of homology.

When the recipient genome contained the mutant allele ade8-18, the rate leading to white (ade8-18) excision products is about four times the rate to red, i.e. 7.3×10^{-5}/division vs 1.9×10^{-5}/division. A representative set of data is displayed as Table 2. However, when the recipient genome contained the mutant ade8-10 in place of ade8-18, the corresponding rates to white and red excision products were essentially equal--i.e. 3×10^{-5}/division vs 2.5×10^{-5}/division. We may note that ade8-10 is located within the BamHI-SalI section. Thus, the ade8-10 site divides the 4 kb insert nearly equally, whereas the ade8-18 mutant divides the same fragment disproportionately.

Given an integration event of the type ....ade8-URA3-TRP1-ADE8....., a crossover at a or b would generate red or white excision products respectively as shown in Fig. 4. The ratio of red to white excision products is approximately equal to the ratio of the sequence homology lengths in region b compared to region a. Illustrative data for the 4 kb fragment integrated into recipients carrying ade8-18 and ade8-10 are given in Tables 2 and 3.

In addition to varying the position of the mutant site achieved by utilizing different alleles, we may also vary the extent of homology by choosing appropriate subclones of the original 4 kb BamHI fragment. As given in Fig. 2, an EcoR1 site allows us to isolate a YRp17 plasmid containing a 2.5kb insert that confers ADE8 function. Non-tandem duplications derived from this vector were studied with respect to spontaneous mitotic excision. Integrations into haploid recipient strains carrying ade8-18 and ade8-10 were analyzed, and the data is presented in Table 4. In this instance, the ade8-18 site divides the 2.5 kb insert into nearly equal segments. Thus, our model predicts that the ratio of (rate to reds)/(rate to whites) would be unity. The overall total excision rate for this smaller fragment would be little more than half the rate obtained with the 4 kb fragment. Also, reductions in rate

TABLE 2. Plasmid Excision Rate: 4kb Insert Integrated at *ade8-18*

Colony	Units/ Colony	Units Plated	trp^-, ura^-		Rate (excisions/cell/div)	
			White	Red	White	Red
1	1.43×10^{7}	2808	2	2	4.3×10^{-5}	4.3×10^{-5}
2	2.46×10^{7}	2357	15	1	3.8×10^{-4}	2.5×10^{-5}
3	1.44×10^{7}	1870	2	0	6.7×10^{-5}	—
4	1.40×10^{7}	1913	0	0	—	—
5	1.14×10^{7}	2341	3	4	8.0×10^{-5}	1.1×10^{-4}
6	1.41×10^{7}	2176	2	0	5.6×10^{-5}	—
7	1.12×10^{7}	2153	4	0	1.2×10^{-4}	—
8	1.40×10^{7}	2626	3	0	6.7×10^{-5}	—
9	1.51×10^{7}	2084	5	0	1.5×10^{-4}	—
10	1.49×10^{7}	2236	50	0	1.3×10^{-3}	—
11	1.38×10^{7}	2528	3	1	7.3×10^{-5}	2.4×10^{-5}
Total	1.62×10^{8}	25092	89	8		

Median Rate to *ade8-18*,*trp*$^-$,*ura*$^-$ from colony #11: 7.3×10^{-5} excisions/cell/division

Mean Rate to *ADE8*$^+$,*trp*$^-$,*ura*$^-$ from total data: 1.9×10^{-5} excisions/cell/division

Calculated from Rate= ((.4343)x(freq. *trp*$^-$,*ura*$^-$))/$\log_{10}(N_{final\ cells})$

TABLE 3. Plasmid Excision Rates

Size of Insert	Integration Site	Rate* to mutant	Rate* to wild
4 kb	*ade8-18*	7.3×10^{-5}	1.9×10^{-5}
4 kb	*ade8-10*	3.0×10^{-5}	2.5×10^{-5}
4 kb	*trp1-1*	7.6×10^{-5}	8.1×10^{-5}
2.5 kb	*ade8-18*	4.1×10^{-5}	3.4×10^{-6}

*Rate = excisions/cell/division

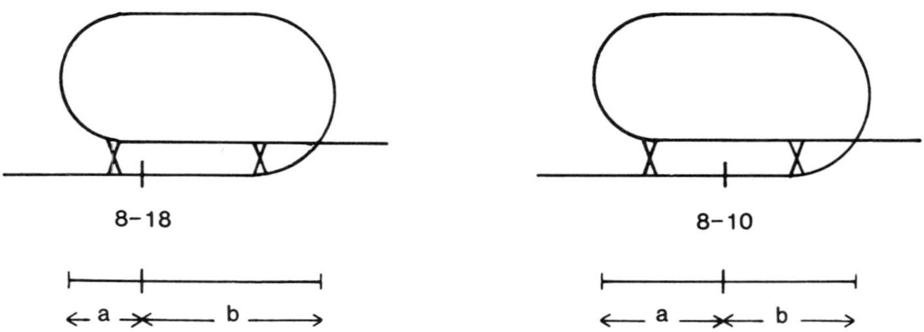

Fig. 4. Intrachromosomal excision events relative to mutant sites in *ADE8*.

TABLE 4. Gene Conversion and PMS in Hybrids With 2,3, and 4 Adenine8 Sequences

Sequence Array	Strain	Total Tetrads	$4^+:4^-$	$6^+:2^-$	$2^+:6^-$	$5^+:3^-$	$3^+:5^-$	$7^+:1^-$ & ab. $6^+:2^-$	%BCF*	%PMS/BCF
+ / −	Bam 5	622	560	15	7	21	18	1	10.0	64
− + / −	Bam 9	515	493	8	3	7	4	0	4.3	48
− − / +	DK101	582	547	6	9	5	15	0	6.4	57
− + / − −	TC140	554	521	5	15	5	8	0	6.0	39
+ + / −	Bam 32	640	583	33	0	24	0	2	9.3	44
+ + / − −	TC136	820	692	110	0	18	0	0	15.6	14
			$8^+:0^-$	$6^+:2^-$	$7^+:1^-$					
− + / − +	TC148	174	168	2	4				3.4	67

*%BCF, the Basic Conversion Frequency, is defined as $((6^+:2^- + 2^+:6^- + 5^+:3^- + 3^+:5^-)/\text{Total Tetrads}) \times 100\%$

would arise at the expense of ade8 excision products, since the smaller fragment was generated by removing 1.5 kb of DNA from the b region of the original 4kb segment. To date, all of our data on mitotic excision in non-tandem duplications is consistent with the simple model presented above.

Finally, we may inquire about mitotic excision s as a function of where the integration site is located within the genome. Three integration events were detected in the vicinity of trpl, a centromere-proximal marker whose extent of sequence homology corresponds to a 1.4 kb TRP1 fragment carried in the YRp17 vector. On the meiotic map (26,27) trpl is located about 0.5 cM from the centromere while ade8 is removed about 150cM distally on the same arm of chromosome IV. A single integrant at trpl was analyzed in detail. The non-tandem duplication has the structure cen- trpl-ADE8-URA3-TRP1.......ade8-18... ..ade2. The total excision rate approximates that observed for excision of the same plasmid when it is incorporated into the genome at ade8-18. Also, the TRP1 ura3 ade8-18 excisions were equal to the trpl ura3 ade8-18 excisions. Thus, the rates of spontaneous mitotic excision for plasmid sequences introduced into the chromosome by homologous recombination are not markedly affected by proximity to a genomic centromere. In summary, it is clear that non-tandem duplications that arise by the integration of cloning vehicles into the genome provide us with a sensitive device for estimating several important parameters of recombination. Furthermore, as presently constituted or coupled to selective DNA construction, the system could serve to identify and isolate DNA sequences that regulate mitotic recombination. As yet, we have not varied the distance between the duplicated sequences, though this is readily achieved by recombinant DNA *in vitro* procedures.

Meiotic gene conversion and postmeiotic segregation in non-tandem duplications.

Most of the independent integration events (28/32) were heterozygous for the plasmid markers, i.e. URA3, TRP1 and ADE8 (see Fig. 1). 515 complete four-spored tetrads from an integrant at ade8-18 were analyzed. This culture is listed as Bam9. In all, twenty-two conversion-like events were detected via the plate dissection-replica transfer method. This reflects a basic conversion frequency of 4.3%--a value equal to about half that of the standard isogenic heterozygote Bam5, where 622 analyzed tetrads yielded a basic conversion frequency of 10%. However, the % PMS/BCF values, 48% and 64%, for Bam9 and Bam5 respectively, are essentially equivalent. Thus, we presume that the presence of the integrated plasmid, yielding a non-tandem duplication, reduces the frequency of heterozygous pairings. Pairings are of two types i.e., +/- or-/- and if these occur with equal liklihood and the repair parameter (50%) is uncharged, our predicted value for %BCF and % PMS/BCF would concur with those observed (see Table 4). Clearly a, a novel prediction is suggested by these findings. If the non-tandem duplications were iterated as +----- we would predict a reduction in % BCF proportional to the number of iterated ade8-18 alleles, but with no change in the % PMS/BCF ratio. Recombinant DNA constructions of this sort have been reported (23).

The availability of non-tandem duplication, such as ADE8-URA3-TRP1-ade8-18 as well as the corresponding homosequential forms ADE8-URA3- TRP1-ADE8 and ade8-18-URA3-TRP1-ade8-18 obtained via intrachromosomal information transfer, allow us to generate fifteen unique hybrids that carry two, three or four ade8 alleles per diploid. Among these, ten contain dissimilar homologous chromosomes. Data pertaining to five situations is given in Table 4. We examine first DK101, a diploid with the following genotype: ade8-18 URA3 TRP1 ade8-18/ADE8.

The meiotic behavior of this hybrid, containing two alleles and one allele on respective homologues, may be contrasted with Bam5. This isogenic control yielded data that is statistically homogeneous with our previous analysis of 15,480 tetrads (16,17). Both samples display a slight but significant disparity relative to the strand on which intragenic recombination is initiated. This may be expressed by the dissymmetry coefficient of $(6^+:2^- + 5^+:3^-)/(2^+:6^- + 3^+:5^-)$ which is 1.36 in the 15,480 tetrad sample and 1.5 in Bam5. In DK101 the dissymmetry coefficient falls to 0.46 i.e., 11/24. How can these findings be rationalized in terms of a simple testable molecular model?

We may presume that the dissymmetry coefficients characterizing the control data reflect a higher liklihood of conversional initiations involving the strand bearing the wild type allele as given in the Aviemore model (24). We may note that in the total control data, there are 396 and 312 PMS events of the $5^+:3^-$ and $3^+:5^-$ varieties respectively. Taken at face value, these data indicate that the disparity arises prior to the occurrence of correctional repair on the heteroduplex. It is only subsequent to this repair that gene conversion or restoration are established. Given the above, we emphasize that one strand in DK101 bears two ade8-18 sequences while the homologue bears only a single ADE8 sequence. Thus, the observed dissymmetry coefficients can be accounted for as follows: Since DK101 contains two mutant alleles on one strand and a wild type allele on the other, we may assign to these initiation likelihood values derived from the control data, i.e. 0.423 x 2 and 0.527 respectively. Normalizing these to unity, we obtain 0.595 for the strand with two ade8-18 alleles and 0.405 for the wild type strand. Thus, with these derived initiation values, we predict 0.405 x 35 = 14 events of the $6^+:2^- + 3^+:5^-$ type and 0.595 x 35 = 21 events of the $2^+:6^- + 3^+:5^-$ type. The actual number of observed events corresponding to these classes are 11 and 24--an acceptable fit between the observed data and hypothesis. In summary, each sequence retains its normal properties in such non-tandem duplications.

A similar analysis applies to TC140. Here, the two homologues contain ADE8-URA3-TRP1-ade8-18 and ade8-18-URA3 TRP1-ade8-18. If misalignment of homologous sequences did not occur, or occurred with a negligible frequency, this diploid should exhibit a dissymmetry coefficient equivalent to the standard value 1.36. The observed value 0.43 (10/23) is significantly lower than the predicted value. Now, if we assume that misalignment at meiotic prophase may occur, we visualize that three different synaptic configurations might occur with equal probability or 0.33; these patterns are:

ADE8 - URA3 - TRP1 - ade8-18
ade8-18-URA3 - TRP1 - ade8-18

                ADE8 - URA3 - TRP1 - ade8-18
ade8-18 - URA3 - TRP1 - ade8-18

ADE8 - URA3 - TRP1 - ade8-18
                ade8-18 - URA3 - TRP1 - ade8-18

It may be noted that the first two configurations can readily yield gene conversions and postmeiotic segregations, while the last configuration can yield only a homoduplex at the paired ade8-18 sites. Alternatively, the unpaired ADE8 segment might behave as wild type sequence heterozygous for a deletion. In the latter instance, no postmeiotic segregations would be generated. Thus, overall we predict a reduction of the basic conversion frequency by a fraction of about 1/3 or to 6.7%-- a value in agreement with the observed 6.0%. Again, it may be noted that in this hybrid, containing a single heterozygous site within otherwise identical non-tandem duplications in each homologue, the respective ade8 sequences retain their individual properties.

In the culture designated Bam32, with 642 total tetrads analyzed, we may observe a situation that is essentially the reverse of DK101. Here, an isosequential non-tandem duplication for the wild type ADE8 sequences is heterozygous for a single copy of the mutant allele ade8-18, i.e., ADE8 URA3 TRP1 ADE8/ade8-18 Several features of the data may be noted. First the $2^{+}:6^{-}$ and $3^{+}:5^{-}$ categories are conspicuously absent. Moreover, the $6^{+}:2^{-}$ and $5^{+}:3^{-}$ classes occur with approximately equal frequency. The wide ratio class contains a single ab $6^{+}:2^{-}$ and one $7^{+}:1^{-}$ or events taken to represent two independent heteroduplexes. In the ab $6^{+}:2^{-}$, two sectored ascosporal colonies accompany two wild type colonies; in the $7^{+}:1^{-}$ a single sectored ascosporal colony accompanies three corresponding wild type colonies. Thus, the ab$6^{+}:2^{-}$ ascus may reflect two separate events that involve all four chromatids and where the heteroduplexes span a single ade8-18 site and these are passed uncorrected into the functional ascospores. Likewise, the $7^{+}:1^{-}$ segregation mirrors two equivalent occurrences and presumably only one of the heteroduplexes is repaired in the direction of wild type. Adding the respective $6^{+}:2^{-}$ and $5^{+}:3^{-}$ segregations generated by the wide ratio asci to the observed $6^{+}:2^{-}$ and $5^{+}:3^{-}$ clones, we obtain 34 and 27 events respectively or a %PMS/BCF of 44.3%, a value roughly comparable to the corresponding ratio in our control sample--52%. The absence of apparent $2^{+}:6^{-}$ and $3^{+}:5^{-}$ segregation asci can be accounted for by simply assuming that homologous pairing and heteroduplex formation spanning the ade8-18 site, with or without correctional repair will be associated with an unaltered ADE8 sequence in cis array.

Accordingly, potential conversions or PMS segregations resulting from events initiating on the ade8-18 strand remain cryptic. Masked by the ADE8 on the same strand, such events therefore appear as ordinary $4^{+}:4^{-}$ segregations.

We may proceed to calculate the expected number of cryptic events and subsequently compare this theoretical value with a value determined from an experimental procedure described below. Using the same rationale for normalization as in DK101 (0.577+0.577+.423=1.577), we obtain 1.154/1.577=0.732 and 0.423/1.577=0.268-- the respective initiation probabilities for the corresponding parental strands ADE8 URA3 TRP1 ADE8 and ade8-18. Because 61 events were observed and since these represent only 73.2% of the total, the actual number of events should be corrected to 83.3. Accordingly, 26.8% of the total events should have initiated from the mutant ade8-18 strand and yielded 22.3 cryptic events to be found among the 583 normal $4^{+}:4^{-}$ segregations.

Now, we may consider the experimental protocol for detecting cryptic events. A total of 167 normal $4^{+}:4^{-}$ meiotic segregations from Bam 32 were tested as follows: the red segregant ascosporal clones were patched to YEP master plates. Individual white colonies from each patch were isolated, purified and tested for tryptophane and uracil requirements as well as complementation by crossing with an ade8-18 tester strain. The tested cultures might be ADE8 -URA3-TRP1-ADE8 , or normal. Alternatively, they might be ade8-18 URA3 - TRP1-ADE8 i.e., conversions, or ade8-18/ADE8-URA3-TRP1-ADE8 i.e. a PMS. In either case, the conversions and the PMS events will generate white colonies that are demonstrably ade8-18 genotypically. Such derivatives might arise as a consequence of intrachromosomal gene conversion with or without associated exchange. We examined 334 red ascosporal colonies from 167 complete asci. Twenty four colonies, each representing a single tetrad, yielded the expected white derivatives. However, in all but two instances, these were URA3 TRP1 and ADE8. Hence, they mirror spontaneous mutantional alterations at unrelated loci that impose blocks in the de novo adenine biosynthetic pathway prior to ade2--another manifestation of the Roman Effect. The two remaining instances were ade8-18-ura3-trp1 and hence reflect the cryptic gene conversion or PMS along with the excision of the original plasmid sequence either subsequently or simultaneously. Finally, we may compare our observed number of two cryptic events with the number expected or 6.3 events (167/583 x 22). We may conclude that cryptic events do in fact occur and these arise in proportions similar to those given above. Hence, the basic conversion frequency of 9.3% has been underestimated by a failure to include twenty two cryptic events. The adjusted value is 12.9% BCF.

The possibilities for the analysis of synaptic misalignments and therefore unequal meiotic crossing-over is especially favored in the diploid TC 136. Here, two different isosequential non-tandem duplications are combined as follows: ADE8-URA3-TRP1-ADE8/ade8-18-URA3-TRP1-ade8-18.

One homologue carries two ADE8 arrays and the other carries two ade8-18 arrays. Otherwise, the homologues are congenic, though each was derived by an intrachromosomal gene conversion of separate integration events.

In all, 820 complete tetrads were analyzed (Table 4). The apparent %BCF was 15.6% with 110 segregations of the $6^+:2^-$ variety and 18 tetrads displayed $5^+:3^-$ patterns. However, $2^+:6^-$ and $3^+:5^-$ asci were absent. What rationale will account for the marked disparity observed in these data as well as the rather high %BCF value. A priori, we might consider that each heterozyous site of the non-tandem duplication might behave independently of the remaining site. Conversions of mutant to wild at either site would yield $6^+:2^-$ tetrads and uncorrected heteroduplexes formed by initiations on the wild type strand would generate $5^+:3^-$ asci. In contrast, $2^+:6^-$ tetrads would require simultaneous conversions of wild to mutant at both sites and PMS events with initiations on the ade8-18 stand would remain cryptic. To the above we must add the contribution of reciprocal exchange between the duplicated sequences. Such events would appear as $6^+:2^-$ segregations superficially, but crypticity tests would reveal that two ascosporal clones were heterosequential. In contrast $6^+:2^-$ tetrads arising from informational transfer alone would contain a single heterosequential spore clone. Further speculation in the absence of adequate crypticity analysis is not warranted. Considered together, the factors discussed above provide a basis for the observed disparity.

A qualitative statement bearing on the question of misalignment during meiotic prophase may be obtained by considering the 174 unselected tetrads for hybrid TC148. This diploid is homozygous for a heterosequential non-tandem duplication. As displayed in Table 4, six aberrant tetrads were identified; two were $6^+:2^-$ and four were $7^+:1^-$. The $6^+:2^-$ asci may have originated as intrachromosomal gene conversions of the wild to the mutant sequence; or they may reflect misalignments associated with a similar conversional direction. The $7^+:1^-$ segregations are all taken to represent PMS events--i.e. misalignments accompanied by information transfer from the mutant sequence to the wild type sequence without correction. Thus, some of the meioses display evidence that is indicative of misalignments or intrachromsomal events.

In conclusion, it is apparent that these studies must be regarded as only partial. Required for a total analysis is a more extensive study of crypticity coupled with a rigorous molecular approach via DNA-DNA hybridization and sequence determinations of the various wild type and mutant alleles.

Literature Cited

1. Abei, R., C. Gysler, P. Munz, and U. Leupold. 1980. Heterologous recombination between redundant tRNA genes in Schizosaccharomyces pombe. II. Events leading from the suppressor active to the suppressor inactive state. Tenth Int. Conf. Yeast Gent. and Mol. Biol. p.167. Louvain-la-Nueve, Belgium.
2. Botstein, D. and R. W. Davis. 1982. Principles and practice of recombinant DNA research with yeast. In The Molecular Biology of the Yeast Saccharomyces. II Metabolism and gene expression (ed. J. Strathern et. al.), Cold Spring Harbor, New York (In press.)
3. Clarke, L. and J. Carbon. 1978. Functional expression of cloned yeast DNA in Escherichia coli. Specific complementation of arginino succinase lyase (arg H) mutations. J. Mol. Biol. 120:517.
4. Egel, R. 1981. Intergenic conversion and reiterated genes. Nature 290:191.
5. Esposito, M. S. 1967. X-ray and meiotic fine structure mapping of the adenine-8 locus in Saccharomyces Cerevisiae. Genetics 58:507
6. Esposito, M. S. 1971 Post-meiotic segregation in Saccharomyces. Mol. Gen. Genet. 111:297.
7. Esposito, M. S. 1978. Evidence that spontaneous mitotic recombination occurs at the two-strand stage. Proc. Natl. Acad. Sci. 75:4436.
8. Esposito, M. S. and R. E. Esposito. 1977. Gene conversion, paramutations and controlling elements: A treasury of exceptions. In Cell Biology, a comparative treatise (ed. L. Goldstein and D. M. Prescott), vol. 1, p. 59. Academic Press, New York.
9. Esposito, M. S. and J. E. Wagstaff. 1981. Mechanisms of mitotic recombination. In The Molecular Biology of the Yeast Saccharomyces (ed. J. Strathern, et. al.) Cold Spring Harbor, New York (In press.)
10. Fogel, S. and D. D. Hurst. 1967. Meiotic gene conversion in yeast tetrads and the theory of recombination. Genetics 57:455.
11. Fogel, S. and R. K. Mortimer. 1968. Meiotic gene conversion of nonsense mutations in yeast tetrads. Proc. Int. Congr. Gent. 1:6.
12. Fogel, S. and R. K. Mortimer. 1969. Informational transfer in meiotic gene conversion. Proc. Natl. Acad. Sci. 62:96.
13. Fogel, S. and R. K. Mortimer. 1970. Fidelity of gene conversion in yeast. Mol. Gen. Genet. 109:177.
14. Fogel, S. and R. K. Mortimer. 1971. Recombination in yeast. Annu. Rev. Genet. 5:219.
15. Fogel, S., D. D. Hurst, and R. K. Mortimer. 1971. Gene conversion in unselected tetrads from multipoint crosses. Stadler Genet. Symp. 2:89.
16. Fogel, S., R. K. Mortimer, and K. Lusnak. 1981. In The Molecular Biology of the Yeast Saccharomyces. I Life Cycle and Inheritance (ed. J. Strathern, et. al.), Cold Spring Harbor, New York.
17. Fogel, S., R. K. Mortimer, K. Lusnak, and F. Tavares. 1979. Meiotic gene conversion: A signal of the basic recombinational event in yeast. Cold Spring Harbor Symp. Quant. Biol. 43:1325.
18. Golin, J. E. and M. S. Esposito. 1977. Evidence for joint genic control of spontaneous mutation and genetic recombination in Saccharomyces. Molec. Gen. Genet. 150:127
19. Haber, J. E., D. T. Rogers and J. H. McCusker. 1980. Homothallic conversions of yeast mating-type gene occur by intrachromosomal recombination. Cell 22:227.

20. Hinnen, A., J. Hicks, and G. R. Fink. 1978. Transformation in yeast. Proc. Natl. Acad. Sci. 75:1929.
21. Hurst, D. D., S. Fogel, and R. K. Mortimer. 1972. Conversion associated recombination in yeast. Proc. Natl. Acad. Sci. 69:101.
22. Jackson, J. A., and G. R. Fink. 1981. Gene coversion between duplicated elements in yeast. Nature 292:306.
23. Klein, H. L., and T. D. Petes. 1980. Intrachromosomal gene conversion in yeast. Nature 289:144.
24. Meselson, M. S., and C. M. Radding. 1975. A general model for genetic recombination. Proc. Natl. Acad. Sci. 72:358.
25. Mortimer, R. K., and S. Fogel. 1974. Genetical interference and gene conversion. In Mechanisms in recombination (ed. R. Grell), p. 236. Plenum Press, New York.
26. Mortimer, R. K., and D. C. Hawthorne. 1969. Yeast Genetics. In The yeasts (ed. A. H. Rose and J. S. Harrison), vol. 1, p. 385. Academic Press, New York.
27. Mortimer, R. K. and D. Schild. 1980. Genetic map of Saccharomyces cerevisiae. Microbiol. Rev. 44:519.
28. Radding, C. M. 1973. Molecular mechanisms in genetic recombination. Annu. Rev. Genet. 7:87.
29. Roman, H. 1957. Studies of gene mutation in Saccharomyces. Cold Spring Harbor Symp. Quant. Biol. 21:175.
30. Roman, H. 1980. Recombination in diploid vegetative cells of Saccharomyces cerevisiae. Carlsberg Res. Comm. 45:1315.
31. Scherer, S. and R. W. Davis. 1980. Recombination of dispersed repeated DNA sequences in yeast. Science 209:1380.
32. Stinchcomb, D. T., K. Struhl, S. Scherer, and R. W. Davis. 1979. Isolation and characterization of a yeast chromosomal replicator Nature 282:39.
33. Struhl, K., D. T. Stinchcomb, S. Scherer, and R. W. Davis. 1980. High frequency transformation of yeast: Autonomous replication of DNA molecules. Proc. Natl. Acad. Sci. 76:1035.
34. Szostak, J. W. and R. Wu. 1980. Unequal crossing over in the ribosomal DNA of Saccharomyces cerevisiae. Nature 84:426.
35. Williamson, M. and S. Fogel. 1980. Correction deficient (cor) mutants that alter the frequency and length of correctional tracts in meiotic hybrid DNA. In Tenth International Conference on Yeast Genetics and Molecular Biology, No. 135, p. 44. Louvain-la-Neuve, Belgium.

Part IV

Mating Type: Cell Type Control and Mating Physiology

21. REGULATION OF THE α-SPECIFIC STE3 GENE

George F. Sprague, Jr., David C. Hagen, and Ira Herskowitz†

Institute of Molecular Biology and Department of Biology
University of Oregon, Eugene, Oregon 97403

Rec. Adv. Yeast Mol. Biol. (1982) 1:289-294

The STE3 gene is required uniquely by α cells in order to mate: MATα ste3⁻ cells do not mate, whereas MATa ste3⁻ mate efficiently. Genetic and biochemical evidence that STE3 is regulated by the mating type locus is reviewed here. Genetic interactions between STE3 and MATα2 had suggested that STE3 was regulated by a second function of the α mating type locus, MATα1. The molecular cloning of STE3 DNA allowed regulation of this gene to be examined directly. It was found that STE3 RNA was present in α cells, but not a or a/α cells, and also that STE3 RNA was present in matα2 mutants but absent from matα1 mutants. Thus, MATα1 is required for STE3 expression. In addition, expression of STE3 was regulated during mating with a cells. Specifically, a-factor induced STE3 RNA to a level fourfold above the steady state level seen in α cells.

INTRODUCTION

The mating potential of yeast cells is determined by alleles of the mating type locus, MAT. MATa cells mate efficiently with MATα cells to form MATa/MATα diploid cells, which cannot mate but can be induced to undergo meiosis and sporulation (for review, see Herskowitz and Oshima, 1981). The mating process is complex and involves a number of interactions between a and α cells that lead ultimately to cell and nuclear fusion (for review, see Thorner, 1981). Extracellular pheromones or mating factors produced by a and α cells - a-factor and α-factor, respectively - trigger mating by causing physiological changes in cells of the opposite type. These changes include induction of cell surface components, called agglutination factors, that increase cell-cell adhesion. Since yeast cells are non-motile, the agglutination factors presumably increase the likelihood of the two cell types adhering (and therefore mating) following a chance encounter in nature. a and α-factors also induce arrest of cell division in the G1 phase of the cell cycle prior to the initiation of DNA synthesis, thereby synchronizing the cell cycles of the mating partners. Thus, when cell and nuclear fusion occurs, a diploid cell is formed.

The mating type locus is believed to control the behaviors outlined above through the action of positive and negative regulators encoded by the

---

*Support was obtained from Public Health Service Research Career Development Award AI-00163 and PHS Grant AI-13462 to I.H., from PHS Grant BRSG SO 7RR07080 to the University of Oregon Graduate School, and from PHS Grant GM30027 to G.F.S.

†Present address: Department of Biochemistry and Biophysics, University of California, San Francisco, California 94143

MAT alleles. The targets of these regulators are unlinked genes that give the a, α, and a/α cell types their unique properties. This view of the mating type locus was initially proposed on genetic grounds (MacKay and Manney, 1974b; Strathern, Hicks, and Herskowitz, 1981) and was based on the observation that genes required for mating (STErile genes) are located not only at the mating type locus, but also elsewhere in the genome (MacKay and Manney, 1974a,b; Manney and Woods, 1976; Hartwell, 1980; Sprague, Rine, and Herskowitz, 1981). Among the unlinked STE genes are those that are required for mating by both a and α cells. Other STE genes, however, are necessary in only one cell type and are referred to as a-specific or α-specific STE genes. The role any of the STE genes play in determining the cell type-specific behaviors described above is not known. In any case, the STE genes are likely to be the targets of the MAT-encoded regulators.

Genetic analysis of mating type locus mutations has suggested a specific model for regulation of cell type by MATa and MATα (the α1-α2 hypothesis, Strathern, Hicks and Herskowitz, 1981). According to this model, MATα codes for two regulators, α1 and α2. α1 is a positive regulator that stimulates expression of unlinked α-specific genes such as α-specific STE genes. α2 is a negative regulator that inhibits expression of a-specific genes. MATa encodes one product function, a1, which in conjunction with the α2 product stimulates expression of genes required for sporulation (Figure 1).

Recent studies made possible by the molecular cloning of unlinked genes thought to be regulated by the MAT products have confirmed several features of the model described above and have extended our understanding of the manner by which the MAT-encoded regulators function. Studies with the cloned genes have also provided insight into the mode of action of the pheromones. We will review genetic experiments that identified a particular gene, STE3, as one that was likely to be regulated by MATα. We will also review experiments using a cloned DNA fragment containing STE3. These experiments have demonstrated that RNA production from this locus is controlled by both MATα and by a-factor.

RESULTS AND DISCUSSION

Genetic Evidence that STE3 is Regulated by MATα

According to the α1-α2 hypothesis, matα2 mutants express both α-specific genes, because they contain MATα1 product, and a-specific genes, because they lack the MATα2 product. The expression of α-specific and a-specific genes in the same cell is thought to result in an antagonism between those gene products thereby causing the mating defect characteristic of matα2 mutants. For example, gene products that normally interact during mating may interact prematurely within matα2 mutant cells. This view leads to the prediction that matα2 mutants that carry additional defects in α-specific genes could have the mating phenotype of a cells. This expectation has been fulfilled for two classes of double mutant strains. First, matα2 matα1 double mutants mate as a because inactivation of the MATα1 positive regulator prevents expression of all α-specific genes (Strathern, Hicks, and Herskowitz, 1981; Sprague, Rine, and Herskowitz, 1981). Second, matα2 ste3 double mutants mate

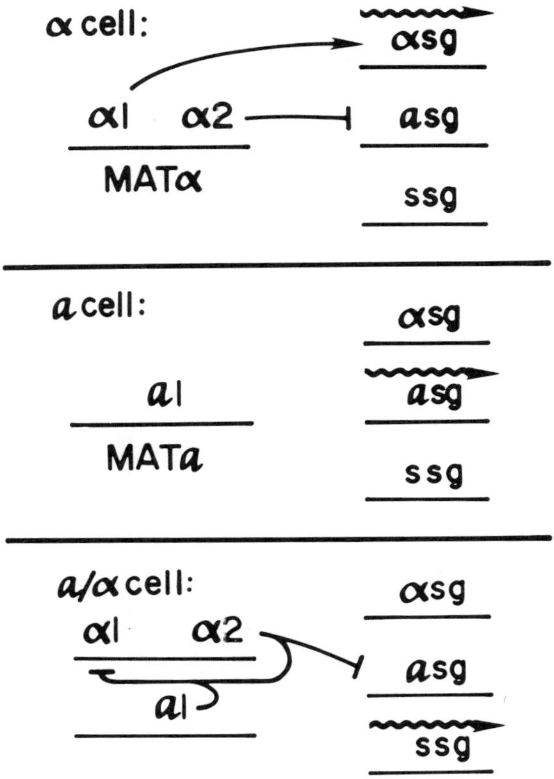

Figure 1. The α1-α2 hypothesis for control of cell type. Expression of α-specific (αsg), a-specific (asg) and sporulation-specific (ssg) is shown for α, a, and a/α cells. A wavy line indicates gene expression; a line with an arrowhead indicates stimulation of expression of those genes; a line with a terminal bar indicates inhibition of expression of those genes. In an α cell expression αsg is stimulated by α1 product and expression of asg is inhibited by α2 product. In an a cell αsg are not expressed and asg are expressed due to the absence of α1 and α2 products. In an a/α cell, the combined action of α2 and a1 products inhibits expression of α1 (Klar et al., 1981; Nasmyth et al., 1981) and therefore αsg are not expressed. α2 also inhibits expression of asg. The mechanism by which ssg is controlled is not clear. One possibility is that the RME1 gene product inhibits expression of ssg and that RME1 expression is inhibited by α2-a1 products in a/α cells (Rine, Sprague, and Herskowitz, 1981). The figure is reproduced from Sprague, Rine, and Herskowitz (1981).

as a. In this case a particular unlinked α-specific gene has been inactivated. The similarity of the mating properties of matα2 mutants defective in either the positive regulator or in a putative regulated function, suggests that STE3 is in fact regulated by the MATα1 product.

Although the function of STE3 product in mating is not known, the genetic analysis suggests several possibilities. STE3 could be a positive regulator of other α-specific genes. Not all α-specific genes can be controlled by STE3, however, since ste3 mutants produce α-factor. Thus, MATα1 may control some α-specific genes directly and others indirectly via stimulation of STE3 expression. A second possibility is that STE3 may negatively regulate some a-specific genes and MATα2 others. When both negative regulators are inactivated, all a-specific genes are expressed and the cell mates as a. Finally, STE3 product may function directly in conjugation. In this case, removal of the STE3 product by mutation relieves antagonism between that product and an important a-specific function.

MATα1 Controls RNA Production from STE3

To determine whether STE3 is indeed controlled by MATα1 and the level at which control is exerted, we have cloned a DNA fragment containing the STE3 gene and used the clone to assay for STE3 RNA in a, α, and a/α cells (Sprague, Jensen, and Herskowitz, submitted). The STE3 gene was isolated as a DNA fragment that complemented the mating defect of ste3 mutants. The fragment was shown subsequently to indeed contain STE3 because it allowed plasmids to integrate at the chromosome STE3 locus.

The STE3-containing fragment hybridizes to a single RNA species from α cells and this species is absent from a and a/α cells. Since only one RNA species hybridizes to the fragment which is capable of complementing ste3 mutations, it is likely that that RNA is the STE3 transcript. When RNA from matα1 and matα2 mutants was examined, STE3 RNA was present in matα2 but not in matα1 mutants. These results indicate that the MATα1 product stimulates RNA production from the STE3 gene. Whether α1 controls synthesis or stability of STE3 RNA is not known. However, since STE3 RNA has not been detected in a cells, it seems more likely that α1 controls transcription rather than message stability.

The experiments described above demonstrate that α1 is necessary for STE3 expression. Is α1 sufficient for expression? That is, would the presence of α1 product in a cells cause STE3 RNA production or do other controls preclude stimulation by α1? To address this question, a plasmid containing the α1 gene under control of yeast alcohol dehydrogenase promoter (constructed by G. Ammerer) was introduced into a cells and STE3 RNA assayed. These cells produced STE3 RNA (G. Ammerer and G. Sprague, unpublished observations), indicating that α1 is sufficient for STE3 expression in a cells.

Induction of STE3 by a-factor

As noted above, a-factor causes a variety of physiological changes in α cells that presumably trigger mating with a cells. The mechanism by which these changes occur is unknown, but one possibility is that a-factor alters expression of genes involved in mating and in progression through the cell cycle. Since expression of STE3 is limited to α cells, we have examined whether expression is affected by a-factor. When the levels of STE3 RNA were assayed after addition of a-factor to α cells, a rapid increase in STE3 RNA to a level fourfold greater than found in untreated cells was observed (D. Hagen and G. Sprague, manuscript in preparation). Inhibition of protein synthesis (by cycloheximide) did not prevent induction of STE3 RNA. Thus, it seems likely that a-factor acts by changing the activity of a pre-existing cellular component that can affect STE3 RNA production. Whether a-factor causes an increase in STE3 transcription or in STE3 RNA stability cannot be determined from these data.

In summary, STE3 gene expresssion is controlled in two ways. First, the MATα1 gene product is required for production of STE3 RNA. Second, during mating expression of STE3 is induced by the pheromone produced by a cells. It is likely that STE3 is representative of a class of genes that is regulated in this manner.

LITERATURE CITED

1. Hartwell, L. 1980. Mutants of Saccharomyces cerevisiae unresponsive to cell division control by polypeptide mating hormone. J. Cell. Biol. 85:811-822.

2. Herskowitz, I. and Oshima, Y. 1981. Control of cell type in Saccharomyces cerevisiae: Mating type and mating-type interconversion, p. 181-209. In: J.N. Strathern, E.W. Jones, and J.R. Broach (eds.) The Molecular Biology of the Yeast Saccharomyces: Life Cycle and Inheritance. Cold Spring Harbor Laboratory, Cold Spring Harbor, N.Y.

3. Klar, A.J.S., Strathern, J.N., Broach, J.R., and Hicks, J.B. 1981. Regulation of transcription in expressed and unexpressed mating type cassettes of yeast. Nature 289:239-244.

4. MacKay, V.L. and Manney, T.R. 1974. Mutations affecting sexual conjugation and related processes in Saccharomyces cerevisiae. I. Isolation and phenotypic characterization of nonmating mutants. Genetics 76:255-271.

5. MacKay, V.L. and Manney, T.R. 1974. Mutations affecting sexual conjugation and related processes in Saccharomyces cerevisiae. II. Genetic analysis of nonmating mutants. Genetics 76:272-288.

6. Manney, T.R. and Woods, V. 1976. Mutants of Saccharomyces cerevisiae resistant to the α mating-type factor. Genetics 82:639-644.

7. Nasmyth, K.A., Tatchell, K., Hall, B.D., Astell, C., and Smith, M. 1981. A position effect in the control of transcription at yeast mating type loci. Nature 289:244-250.

8. Rine, J., Sprague, G.F., Jr., and Herskowitz, I. 1981. rmel mutation of Saccharomyces cerevisiae: Map position and bypass of mating type locus control of sporulation. Molec. Cell. Biol. 1:958-960.

9. Sprague, G.F., Jr., Rine, J. and Herskowitz, I. 1981. Control of yeast cell type by the mating type locus. II. Genetic interactions between MATα and unlinked α-specific STE genes. J. Mol. Biol. 153:323-335.

10. Strathern, J., Hicks, J. and Herskowitz, I. 1981. Control of cell type in yeast by the mating type locus: The α1-α2 hypothesis. J. Mol. Biol. 147:357-372.

11. Thorner, J. 1981. Pheromonal regulation of development in Saccharomyces cerevisiae, p. 143-180. In: J.N. Strathern, E.W. Jones, and J.R. Broach (eds.) The Molecular Biology of the Yeast Saccharomyces: Life Cycle and Inheritance. Cold Spring Harbor Laboratory, Cold Spring Harbor, N.Y.

22. BIOSYNTHESIS AND MODE OF ACTION OF YEAST α-FACTOR MATING PHEROMONE: A MODEL EUKARYOTIC HORMONE*

Debra Barnes, Lindley Blair, Anthony Brake,+ Monica Church, David Julius, Riyo Kunisawa, Janet Lotko, Gary Stetler, and Jeremy Thorner

Department of Microbiology and Immunology
University of California
Berkeley, California 94720

Rec. Adv. Yeast Mol. Biol. (1982) 1:295-305

Haploid cells of the α mating type of the yeast *Saccharomyces cerevisiae* secrete into the culture medium a linear, 13 amino acid-long oligopeptide, called α-factor. This molecule acts as a mating pheromone to prime haploid cells of the opposite mating type, *a* cells, for conjugation with α cell partners. The α pheromone is initially synthesized as a larger precursor protein, prepro-α-factor, which is processed by both endoproteolytic and exoproteolytic cleavages. The final maturation process required for the synthesis of native α-factor is the removal of additional N-terminal residues, which have the repeating structure -Glu-Ala- (or -Asp-Ala-). This step is accomplished through the action of a heat-stable membrane-bound dipeptidyl aminopeptidase. This enzyme is the product of the *STE13* gene. --- A generally applicable method for identifying and isolating genes whose transcription is differentially regulated under two different conditions was developed. This technique is based on the incorporation of the analog 4-thiouridine into nascent RNA. Using this procedure, cloned yeast DNA segments were found whose transcription in *a* cells *in vivo* is apparently modulated in a dramatic fashion within 15 minutes after exposure to α-factor. Three classes of such genes have been identified: genes expressed in vegetatively-growing cells which are no longer transcribed ("turn-off" genes); genes whose expression is increased 10- to 20-fold ("turn-up" genes); and genes expressed only after α-factor administration which are presumably specific to the conjugation process ("turn-on" genes).

INTRODUCTION

This laboratory is investigating the molecular mechanisms that underlie developmental processes in eukaryotic cells that involve control by intercellular signalling. This work addresses three primary questions: (a) How are genes regulated, and what functions are required, such that a given cell type is the producer of a particular extracellular signalling molecule? (b) How are genes regulated, and what functions are necessary, for a cell to have the capacity to respond to the signal molecule released by another cell type? (c) By what biochemical steps does the presence of the signal molecule elicit in its target cell the changes in cell physiology and/or gene expression which constitute its characteristic pattern of response? The system in which these problems are being studied is a unicellular eukaryotic microorganism, *Saccharomyces cerevisiae* (baker's yeast). Despite this deceptively simple lifestyle, this organism exists

* Supported by NIH Research Grant GM21841 to JT, by American Cancer Society Postdoctoral Fellowships 1777 to LB and 1882 to GS, by Damon Runyon-Walter Winchell Cancer Fund Postdoctoral Fellowship 315 to AB, and by USPHS Predoctoral Traineeships GM07232 to DB, MC, DJ, and JL.

+ Present Address: Chiron Corporation, Emeryville, California 94608.

in three distinct cell types which possess markedly different developmental capabilities. There are two haploid cell types, designated a cells and α cells, which are able to fuse ("mate") with one another, to form a third cell type, a/α diploid cells. The ability of the haploids to mate can be considered a developmental option, because it is an alternative to their simple vegetative multiplication. The initial events in this conjugation process are triggered by the reciprocal exchange between the cells of extracellular oligopeptide pheromones. The α haploids secrete into their culture medium a linear, 13 amino acid-long molecule, called α-factor, which acts only on the a cells. The a cells release a different linear, 11 amino acid-long molecule, called a-factor, which affects only the α cells. Diploid (a/α) cells neither produce nor respond to these pheromones. On the other hand, diploid cells can undergo meiosis and sporulation when starved for a nitrogen source, a developmental option which the haploid cells lack. Both genetic studies of the control of cell type in yeast (15,27) and biochemical studies of the mating process (32,33) have been summarized in recent extensive reviews, and references to the pertinent work and related surveys of the literature may be found therein.

In this laboratory, the α-factor pheromone molecule has been purified, completely characterized, and even chemically synthesized (8,10). Furthermore, yeast cells are eukaryotic cells especially amenable to other kinds of biochemical and genetic analyses, both of the traditional sort and of the kind involving recombinant DNA methodology, as is documented by the results being presented at this Workshop. For these reasons, yeast has proven to be a particularly tractable system in which to examine the fundamental questions posed above. Specifically, how do α cells synthesize, process, and secrete α-factor? How do a cells recognize the presence of α-factor and transduce this information? What biochemical events occur in a cells following their exposure to α-factor to initiate their mating "program" (and to prevent continued mitotic cell cycles until the mating process is completed)?

This paper describes recent work which examines the mechanism of biosynthesis of α-factor by α cells and reports the discovery that exposure of a cells to α-factor alters their pattern of gene expression.

MATERIALS AND METHODS

Yeast Strains, Plasmids, and Growth Conditions

XT1172-S245c (19) and A2S3b (an isogenic derivative carrying the ste13-1 mutation [28]) were obtained from Dr. Jasper Rine. Derivatives of these strains suitable for transformation with yeast plasmid vectors were prepared through appropriate crosses with AB35-14A (MATα ura3-52 leu2-3,112) and AB35-13D (MATa ace2-1 his4-580 leu2-3,112 ura3-52 trp1-289), which were constructed in this laboratory and which display a reproducibly high efficiency of transformation. Strain XS3-6B (MATa cry1 ura1 met2 lys2 his4-580 SUP4-3) was prepared by crossing 381G and 382-31 (13), provided by Dr. Leland Hartwell. Tester strains for the detection of α-factor carrying the markers sst1-2 and sst2-4, which make a cells at least one- and two-orders of magnitude more sensitive, respectively, to the effects of the pheromone (6,7), were provided by Dr. Russell Chan. Vectors and plasmids were obtained from the following individuals: YEp13 (3) from Dr. James Hicks; λ1059 (16) from Dr. Jonathan Karn; YEp24 (11) from Dr. David Botstein; p69A, isolated from a library of yeast DNA inserts in YEp13 constructed by Dr. Kim Nasmyth (21), from Drs. Janet Kurjan and Ira

Herskowitz; YEp13-13-3 and YEp13-13-4 from Dr. George Sprague, Jr. Strains were grown unselectively in a rich broth (YPD) or in a defined minimal medium (SD) supplemented with required nutrients. Plasmids were maintained in transformed yeast strains by growth in a selective (e.g. -Leu), but otherwise complete, minimal medium (SC). YPD, SD, and SC were prepared as described (26).

Other Procedures

Purification of α-factor and related peptides was performed by previously described methods (8,9,10). Analysis of α-factor-related antigens was accomplished by radioimmunoassay using anti-α-factor antibodies prepared in this laboratory and ^{125}I-α-factor (Y. Jones-Brown and J. Thorner, data to be published). Amino acid composition and sequence analyses were performed by standard methods. Genetic crosses (26), transformations of yeast and bacterial cells with cloned genes (1), preparations of plasmid and phage DNAs (25), transfer of phage plaques to nitrocellulose filter disks (2,34), preparation of yeast total and polyA+ RNA (29), blotting of DNA and RNA to nitrocellulose filter paper (35), nucleic acid hybridizations (25,35), restriction endonuclease digestions of DNA (25), purifications of DNA fragments from agarose gels (35), ligations (35), and other techniques for the construction of plasmids and recombinant phage were all performed by minor modifications of published procedures.

RESULTS AND DISCUSSION

Processing of α-Factor Pheromone Precursor Requires A Membrane-Bound Dipeptidyl Aminopeptidase

Detection of α-factor-deficient mutants. In order to define the steps involved in the synthesis, processing, and secretion of α-factor, mutants were identified which are defective at various stages of the pathway leading to the production of the mature, biologically active pheromone. The method devised to screen for α-factor-deficient mutants utilizes a "halo assay". Patches or colonies of α-factor-producing cells cause growth arrest of sensitive *a* cells in the G1 phase of the cell cycle (4). Hence, in a surrounding lawn of *a* cells, a clear zone will develop due to the lack of proliferation of the *a* cells in the area immediately around the α-factor-producing cells. The size of the halo is proportional to the amount of α-factor released. In order to make the halo assay sufficiently sensitive, *a* tester strains are used which carry an *sst1* mutation (which makes *a* cells super-sensitive to the pheromone because they are unable to proteolytically degrade it [6,7,8,9]) or an *sst2* mutation (which makes *a* cells hypersensitive to α-factor for unknown reasons [6,7]). Wild-type *α* strains produce a readily detectable halo on both *sst1* and *sst2* tester lawns, as expected. *MATα* strains carrying a *kex2* mutation, in contrast, do not produce a halo on either tester. Indeed, as judged by radioimmunoassay, the amount of α-factor cross-reacting material released by α *kex2* strains is undetectable (Y. Jones-Brown, D. Julius, L. Blair, and J. Thorner, data to be published). This finding was not unexpected since *kex2* mutants are unable to release another secreted protein, "killer" toxin (18), and have been reported to show perturbations in extracellular glycoproteins of the cell wall (23). Quite surprisingly, however, haploid cells of the α mating type of another yeast species, *Saccharomyces kluyveri*, were able to produce a large halo in the *S. cerevisiae a sst2* lawn, but gave no halo at all in the *a sst1* lawn. Although *S. kluyveri* α cells do not mate with *S. cerevisiae a* cells, they do produce an α-factor-related pheromone (20) which differs from that of S. cerevisiae by only three amino acid substitutions within an otherwise completely homologous molecule (24). This result suggested that the sst2 mutation not only causes a cells to be much more sensitive to α-factor, but also causes a cells to relax the specificity of their response. It occurred to us, therefore, that a differential halo test (no halo on the a sst1 lawn; but a pronounced halo on the a sst2 lawn) might be an indication that certain apparently α-factor-deficient S. cerevisiae α cell mutants were actually releasing forms of α-factor related to, but chemically distinct from, normal intact α-factor. Such altered peptides could represent alterations in the α-factor structural gene itself, or incompletely or incorrectly processed molecules.

That this supposition was, in fact, true was demonstrated by the observation that MATα strains bearing the α cell-specific ste13-1 mutation yielded no halo in lawns of wild-type a or a sst1 cells, but gave a large halo in the a sst2 tester lawn. Because a sst2 cells are at least an order of magnitude more sensitive to α-factor than a sst1 cells, the possibility existed that that the failure of α ste13 mutants to produce a halo in the a sst1 lawn was merely due to a defect leading to decreased production of otherwise normal α-factor rather than due to a defect resulting in the production of an altered form of α-factor. However, analysis of the amount of α-factor-related peptides secreted by the α ste13 cells by radioimmunoassay demonstrated that they produced as much or more α-factor cross-reacting material than the isogenic parent strain.

MATα ste13 mutants release improperly processed forms of α-factor. To prove that altered peptides were produced by α ste13 cells, these molecules were purified to homogeneity from the cell-free supernatant fluid of cultures of the mutant cells using modifications of the procedures devised in our laboratory for purifying normal α-factor (9,10). The behavior during purification of the altered α-factor molecules secreted from α ste13 cells, dubbed "α-factor*", suggested that the peptides were a collection of molecules which all were larger and more hydrophilic than the normal pheromone. The different α-factor* molecules were separated by high pressure liquid chromatography and completely characterized. Amino acid composition and sequence analysis confirmed that the bulk of the α-factor* peptides differ from normal α-factor by the presence of four extra residues at the N-terminus and have the structures:

H_2N-GluAlaGluAlaTrpHisTrpLeuGlnLeuLysProGlyGlnProMetTyr-COOH

H_2N-AspAlaGluAlaTrpHisTrpLeuGlnLeuLysProGlyGlnProMetTyr-COOH

The extra residues present in these α-factor* peptides are completely consistent with those predicted for the primary translation product, as determined by nucleotide sequence analysis of the cloned α-factor structural gene (J. Kurjan and I. Herskowitz, personal communication). Hence, the isolation and characterization of the α-factor* peptides confirms that the α pheromone is produced by processing of a larger precursor polypeptide.

STE13 is the structural gene for a heat-stable membrane-bound dipeptidyl aminopeptidase. The nature of the residues not removed from the collection of α-factor* peptides suggested that ste13 mutants might be defective in an enzymatic activity specific for the excision of a repeating -X-Ala- sequence.

Such enzymes, called dipeptidyl aminopeptidases, have been described (30) and have specificity for cleaving on the carboxyl side of Ala (or Pro) in repeating sequences of this type. Furthermore, dipeptidyl aminopeptidases have been implicated, at least by *in vitro* experiments, in the processing of other secreted peptides, for example the bee venom peptide mellitin (17). Because dipeptidyl aminopeptidase activities also hydrolyze efficiently -X-Pro-, we have utilized Ala-Pro-*p*-nitroanilide as a substrate for assaying this enyzme in various subcellular fractions of yeast cells. This is a particularly useful synthetic substrate since there are no activities which we have been able to detect in yeast extracts which can hydrolyze just Pro-*p*-nitroanilide; therefore, hydrolysis of Ala-Pro-*p*-nitrolanilide must reflect exclusively the action of a dipeptidyl aminopeptidase, and not non-specific aminopeptidase or carboxypeptidase digestion. Indeed, Ala-Pro is released as a unit from this substrate by activities in yeast extracts, as determined by thin layer chromatography of the product.

Using the synthetic substrate, we have detected two dipeptidyl aminopeptidase activities in normal yeast cells. Both of these enzymes have their highest specific activities in the particulate material, and preliminary subcellular fractionation indicates that the membranes with which these activites are associated are endoplasmic reticulum- or Golgi-derived (D. Julius, T. Etchevery, W. Hansen, and R. Schekman, unpublished results). One of these dipeptidyl aminopeptidase activities accounts for 30-50% of the total and is quite heat-stable, surviving a 60°C treatment for 15 minutes. The other enzyme constitutes the remainder of the activity and is completely inactivated by such a heat treatment.

In *ste13* mutants, there is less total dipeptidyl aminopeptidase activity and all *the* activity that is present is inactivated by a 60°C heat treatment. This result indicates that the heat-stable activity is missing in *ste13* cells. In fact, the loss of activity in the *ste13* mutant is strictly *additive* with the amount of activity remaining in heat-treated wild-type extracts. Furthermore, a cloned fragment of yeast DNA which was selected to complement the *ste13* mutation causes normal α cells to overproduce a heat-stable dipeptidyl aminopeptidase activity three- to five-fold, depending on the particular transformant and preparation examined. This particular cloned segment harbors the *STE13* gene, as determined genetically, since it integrates at that locus and completely suppresses all the phenotypes of *ste13* mutants when integrated, i.e. as a single copy (G. Sprague, personal communication; A. Brake and L. Blair, unpublished results). Another cloned fragment which only weakly suppresses the *ste13* mutation was also isolated (G. Sprague, personal communication). Normal *α* cells carrying this second clone also overproduce dipeptidyl aminopeptidase activity nine- to fifteen-fold; however, all the activity detected is heat-labile. Thus, the two cloned segments appear to code for the two different types of dipeptidyl aminopeptidase enzymes.

We have been able to assay dipeptidyl aminopeptidase activity in whole cells permeabilized by treatment with the non-ionic detergent Brij 58 following one cycle of freeze-thaw. This has made it possible to analyze the segregation of the heat-stable enzyme deficiency in tetrads. For this purpose, MATα/MATα *ste13/STE13 rme/rme* diploids were constructed and sporulated, so that *all* the *meiotic* products would be α cells in order to be able to score all the phenotypes of the *ste13* defect. In 115 out of 122 tetrads examined, the inability to produce a *halo* in an *a sst1* lawn segregated 2:2, and co-segregated with an inability to mate with wild-type a cells, as expected. Five of these tetrads were chosen at random for enzyme assay. Small cultures of each spore were grown to late exponential-early stationary phase, permeabilized by the procedure just described, and the content of heat-stable dipeptidyl aminopeptidase examined. In all five tetrads, the lack of the heat-stable activity co-segregated with the other phenotypes of the ste13 mutation (sterility and α-factor-deficiency). Finally, the STE13 gene almost certainly codes for a protein, since we have recently isolated a nonsense (ochre) mutation in this locus. Taken together, these results indicate that the STE13 locus is probably the structural gene for the heat-stable membrane-bound dipeptidyl aminopeptidase.

Because quantitative bioassays have shown that α-factor* has a specific biological activity about two orders of magnitude lower than that of wild-type α-factor and because the mating-deficiency of α ste13 mutants can be completely overcome either by adding authentic α-factor, or by having normal α cells present, or by using supersensitive a cells as the mating partners (R. Chan, L. Melnick, L. Blair, and J. Thorner, data to be published), the "sterility" of α ste13 cells is due only to their failure to secrete a pheromone of sufficient biological potency. Thus, our results represent the first instance in which the biochemical basis for the mating defect of a ste mutation has been elucidated. A direct corollary of these findings is that α-factor is required for the mating process to occur at normal efficiency.

Dipeptidyl aminopeptidase action is rate-limiting for the production of α-factor even in normal α cells. As another test of the conclusion that the ste13 lesion disrupts proper processing of α-factor precursor, α ste13 mutants were transformed with a multi-copy plasmid carrying the α-factor structural gene. Although the transformed mutants overproduced α-factor-related peptides about five-fold, as detected by the radioimmunoassay, the material produced was still only able to produce a halo on a sst2 tester cells, indicating that only α-factor* was being secreted (which was confirmed biochemically). Thus, as expected, the ste13 lesion was epistatic to having many intact copies of the normal pheromone structural gene. As a control for this experiment, however, normal α cells were also transformed with the cloned α-factor gene. Although a somewhat larger halo was observed in lawns of a sst1 testers and increased biological activity was detectable by quantitative bioassay as well, the level of increase was only a few fold, despite the fact that radioimmunoassay indicated that the transformed normal cells secreted 20- to 30-times as much α-factor-related cross-reacting material as normal α cells transformed with just the vector (YEp13). Remarkably, dramatic overproduction of biological activity by the transformants was only observed using an a sst2 tester lawn. These results indicated that transformed wild-type α cells were producing mainly α-factor*-like peptides. These observations suggest that when the α-factor precursor is made at high levels, processing of all the molecules by dipeptidyl aminopeptidase cannot keep pace with their production (which was also confirmed by biochemical analysis of the peptides).

If the processing step carried out by dipeptidyl aminopeptidase is rate-limiting for α-factor maturation in normal cells, then it would be expected that normal α cells transformed with a composite plasmid carrying both the STE13 gene and the α-factor structural gene would overproduce mature α-factor molecules, rather than α-factor*. Such a composite plasmid was constructed by inserting an Eco RI fragment carrying the entire α-factor gene into

YEp13-13-3 that had been partially digested with Eco RI. Normal α cells transformed with the composite plasmid produce halos of equal size on both a sst1 and a sst2 lawns. Moreover, these halos are distinctly larger than those produced by strains carrying the vector containing the α-factor gene alone. These results indicate that the bulk of the α-factor produced by the cells transformed with the composite plasmid has been fully processed to its mature form.

These observations, combined with the phenotypes of the ste13 mutation and the characterization of the product of the STE13 gene, provide the first *in vivo* evidence that a dipeptidyl aminopeptidase has an essential role in the processing of the precursor form of a secreted hormone.

Hormone-Responsive Genes in Yeast: α-Factor Changes the Pattern of Gene Expression in a Cells

A general method for identifying transcripts that are differentially expressed under two separate conditions. In order to determine if exposure of a cells to α-factor changes their pattern of gene expression, it was essential to be able to isolate the newly-synthesized mRNA from pheromone-treated cells and from control cells. This was accomplished by "tagging" these RNA molecules by incorporating 4-thiouridine, in our case during a 30 minute pulse of a cells in either the presence or absence of 50 units/ml of α-factor. To enhance the efficiency of labelling, uracil auxotrophs were used. The 4-thiouridine in nascent RNA allows the purification of the newly-made polyA+ mRNA species by chromatography on oligo dT-cellulose followed by chromatography on Hg^{++}-agarose. Because of tautomeric shift (shown below),

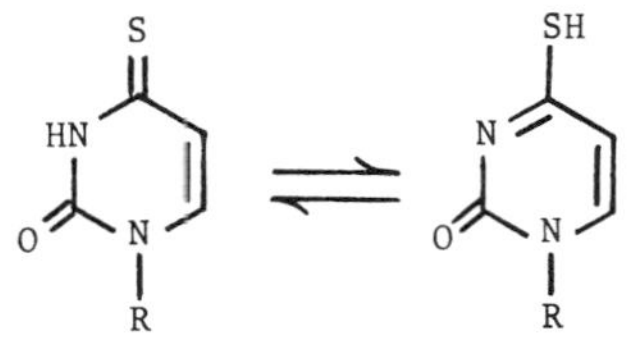

Figure 1. Structure of 4-thiouridine. The sulfur atom bonded to the pyrimidine ring has a significant degree of sulfhydryl character, as judged by its ability to react with thiol-directed reagents (5).

the sulfur atom in 4-thiouridine behaves like a sulfhydryl group and will react reversibly with mercuric ions. The necessity for obtaining such highly enriched and specific mRNA fractions was to be able to use them as templates for the synthesis of very selective cDNA probes of high specific radioactivity. Such cDNAs could then be used to screen libraries of yeast DNA segments cloned in bacteriophage λ or plasmid vectors by differential plaque or colony hybridization.

In theory, of course, incorporation of 4-thiouridine into nascent RNA could be accomplished under any conditions imposed by an experimenter, for example shifts of temperature-sensitive mutants from one temperature to another, transfer of cells from one carbon source to another, or following heat shock. Hence, the method we have devised represents a generally applicable procedure for identifying and isolating genes whose transcription is differentially regulated under any two different conditions. The specific procedure which we followed to prepare cDNA probes for detecting transcripts modulated by α-factor administration is outlined in the diagram below.

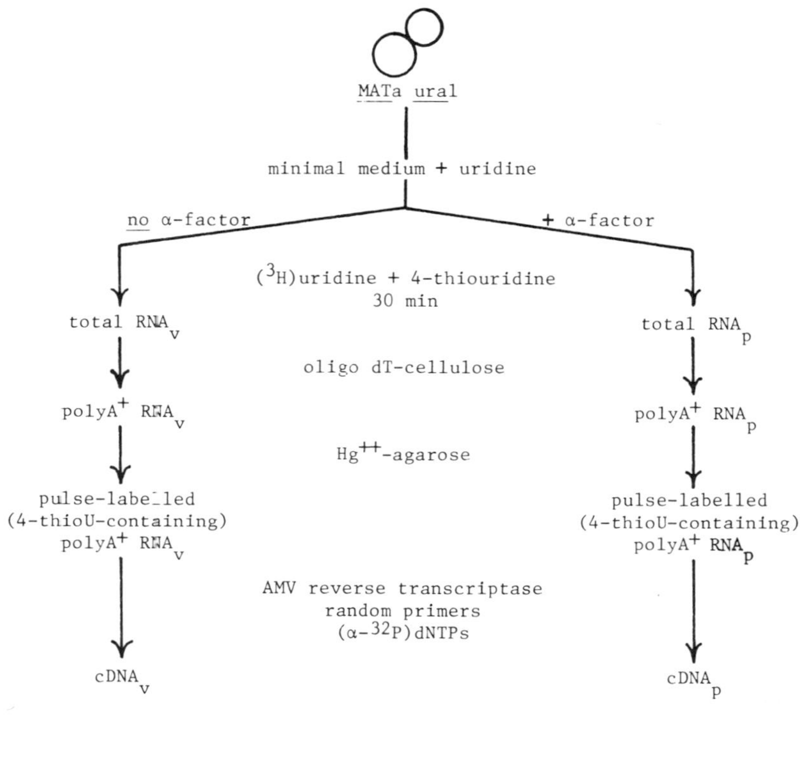

Figure 2. Preparation of cDNA probes for the detection of α-factor-controlled transcription units. XS3-6B (a ura1) was grown in minimal medium + uridine until mid-exponential phase. The cells were collected by brief centrifugation, washed in minimal medium without uridine, and resuspended in minimal medium containing 0.15 mM 4-thiouridine and 5 x 10^4 cpm/ml [6-^{3}H]uridine (2 nM). After 10 minute pre-incubation to allow equilibration of the pyrimidine nucleotide pool, 50 units/ml of α-factor in 0.1 M Na-acetate (pH 6) were added to one half of the culture; the other half received an equivalent volume of 0.1 M Na-acetate. After 30 more minutes of labelling, cells were rapidly chilled over crushed ice

and total cellular RNA was prepared by standard methods (phenol extraction). The total polyA+ fraction was selected by chromatography on oligo dT-cellulose. The newly-synthesized (thiouridine-containing) polyA+ RNA was then selected by its binding to Hg^{++}-agarose and eluted with 2-mercaptoethanol. Such purified RNAs were used as the substrates for preparing [^{32}P]cDNA copies using [α-^{32}P]dNTPs, AMV reverse transcriptase, and random calf thymus DNA primers.

As described in Figure 2, radioactive cDNAs, which were representative of the nascent mRNAs synthesized both in vegetatively-growing a cells and in a cells during the first 30 minutes after their exposure to α-factor, were prepared. These probes were used to screen replicate nitrocellulose filters to which a library of yeast DNA segments had been transferred (2,34) in the form of plaques in the bacteriophage λ vector, λ1059. The library was constructed using large fragments of yeast DNA prepared by sucrose gradient sedimentation of yeast genomic DNA partially digested with the restriction endonuclease Sau3a. The average insert size of the library was 16.5 kb. Two types of differential plaque hybridizations (29) were performed using the phage libraries.

First, duplicate filters were hybridized either with the cDNA probe made from the nascent RNA of α-factor-treated cells ($cDNA_p$) or with the cDNA probe made from the nascent RNA of vegetatively-growing cells ($cDNA_v$). Following hybridization and washing, autoradiograms of the two sets of filters were compared to each other. Putative plaques which displayed apparently clear-cut differences in their degree of labelling by the two different probes were picked for further testing. Due to the fact that the average insert size in the phages being screened was large enough to potentially code for several genes, an inherent problem with this method might be that the presence of a gene whose transcription is responsive to α-factor may be masked during the screening if it is surrounded by several genes which are transcriptionally active continuously. In an attempt to minimize this difficulty, a second set of differential plaque hybridizations was performed in which a 3,000-fold excess of unlabelled competitor polyA+ RNA, isolated from vegetative a cells, was added along with the $cDNA_p$. Additional phages were obtained by this second screening procedure. All the candidate phages were plaque-purified and re-tested three times. Out of over 20,000 plaques screened, only 5 recombinant phage were found that reproducibly behaved as if they carried genes whose transcription was affected by the exposure of a cells to α-factor. These phages were grown up on a large scale and their DNAs were purified.

In order to prove that each of these recombinant phage DNAs carried a yeast DNA sequence which was differentially transcribed in vegetative a cells and in α-factor-treated a cells, radioactive probes were prepared by nick-translation of each whole phage DNA. These radioactive probes were used to examine the spectrum of complementary RNA molecules and the kinetics of their appearance in yeast cells in the following way. *MATa sst1* cells were grown in minimal medium to mid-exponential phase. At time zero, α-factor was added at a final concentration of 10 units/ml and a sample of the culture was immediately withdrawn and chilled on ice. At various times after addition of the pheromone, usually at intervals of 15 minutes, samples of the culture were also withdrawn. PolyA+ mRNA was then prepared from the culture samples taken at each of the time points and subjected to agarose gel electrophoresis. The separated RNA molecules were transferred to nitrocellulose paper by the blotting procedure of Thomas (31). The radioactive recombinant phage DNAs then were used as hybridization probes.

This type of analysis confirmed that the recombinant phage identified by the screening procedure we devised did indeed carry DNA fragments whose transcription appears to be modulated by α-factor. Three different patterns of response are represented in the collection of sequences we have identified.

At least one insert contains a sequence which does not seem to be transcribed after *a* cells are exposed to α-factor. After α-factor administration, this transcript decays with a half-life of about 30 minutes. We have called this class of sequences "turn-off" genes. Such genes probably represent functions specific to vegetatively-growing cells. Such a transcript might code for a product which is required in cell cycle initiation, for example.

Each of at least two other inserts contains a sequence which is apparently transcribed at a low basal rate in vegetative *a* cells; yet, these transcripts are present in ten- to twenty-fold higher abundance after *a* cells are exposed to α-factor, as determined by densitometer scans of the autoradiograms. We have termed this class of sequences "turn-up" genes. Such genes might code for functions which may need to be present at elevated levels in conjugating cells, like cell wall biosynthetic enzymes.

At least one insert codes for three different transcripts which are undetectable in vegetative *a* cells, but which are present in *a* cells after exposure to α-factor. We have designated this class of sequences "turn-on" genes. This last class presumably represents genes coding for products uniquely involved in the mating process, like the cell surface agglutinin molecules required for the adhesion of *a* cells to α cells which is known to be an inducible function (12).

The changes observed for all three of these classes of transcripts are detectable within 10-15 minutes after administration of α-factor. Nonetheless, additional experiments will have to be done to determine if these effects are truly specific responses to α-factor treatment *per se* or are also elicited by other treatments or conditions which arrest yeast cells in the G1 phase of the cell division cycle. In this regard, however, based on its length (~1500 nucleotides), the "turn-off" transcript identified in this work is not the transcript of either the histone H2A or H2B genes or the *CDC28* gene, all of whose mRNAs are known to be present at greatly reduced levels in G1-arrested cells (14,22; S. Reed, personal communication).

In any event, it seems that α-factor treatment may alter the pattern of gene expression in *a* cells in a very specific fashion. The availability of the cloned genes should permit us to determine the functions coded for by these regions and their role in the mating process, as well as the mechanism by which transcription of these sequences is controlled by the pheromone.

LITERATURE CITED

1. Beggs, J. D. 1978. *Nature* 275: 104-109.
2. Benton, W. D. and Davis, R. W. 1977. *Science* 196: 180-182.
3. Broach, J. R., Strathern, J. N., and Hicks, J. B. 1979. *Gene* 8: 121-133.
4. Bücking-Throm, E., Duntze, W., Hartwell, L. H., and Manney, T. R. 1973. *Exp. Cell Res.* 76: 99-110.

5. Carbon, J. and David, H. 1968. Biochemistry 7: 3851-3858.
6. Chan, R. K. and Otte, C. 1982. Molec. Cellul. Biol. 2: 11-20.
7. Chan, R. K. and Otte, C. 1982. Molec. Cellul. Biol. 2: 21-29.
8. Ciejek, E. M. 1980. Ph.D. Thesis, University of California, Berkeley, CA.
9. Ciejek, E. and Thorner, J. 1979. Cell 18: 623635.
10. Ciejek, E., Thorner, J., and Geier, M. 1977. Biochem. Biophys. Res. Commun. 78: 952-961.
11. Fasiolo, F., Bonnet, J., and La Croute, F. 1981. J. Biol. Chem. 256: 2324-2328.
12. Fehrenbacher, G., Perry, K. and Thorner, J. 1978. J. Bacteriol. 134: 893-901.
13. Hartwell, L. H. 1980. J. Cell Biol. 85: 811-822.
14. Hereford, L. M., Osley, M. A., Ludwig, J. R., and McLaughlin, C. S. 1981 Cell 24: 367-375.
15. Herskowitz, I. and Oshima, Y. 1981. In Molecular Biology of the Yeast Saccharomyces: Life Cycle and Inheritance (Strathern, J. N., Jones, E. W., and Broach, J. R, Eds.) Cold Spring Harbor Laboratory, NY, pp. 181-209.
16. Karn, J., Brenner, S., Barnett, L., and Cesareni, G. 1980. Proc. Nat. Acad. Sci., USA 77: 5172-5176.
17. Kreil, G., Haiml, L. and Suchanek, G. 1980. Eur. J. Biochem. 111: 49-58.
18. Leibowitz, M. J. and Wickner, R. B. 1976. Proc. Nat. Acad. Sci., USA 73: 2061-2065.
19. MacKay, V. L. and Manney, T. R. 1974. Genetics 76: 255-271.
20. McCullough, J. and Herskowitz, I. 1979. J. Bacteriol. 138: 146-154.
21. Nasmyth, K. A. and Tatchell, K. 1980. Cell 19: 753-764.
22. Reed, S. I., Ferguson, J., and Groppe, J. C. 1982. Molec. Cellul. Biol. 2: 412-425.
23. Rogers, D. T., Saville, D., and Bussey, H. 1979. Biochem. Biophys. Res. Commun. 90: 187-193.
24. Sakurai, A., Sato, Y., Park, K. H., Takahashi, N., Yanagishima, N., and Banno, I. 1980. Agric. Biol. Chem. 44: 1451-1453.
25. Schleif, R. F. and Wensink, P. C. 1981. Practical Methods in Molecular Biology, Springer-Verlag, NY, 220pp.
26. Sherman, F., Fink, G. R., and Lawrence, C. 1979. Methods in Yeast Genetics: A Laboratory Manual, Cold Spring Harbor Laboratory, NY, 98pp.
27. Sprague, G. F., Jr., Blair, L. C., and Thorner, J. 1982. Ann. Rev. Microbiol. 36: in press.
28. Sprague, G. F., Jr., Rine, J. and Herskowitz, I. 1981. J. Molec. Biol. 153: 323-335.
29. St. John, T. P. and Davis, R. W. 1979. Cell 16: 443-452.
30. Suarez-Rendueles, M. P., Schwencke, J., Garcia-Alvarez, N. and Gascon, S. 1981. FEBS Letts. 131: 296-300.
31. Thomas, P. S. 1980. Proc. Nat. Acad. Sci., USA 77: 5201-5205.
32. Thorner, J. 1980. In The Molecular Genetics of Development (Leighton, T. J. and Loomis, W. F., Eds.) Academic Press, NY, pp. 119-178.
33. Thorner, J. 1981. In Molecular Biology of the Yeast Saccharomyces: Life Cycle and Inheritance (Strathern, J. N., Jones, E. W., and Broach, J. R., Eds.) Cold Spring Harbor Laboratory, NY, pp. 143-180.
34. Woo, S. L. C. 1979. Mets. Enzymol. 68: 384-395.
35. Wu, R. (Ed.) 1979. Methods in Enzymology, Volume 68, Academic Press, NY, 555pp.

23. PREPARATION OF PRODUCT-SPECIFIC ANTISERA BY GENE FUSION: ANTIBODIES SPECIFIC FOR THE PRODUCT OF THE YEAST CELL CYCLE GENE CDC28*

Steven I. Reed

Biochemistry and Molecular Biology Section
Department of Biological Sciences
University of California
Santa Barbara, California 93106

Rec. Adv. Yeast Mol. Biol. (1982) 1:306-319

Antisera with specificity for the product of a yeast cell division cycle (cdc) gene were prepared by immunizing rabbits to a novel hybrid polypeptide. A segment of the yeast gene CDC28 was fused to the E. coli lacZ gene, which encodes β-galactosidase, by insertion of yeast sequences into the plasmid pBGF1. pBGF1 contains the lac promotor-operator and most of the lacZ gene. An EcoRI site 16 codons upstream from the carboxyterminus of the β-galactosidase coding region served as a convenient splicing site for the heterologous sequences. To insure that an open reading frame be maintained between the two gene segments for some portion of the fusions, the CDC28-encoding segments were first subjected to limited digestion with nuclease BAL-31 to produce random junction points. A hybrid polypeptide encoded by such a continuous open reading frame was purified from E. coli by preparative SDS polyacrylamide gel electrophoresis and used to immunize rabbits. The resulting antisera were shown to have specificity for CDC28 gene product synthesized by cell free translation of yeast mRNA.

INTRODUCTION

Mutational analysis of a cellular process permits the identification of relevant genetic loci. Using genetic methods, investigators can then infer functions and patterns of interaction of the products of these loci, yet, most often, the molecular nature of these functions and interactions has remained elusive, particularly if the process under scrutiny is complex. Study of the yeast cell division cycle has followed such a pattern. Extensive mutational analyses have been performed in Saccharomyces on the progression through the cell division cycle and on its control (7,8,18,19). Subsequent physiological studies using cell division cycle (cdc) mutants (6,9) have established a fundamental and unifying model (7,18) for the organization of the cell cycle. In addition, pseudoreversion studies have suggested interactions between the products of a number of genes (14; our unpublished results). Yet, the elucidation of cell division in molecular terms has been hampered by a lack of information concerning the physical and biochemical properties of the products of the cdc genes. Conventional genetic methods unfortunately offer no strategy for proceeding from mutational definition of genes to identification of the products they encode and determination of the precise physiological roles that these products perform. As a result, in yeast, only a handful of the cdc loci have been matched with protein products (18) of known physiological function.

*Supported by US Public Health Service Grant 5 R01 GM28005-02.

Recombinant DNA methodologies allow the routine isolation of yeast genes on recombinant plasmids by complementation of mutants (1,3,10,15). While physical isolation of genes renders the analysis of gene structure and regulation accessible to investigation by biochemical methods, it also provides a direct approach to the biochemical analysis of encoded products. We have developed a method whereby a cloned gene can be used to prepare antisera which are specific for the product it encodes. The method entails the fusion of a segment of the gene to the lacZ gene of E. coli near the carboxyterminal end of its coding region. Gene fusions are screened for the maintenance of a continuous reading frame on the basis of their ability to produce a hybrid polypeptide of the predicted size. Such polypeptides are then isolated and used to prepare antisera which are expected to have some specificity for the product encoded by the native undisrupted gene. We have successfully employed this strategy for the yeast cell division cycle gene CDC28 (8,15,19,20) and have prepared antisera which can precipitate CDC28 product synthesized from yeast mRNA by cell-free translation.

MATERIALS AND METHODS

Organisms, DNAs, Enzymes, Plasmid Preparations, Transformations

The following strains were used: yeast, JF210-92 (MATα, trp1, leu2) containing the plasmid YRp7(CDC28.4)HCN (12); Escherichia coli, SF8(C600, hsdM, hsdR, recBC, lop11). Plasmid pBR322 has been described elsewhere (2). Plasmid pBGF1 was constructed by transferring a (5.7 kpb) HindIII-EcoRI fragment from the plasmid pBGP120 (17) into pBR322 from which the original HindIII-EcoRI segment had been removed. pBGF1 contains the lac promotor-operator and the portion of the lacZ gene extending to the EcoRI site 16 codons upstream from the carboxyterminus of the β-galactosidase coding region. Plasmid pRC1(CDC28.5) was constructed by removing from pRC1(CDC28.2), a plasmid described previously (4), a SalI-PvuII fragment containing the CDC28 coding region and the 3' intragenic spacer region (20) and inserting it into pRC1 (4) which had been cleaved with SalI and SmaI. By this construction, the PvuII terminus of the fragment and the SmaI terminus of the vector are fused by a blunt-end ligation into a novel hybrid junction (see Fig. 2). Plasmid preparations from E. coli as well as transformation were performed as described previously (4). Small scale plasmid preparations from E. coli (minilysates) were by the method of Holmes and Quigley (11). All restriction enzymes, T4 DNA ligase, and nuclease-free bovine serum albumin were purchased from New England Biolabs, Bethesda Research Laboratories, or purified as part of the undergraduate biochemistry laboratory course at the University of California, Santa Barbara by published procedures and were used under conditions suggested by the manufacturers. All yeast and bacterial culture media have been described (4).

BAL-31 Deletions

Deletions were prepared by digesting 2 micrograms of purified fragment DNA with 1.6 units of BAL-31 (Bethesda Research Laboratories). Reaction conditions were 12 mM $CaCl_2$, 12 mM $MgCl_2$, 20 mM Tris HCl, pH 8.0, 1 mM EDTA in 100 milliliters. 30 microliter aliquots were removed at 3.0, 3.5 and 4.0 minutes and digestion was stopped by extraction with PCI [phenol: chloroform: isoamyl alcohol (50: 50: 1)]. Aqueous phases from the three time points were pooled, ethanol precipitated and the pellet dissolved in 11 microliters of water.

Addition of EcoRI Linker Sequences and Cloning into pBR322

The redissolved mixture of deleted fragments was adjusted to the following conditions in 50 microliters: 66 mM Tris HCl, pH 7.6, 6.6 mM $MgCl_2$, 10 mM dithiothreitol, 0.2 mM ATP, 100 microgram/milliliter bovine serum albumin, 0.1 unit of phosphorylated *Eco*RI linker (Collaborative Research) and 4 units of T4 DNA ligase (Bethesda Research Laboratories). The reaction was allowed to proceed for 16 hours at 14°C. After extraction with PCI and ethanol precipitation, the fragment preparation was cleaved with an excess of *Eco*RI (10 New England Biolab units/microgram of fragment). Samples were extracted once more with PCI, ethanol precipitated and added to 4 micrograms of *Eco*RI-cleaved pBR322 DNA which had been subsequently digested with bacterial alkaline phosphatase (Bethesda Research Laboratories). Phosphatase treatment was done at a ratio of 100 units/microgram of pBR322 DNA for 5 hours at 40°C in 10 mM NaCl, 10 mM Tris HCl, pH 8.0 and 1 mM EDTA followed by extraction with PCI and ethanol precipitation.

Fragments with attached *Eco*RI termini and *Eco*RI-cleaved pBR322 DNA were reacted in 40 microliters with 1 unit of T4 DNA ligase for 16 hours at 23°C. The reaction was then extracted with PCI, ethanol precipitated and redissolved in 100 microliters of 10 mM Tris HCl, pH 8.0, 1 mM EDTA. 25 microliters was used to transform *E. coli* cells and colonies containing recombinant plasmids were detected by colony hybridization.

Colony Hybridizations

Colonies containing recombinant plasmids were detected by a modification of the method of Grunstein and Hogness (57). Plates containing from 50 to 200 transformant colonies were replica-plated in duplicate. A disc of Whatman 3MM paper was carefully layered on one of each duplicate pair after incubation for 5 hours at 37°C. After 3 additional hours at 37° the disc was removed and placed colony-side-down on an LB plate containing 250 micrograms per milliliter chloramphenicol for 12 to 16 hours. Filters were removed and prepared for hybridization by incubation for 7 minutes in 0.5 M NaOH followed by two five-minute incubations in 1 M Tris HCl, pH 7.2, and a final incubation in 0.5 M Tris HCl, pH 7.2, 1.5 M NaCl for five minutes. Filters were then rinsed several times with 95% ethanol and dried under a heat lamp for 10 minutes. They were then stacked in a 100 mm plastic petri dish and 3 ml of buffer containing 0.5 M NaCl, 0.1 M sodium phosphate, pH 7.0, 6 mM EDTA and 1% sodium lauryl sulfate (SDS) and 200,000 cpm of nick-translated probe (2.8 kbp fragment in Fig. 2; 2-5 x 10^7 cpm/microgram) were added per filter. Nick translations were performed as has been described previously (21). The hybridization reaction was sealed by laying a sheet of Saran Wrap (Dow Chemical) on the solution and crimping it down by replacing the lid of the petri dish. The reaction was allowed to proceed for 12-16 hours in a 65° incubator. Filters were then washed three times for 15 minutes in 5 X SSC, 0.2% SDS at 65°C and twice in 2 X SSC at 23° (SSC = 0.15 M sodium chloride, 0.015 sodium citrate). They were then dried for 15 minutes under a heat lamp and exposed to Kodak XAR-5 film in the presence of a Cronex Lightening Plus intensifying screen for 8-20 hours at -75°C. Colonies which gave a hybridization signal were then located on the duplicate replica-plate for further analysis.

Restriction Analysis of pBR322 Clones and Construction of β-galactosidase Fusions

Approximate endpoints of deleted fragments were determined by digestion of recombinant plasmids, prepared as mini-lysates with EcoRI and SacI followed by electrophoresis on 1% agarose gels. The deleted fragments chosen for further study are pictured schematically in Figure 2. These were purified by digesting 100 micrograms of purified plasmid with EcoRI, followed by separation on a 4.5% polyacrylamide gel. Fragments were removed from the gel matrix as described by Maxam and Gilbert (13) and quantitated by agarose gel electrophoresis against known DNA fragment standards. Purified fragments were then mixed with EcoRI-cleaved pBGF1 DNA at an equimolar ratio and a final DNA concentration of 200 micrograms per milliliter and incubated with T4 DNA ligase according to manufacturer's specifications. DNA was then transformed to *E. coli*, colonies containing recombinant plasmids detected by colony hybridization and orientation of fragments determined by restriction analysis of mini-lysates.

Screening for Fusion Polypeptides

Transformants containing gene fusions were inoculated into M9 medium supplemented with casamino acids, tryptophan, thymine, 0.2% glycerol and 20 micrograms per milliliter ampicillin (Bristol Laboratories) and grown overnight to stationary phase at 34°C. One milliliter of this culture was then inoculated into nine milliliters of the same medium without ampicillin and incubated at 34°C for 100 minutes. At that time, the culture was made 50 micrograms per milliliter in isopropyl thiogalactoside. After 20 additional minutes of incubation, the cultures were made 10 mM in sodium azide and rapidly cooled to 0°C, on ice. Cells were harvested by centrifugation in a clinical centrifuge at 0° and pellets resuspended in 100 microliters of SDS polyacrylamide gel sample buffer (22). After transfer to a 1.5 milliliter Eppendorf type tube, samples were subjected to sonic disruption using a needle probe of a Braunsonic 1510 sonic disruption device for two 10-second bursts with care taken to avoid foaming. Samples were then placed in a boiling water bath for two minutes and centrifuged in an Eppendorf microfuge for 5 minutes to remove debris. 10 microliter aliquots were then analyzed by electrophoresis on 7.5% SDS polyacrylamide gels (22). Protein bands were detected by staining with Coomassie Brilliant Blue R.

Immunization of Rabbits

Bacterial lysate was prepared as described above except that the procedure was scaled up by a factor of 100. Considerably more extensive sonic disruption was required to break cells and fragment nucleic acid due to the increased volume of the cell suspension (5-10 20-second bursts). 400 microliters were loaded per 1.3 mm thick preparative 7.5% SDS slab gel. Samples were stacked at 2.5 milliamp per gel and then run at 7.5 milliamp until twice the time required for the tracking dye (bromphenol blue) to migrate to the end of the gel had elapsed. The gel was stained in Coomassie Blue [0.25% in ethanol: acetic acid: water (45: 10: 45)] for 30 minutes and then destained in water with several changes until bands could be observed with the aid of a light box. Bands were excised and prepared for immunization of rabbits as described by Tjian et al. (24), except that boosts were performed without adjuvant.

Detection of Anti-P28 Activity

CDC28 mRNA was purified and translated using a rabbit reticulocyte lysate as has been described previously (20). For immunoprecipitation, 25-microliter translation reactions were diluted with 40 microliters of 20 milligram per milliliter bovine serum albumin (Sigma, Fraction V) and 6.5 microliters of 10% SDS. Samples were then boiled for 2 minutes followed by cooling to 0° on ice. Four 16-microliter aliquots were then transferred to separate tubes and each was brought to 1% Triton X-100, 50 mM NaCl, 20 mM Tris HCl, pH 7.5 in a final volume of 64 microliters. 15 microliters of either pre-immune or immune serum were then added and incubated on ice for 16 to 20 hours. 25 microliters of Staphylococcus aureus (Staph A) cells prepared according to Kessler (12) and resuspended in 1% Triton X-100, 50 mM NaCl, 20 mM Tris HCl, pH 7.5, 2 mg per milliliter bovine serum albumin were added and incubated for an additional 30 minutes on ice. Three washes in the same buffer were carried out at 4° by centrifugation in an Eppendorf microfuge followed by a final wash in 10 mM Tris HCl, pH 7.5. Staph A pellets were then resuspended in SDS polyacrylamide gel sample buffer (24) (25 microliters), boiled for 2 minutes and recentrifuged to remove insoluble material. Supernatants were electrophoresed and ^{35}S-labelled proteins autofluorographed as has been described previously (20).

RESULTS

Strategy

The strategy employed for the preparation of gene product-specific antisera is outlined in Figure 1. β-galactosidase from E. coli is used as a vehicle for presentation of the CDC28 polypeptide as an antigen. The hybrid polypeptide is produced in vivo as the result of a gene fusion constructed in vitro. Fragments deleted randomly into the 5' region of the CDC28 gene produced by reaction with the double strand nuclease BAL-31 serve as potential donors of CDC28 product antigenic determinants [Fig. 1 (a)]. These gene segments are linked to the carboxyterminal region of lacZ via synthetic EcoRI linkers [Fig. 1 (b,c)]. The site of gene fusions is a naturally occurring EcoRI site sixteen codons upstream from the carboxyterminus of the β-galactosidase coding region in plasmid pBGF1 (Fig. 1). Fusions which maintain the reading frame of lacZ through the CDC28 segment are detected by the appearance of a predicted fusion polypeptide upon electrophoresis of bacterial lysates [Fig. 1 (d)]. The hybrid polypeptide is then partially purified by SDS polyacrylamide gel electrophoresis and used to immunize rabbits [Fig. 1 (e,f)]. Serum is finally tested for specificity by demanding that it precipitate CDC28 product translated from yeast mRNA using a rabbit reticulocyte cell free lysate [Fig. 1 (g)].

Production of 5' Deleted CDC28 Gene Segments

Deletions extending into the 5' coding region of CDC28 were produced by treatment of a 2.8 kbp DNA fragment containing the entire gene with the exonuclease BAL-31 (Fig. 2). Since the 5' end of the coding region was estimated to be approximately 400 base pairs away from one end of the fragment, the reaction conditions were calibrated to remove about 1000 base pairs or 500 base pairs from each end, as determined by agarose gel electrophoresis (data not shown). EcoRI linker sequences were attached to the digested fragment DNA,

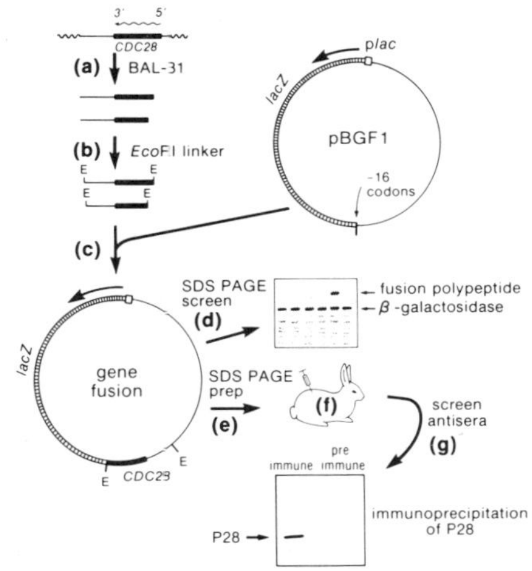

Figure 1

Strategy for production of product-specific antisera. The transcriptional orientation of the CDC28 gene was determined using T4 DNA polymerase according to the method of O'Farrell (16). A 2.8 kbp fragment containing the CDC28 coding region, a small 5' flanking segment and the entire intergenic spacer region on the 3' side, as well as some vector sequences on both sides, served as the initial substrate for gene fusion. This fragment was digested with a pre-calibrated amount of the double-strand exonuclease BAL-31 (a) to give deletions into the 5' end of the CDC28 coding region. EcoRI linker sequences were added to the blunt ends of the digested fragments (b) allowing insertion into plasmid pBGF1 (c). pBGF1 contains the lac promotor-operator as well as the lacZ gene encoding β-galactosidase except for the 16 carboxyterminal codons downstream from the EcoRI site (designated E). Plasmids containing CDC28 sequences were detected by colony hybridization and gene fusions were screened by restriction analysis. Of these, fusions which maintained a continuous open reading frame were detected by analysis of E. coli transformant protein extracts by SDS polyacrylamide gel electrophoresis (d). In-frame fusions are expected to produce a stainable band of lower mobility than that of β-galactosidase (d). Preparative quantities of the fusion polypeptide were then prepared (e) by scaling up the electrophoresis procedure. Rabbits were immunized (f) using the material excised from polyacrylamide gels. Finally, antisera were screened for ability to precipitate CDC28 product (P28) synthesized from yeast mRNA by cell-free translation (g).

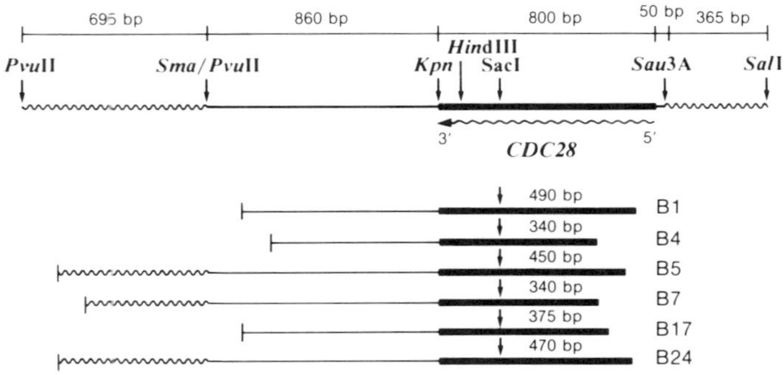

Figure 2

BAL-31 deletions into the CDC28 coding region. A 2.8 kbp restriction fragment obtained by digestion of pRC1(CDC28.5) with SalI and PvuII was subjected to limited treatment with the double strand nuclease BAL-31. Deleted fragments were cloned into pBR322 using synthetic EcoRI linkers and deletion endpoints were localized by digestion of recombinant plasmids with EcoRI and SacI. Based on this analysis, six deletions designated B1, B4, B7, B17, and B24, were chosen to construct gene fusions. The approximate distance from the SacI site within the CDC28 coding region to the deletion endpoint near the 5' end of the coding region is given for each. The location of the SacI site is indicated by an arrow. (▬▬▬), CDC28 mRNA coding sequence, (———) yeast genomic non-coding sequence, (〰〰) vector sequence. (◄〰), indicates orientation of CDC28 transcript.

permitting insertion at the EcoRI site of pBR322. Recombinant plasmids were detected by colony hybridization and insert endpoints mapped by digestion with EcoRI and SacI. The shorter EcoRI-SacI fragment was assumed to define the 5' endpoint of the deleted gene. Using this assumption, the structures of six deleted CDC28 genes obtained are shown in Figure 2. These six, of 24 originally chosen for analysis, constitute the 5'-deleted segments retaining the largest portions of the CDC28 coding region.

Construction of LacZ Fusions

Fragments containing deleted CDC28 genes were isolated by cleaving the six pBR322 recombinants described above with EcoRI followed by electrophoresis in 4.5% polyacrylamide gels and purification. The purified fragments were then inserted into the EcoRI site of plasmid pBGF1 which defines a potential fusion point 16 codons upstream from the carboxyterminal end of the coding sequence of β-galactosidase. The orientation of each of the fragments with respect to the vector was determined by cleavage with SacI, since a SacI site is situated asymmetrically within each of the cloned segments. There are two additional SacI sites in the vector, pBGF1.

Detection of Continuous Open Reading Frames

Assuming that the CDC28 deletion endpoints were truly random, one in three of the gene fusions should have maintained the proper reading frame at the junction between the lacZ and CDC28 moieties. Such fusions, then should result in the production in E. coli of a novel polypeptide containing constituents encoded by both gene segments. The presence of a continuous open reading frame was determined by assaying for an additional major polypeptide in the transformant strains. E. coli strains harboring β-galactosidase fusions with the six deleted fragments shown in Fig. 2 were induced for β-galactosidase using IPTG (isopropyl thiogalactoside), harvested and lysed by sonic disruption directly in SDS polyacrylamide gel electrophoresis sample buffer. Proteins, separated by electrophoresis on a 7.5% SDS polyacrylamide gel, extracted from these strains as well as from a strain harboring the vector, pBGF1, are shown in Figure 3. Although the steady state level of β-galactosidase varies from strain to strain, only one produces an additional polypeptide of the size anticipated. The fusion pBGF1(B24) directs synthesis of a novel polypeptide of approximately 158,000 daltons. The expected contribution of 133,000 daltons of β-galactosidase and 25,000 daltons of CDC28 product is consistent with this apparent size. Although the low steady state level observed (Fig. 3) implies that the polypeptide is unstable in an E. coli cell, the abundance is adequate and the size optimal for electrophoretic purification.

Electrophoretic Purification of Fusion Polypeptide and Immunization of Rabbits

Preparative SDS polyacrylamide gels were used to obtain partially purified fusion polypeptide. As there are few polypeptides encoded by E. coli of molecular weight greater than 150,000 daltons of sufficient abundance to be detected by staining with the Coomassie Blue, a high degree of purification can be achieved in a single electrophoretic step. In order to maximize separation in the high molecular weight region, electrophoresis was performed until most of the proteins (less than 100,000 daltons in molecular weight) were run off of the gel. Under these conditions, the fusion polypeptide could be excised with

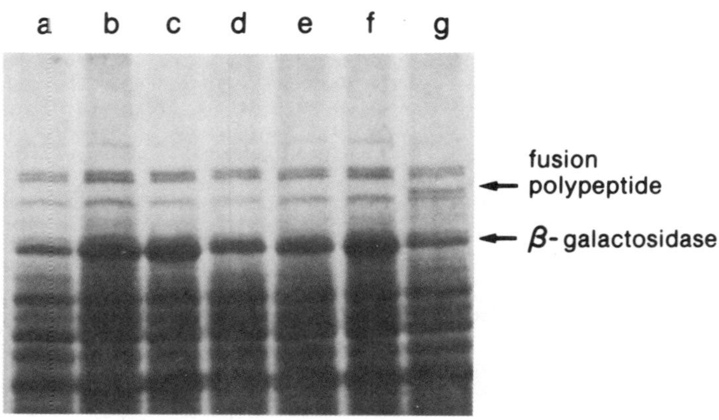

Figure 3

Electrophoretic screening for fusion polypeptides. Small scale lysates were prepared from E. coli transformant cultures and separated on 7.5% polyacrylamide gels in the presence of SDS. Protein bands were detected by staining with Coomassie Brilliant Blue R. Lanes a through g correspond respectively to lysates from E. coli SF8 harboring the vector, pBGF1, and fusions with the six deleted fragments: B1, B4, B5, B7, B17 and B24. The band corresponding to β-galactosidase (135,000 daltons) is indicated as well as the fusion polypeptide encoded by pBGF1(B24) (lane g).

minimal to moderate contamination from the band immediately below and occasional contamination from the band above (see Figure 3), depending on the run.

Excised bands were prepared essentially according to Tjian et al. (24). Polypeptide-containing polyacrylamide slices were lyopholized and then ground to a powder using a mortar and pestle. The powder was then resuspended in 0.1% SDS and injected after emulsification with Freund's complete adjuvant. Boosts were performed at 25 and 35 days subsequent to the initial immunization using the powder resuspended in 0.1% SDS without adjuvant. The amount of polypeptide per initial immunization was estimated at 10-20 micrograms with a similar amount used for each of two boosts. Blood samples for preparation of serum were taken 1 week prior to the initial immunization and 12 days subsequent to the second boost.

Detection of Anti-CDC28 Product (P28) Immunological Activity

Serum collected from two rabbits subjected to the immunization regime described above was initially tested for anti-β-galactosidase activity using an Ouchterlony diffusion assay and found to be positive (data not shown). To determine whether the sera also contained anti-*CDC28* product (P28) activity, hybrid release translation was performed on yeast mRNA. *CDC28* mRNA partially purified by hybrid formation with *CDC28*-containing plasmid bound to DBM paper was translated using a rabbit reticulocyte lysate (20). The translation products directed by partially purified *CDC28* RNA and the equivalent amount of *CDC28*-depleted RNA (unbound fraction from the hybrid selection procedure) are shown in lanes a and b, respectively, of Fig. 4. P28 is seen as a prominent band at 27,000 daltons. Lanes c and d contain immunoprecipitates from a cell-free translation reaction equivalent to lane a where pre-immune serum (lane c) is compared to post-immune serum (lane d). Under the conditions employed, immunoprecipitation of the *CDC28* product is highly specific. Conditions of high stringency were required (see *Materials and Methods*), however, to prevent non-specific precipitation of the *CDC28* polypeptide, only a fraction of which was then precipitated. It is not known whether this inefficiency is due to the inavailability of some of the antigen molecules, the high ionic detergent concentrations that were necessary for specificity, or a low antibody titer. We are presently attempting to resolve this question so that the efficiency of P28 immunoprecipitation may be improved.

DISCUSSION

We present here a practical method for the preparation of antisera with specificity for the product of, in principle, any cloned gene. The procedure utilizes the construction of a hybrid gene consisting of the *lac* operator-promotor region, most of the adjacent *lacZ* gene of *E. coli* and as much as possible of the coding region of the gene of interest. The junction occurs near the carboxyterminus of the β-galactosidase coding region and near the aminoterminus of the component to be spliced. This arrangement has several important advantages. Firstly, expression of the encoded hybrid polypeptide can be efficiently regulated in *E. coli* since the *lac* operator-promotor can be maintained either in the repressed or, in the presence of inducers, derepressed mode to give a high level of transcriptional activity. Both the potential for high levels of expression and the ability to maintain repression, should the fusion product be deleterious to the host, are significant advantages. We, in fact, have observed that induction of the gene fusion which produced a hybrid

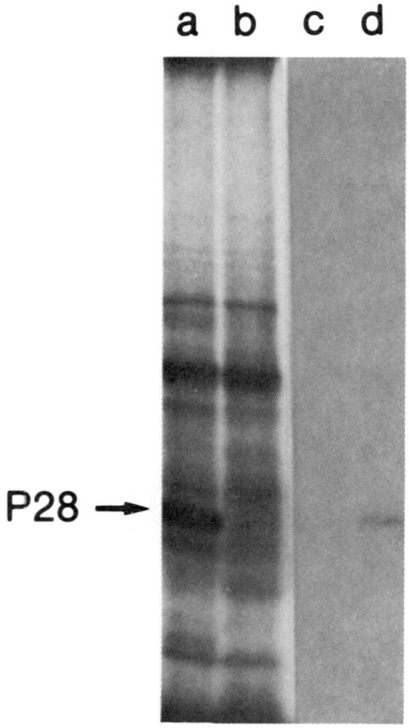

Figure 4

Detection of immunological specificity for the CDC28 product. CDC28 mRNA was partially purified by hybridization to plasmid pBR322(CDC28.1) bound covalently to DBM paper and translated using a rabbit reticulocyte lysate as has been described (20). Incorporation of ^{35}S-methionine followed by autofluorography allows visualization of proteins synthesized. Lane a corresponds to translation products directed by the partially purified CDC28 mRNA whereas lane b corresponds to products directed by an equivalent amount of the unbound fraction removed during the post-hybridization washing procedure, presumably depleted for CDC28 mRNA. The CDC28 product (P28) is seen as a prominent band in land a and represents the only significant difference between lanes a and b. *In vitro* translation products equivalent to those displayed in lane a were allowed to react with pre-immune and post-immune serum from the same rabbit and immune complexes precipitated using *Staph A* as is described in *Materials and Methods*. Pre-immune and post-immune precipitable translation products are seen in lanes c and d, respectively. P28 is precipitated by post- but not pre-immune serum.

polypeptide, pBGF1(B24), for longer than a brief interval, resulted in considerable cell lysis. It was thus important to be able to maintain cultures in the uninduced state except for a short time prior to harvest.

The second advantage of the system concerns the use of β-galactosidase, a large polypeptide, for presentation of the gene product as an antigen. The rationale, albeit unproven, is that the spliced foreign polypeptide will be protected from the potentially hostile milieu of the bacterial cell by the large carrier polypeptide. We have constructed a continuous reading frame fusion using the same CDC28 segment, B24, and a considerably smaller segment of the lacZ gene and were unable to detect a hybrid polypeptide at significant steady state levels (unpublished results), supporting this premise. Furthermore, the large size of the fusion polypeptide anticipated using pBGF1 allows for easy purification. There are few proteins of abundance in *E. coli* of molecular weight greater than that of β-galactosidase (135,000 daltons) so that fusion proteins are easy to detect in whole cell lysates separated by SDS polyacrylamide gel electrophoresis and easy to recover in reasonably pure form by the same method. This simple strategy would be impractical if the fusion protein were significantly smaller.

Immunizations were carried out with relatively small amounts of protein still enmeshed in polyacrylamide matrix. It is not known how much of the polypeptide became available to the rabbits' immune systems. It is suspected, however, based on low titers against β-galactosidase (unpublished observations) that the titers against P28 were also low. In fact the serum from one of two rabbits was only barely capable of a specific immunoprecipitation response above background (data not shown) and the other, although well above background, could only precipitate a fraction of the P28 present in the lysate, possibly, as has been mentioned, due to a low antibody titer. Antisera of higher titer could probably be prepared by using larger amounts of antigen in the immunization procedure. As antigen is easy to purify, such a modification is feasible.

Stringent detergent conditions were required to prevent non-specific precipitation of P28 from cell-free lysates. It was not established whether P28 is insoluble or has a high affinity for Staph A cells or both. Initial boiling of samples in 1% SDS and performance of precipitations in 0.2% SDS, 1% Triton X-100 allowed specific precipitation of P28 by immune sera as can be seen in Fig. 4. It is not known whether these conditions eliminate a subset or majority of potential antibody-antigen interactions resulting in reduced efficiencies of immunoprecipitation. This, however, is not expected to be a problem of general concern for gene products of normal solubility.

In organisms where intensive genetic studies have yielded collections of mutants defective in cellular processes and where genes defined by mutants have been physically isolated using recombinant DNA techniques, methods such as the one reported here should permit extension of studies to the characterization of the structures and, ultimately, of the molecular functions of the encoded gene products. The inability to identify and characterize the products of interesting genes has been a serious impediment to the elucidation, in molecular terms, of important cellular and organismal processes where sophisticated genetic analysis has been possible. Our method also permits the preparation of antisera specific for products of genes expressed at extremely low levels, since expression of antigen is now linked to the lac operator-promotor of *E.*

coli. In principle, then, this method might find application even where the gene product is known but difficult to obtain due to low abundance or intractability to purification.

An alternative method for producing antibodies specific for the products of cloned genes has been described (23,25). It involves the extrapolation of amino acid sequence from the DNA sequence, the selection of likely antigenic runs of amino acids from the sequence and the synthesis of a synthetic peptide for immunization. The disadvantages of this approach are that it involves having sequenced the gene, some knowledge of antigenic amino acid combinations, the ability to construct synthetic peptides, and the limitation of the immunization to a relatively small set of antigenic determinants. Our method requires only prior cloning of the gene, localization and determination of the orientation of the coding sequence and some simple biochemical and microbiological manipulations and, as such, should be of more general application.

I would like to thank Dr. John Sninsky for plasmid pBGP120, Drs. David Asai and Randy Sheckman for invaluable advice concerning immunological techniques, and Ms. Debbie Mustard for typing the manuscript.

LITERATURE CITED

1. Beggs, J. D. (1978) Nature 275, 104-108.

2. Bolivar, F., Rodriguez, R. L., Green, P. J., Betlach, M. C., Heyneker, H. L., Boyer, H. W., Crosa, J. H & Falkow, S. (1977) Gene 2, 95-113.

3. Broach, J. R., Strathern, J. M. & Hicks, J. B. (1979) Gene 8, 121-133.

4. Ferguson, J., Groppe, J. C. & Reed, S. I. (1981) Gene 16, 191-197.

5. Grunstein, M. & Hogness, D. H. (1975) Proc. Natl. Acad. Sci. USA 72, 3961-3965.

6. Hartwell, L. H. (1976) J. Mol. Biol. 104, 803-817.

7. Hartwell, L. H., Culotti, J., Pringle, J. & Reid, B. (1974) Science 183, 46-51.

8. Hartwell, L. H., Mortimer, R. K., Culotti, J. & Culotti, M. (1973) Genetics 74, 267-286.

9. Hereford, L. M. & Hartwell, L. H. (1974) J. Mol. Biol. 84, 445-461.

10. Hinnen, A., Hicks, J. B. & Fink, G. R. (1978) Proc. Natl. Acad. Sci. USA 75, 1929-1933.

11. Holmes, D. & Quigley, M. (1981) Anal. Biochem. 114, 193-197.

12. Kessler, S. W. (1975) J. Immunol. 115, 1617-1624.

13. Maxam, A. & Gilbert, W. (1977) Proc. Natl. Acad. Sci. USA 74, 560-564.

14. Moir, D., Stewart, B.C., Osmond, B. C. & Botstein, D. (1982) Genetics, in press.

15. Nasmyth, K. A. & Reed, S. I. (1980) Proc. Natl. Acad. Sci. USA 77, 2119-2123.

16. O'Farrell, P. (1981) BRL Focus 3, No. 3, 1-3.

17. Polisky, B., Bishop, R. J. & Gelfand, D. H. (1976) Proc. Natl. Acad. Sci. USA 73, 3900-3904.

18. Pringle, J. R. & Hartwell, L. H. (1982) in The Molecular Biology of the Yeast Saccharomyces, J. N. Strathern, E. W. Jones and J. R. Broach, eds. (Cold Spring Harbor Laboratory Press, Cold Spring Harbor, NY), pp. 97-142.

19. Reed, S. I. (1980) Genetics 95, 561-577.

20. Reed, S. I., Ferguson, J. & Groppe, J. C. (1982) Mol. Cell. Biol. 2, 412-425.

21. Rigby, P. W. J., Dieckmann, M., Rhodes, C. & Berg, P. (1977) J. Mol. Biol. 113, 237-251.

22. Studier, F. W. (1973) J. Mol. Biol. 79, 237-248.

23. Sutcliffe, J. G., Shinnick, T. M., Green, N., Liu, F.-T., Niman, H. L. & Lerner, R. A. (1980) Nature (London) 287, 801-805.

24. Tjian, R., Stinchcomb, D. & Losick, R. (1974) J. Biol. Chem. 250, 8824-8828.

25. Walter, G., Scheidtmann, K.-H., Carbone, A., Laudano, A. & Doolittle, R. F. (1980) Proc. Natl. Acad. Sci. USA 77, 5197-5200.

Part V

Mating Type: Control of Silent Cassettes and Mating Interconversion

24. THE MATα2 GENE

Ira Herskowitz

Department of Biochemistry & Biophysics
Division of Genetics
University of California
San Francisco, California 94143

Rec. Adv. Yeast Mol. Biol. (1982) 1:320-331

Mike Esposito told me that one of the purposes of this volume is to provide a record of the state of the field. Rather than restate the contents of the talks from my laboratory by Janet Kurjan (on α-factor) and by Rob Jensen (on control of HO), I thought that I would take this opportunity to review several different aspects of mating physiology, cell type determination, and mating type interconversion from a novel perspective. The full stories on α-factor and on HO are soon to be published (19, 12), and it is more than I can bear to reproduce these schticks in full for this volume. Relevant information on α-factor and HO will, however, be described in the course of this epic.

To provide a record of our thinking about mating physiology, mating type determination, and mating type interconversion, I have chosen to describe much of what we know about the α2 gene of the α mating type locus and the logic that has led to our present position. Equally interestingly (I hope), I shall describe some speculations on the α2 product and its potential role in the switching pattern of homothallic strains.

As will become readily apparent, we are in that marvelous position with respect to α2 of having only a limited amount of concrete biochemical information. Thus, much of what we know and speculate about α2 is based on physiological and genetic observations and consequent deductions.

In our laboratory, we have pursued studies on cell type differentiation and on mating type interconversion in parallel. It has proved to be the case that each of these "different" areas has provided much information that is relevant to the other. Because many of our assays involve the complex processes of mating, pheromone response, and mating type switching, we have inevitably been drawn into several aspects of mating physiology and its genetic control.

THE α1-α2 HYPOTHESIS

Discovery of the MATα2 gene. The story of the MATα2 gene of course begins with the identification of mutants defective in that gene (20). One of the mutants derived from an α strain with a defect in mating (strain VC73) carried a mutation (ste73) with properties that made it difficult to study genetically. Rare diploids formed between VC73 and MATa strains were unable to sporulate; hence it was not possible at that time to determine whether the ste73 mutation was located at the mating type locus. MacKay and Manney however noted that the diploid formed between VC73 and a MATa strain, when mated to a MATα/MATα strain, yielded a tetraploid strain that sporulated. The only simple explanation for these results was that the ste73 mutation was a mutation in MATα that is recessive to wildtype α mating type locus and that the wildtype gene is required in a/α cells for sporulation. All of these deductions have proved to be correct.

Confirmation that ste73 was a mutation of MATα came from mapping analyses, performed in two ways. First, Jim Hicks constructed sporulation-proficient tetraploids similar to those of MacKay and Manney and analyzed the diploid spores produced upon sporulation. This analysis showed that the ste73 mutation was closely linked to the CRY1 locus, which is closely linked to the mating type locus (5, 35). Tetrad analysis of diploids (which is considerably easier to understand!) was performed soon afterwards by Jeff Strathern and yielded similar results (32, 35). The fact that ste73/MATa diploids ordinarily do not sporulate was circumvented by utilizing strains that carry a second mutation, unlinked to the mating type locus, which allowed sporulation. Hopper and Hall (11) identified a mutation (CSP1, at that time thought to be dominant) that allows a/a and α/α strains to sporulate. Strathern thus convinced ste73 strains to sporulate by mating them to a MATa strain carrying CSP1. The tetrads then showed a simple 2 a:2 non-mater segregation, confirming that ste73 mapped to the mating type locus. (We now know that CSP1 is a recessive mutation and is presumably identical to the rme mutation (13). Fortunately the parent of the ste73 mutant and thus VC73 also carried this mutation.) These two analyses confirmed that the ste73 mutation mapped to the α mating type locus. Furthermore, analysis of the segregants showed that a single mutation, ste73, was responsible for the two phenotypes of the original isolate (VC73), inability to mate and inability to sporulate when mated to an a strain. In addition, diploid segregants derived from tetraploids carrying ste73 were produced and could be tested for their phenotype: ste73/MATα strains behaved as α; ste73/MATa strains, as noted above, did not sporulate. [ste73/MATa cells exhibit the same mating properties as ste73 strains (see below) and thus differ from MATa/MATα cells in this respect as well.]

α2 is a negative regulator. Although VC73 was isolated as a "sterile", that is as a mating-defective derivative of an α strain, a close look at its mating behavior showed that it does mate a bit: but it mates as an a (with an efficiency approximately 0.1 to 1% of that by a true a cell) (5, 35, 28). This phenotype of VC73 was the first clue to another function of what we now call the α2 gene. The second clue was Jim Hicks's observation that VC73 also behaved as if it were an a cell in that it degrades α-factor (6). This a cell-specific function, which we termed "Bar", is exhibited by a cells and by VC73 but not by a/α cells or by various other ste mutants derived from α cells. Because the ste73 mutation is recessive, we argued that this mutation inactivates a gene (MATα2) whose product inhibits expression of functions ordinarily expressed only in a cells. In other words, we proposed that α2 is a negative regulator of a-specific genes (asg). It has subsequently been found that strains defective in the α2 gene exhibit several other phenotypes of a cells: response to α-factor (39), production of a-factor (2), and production of the a cell agglutination substance (40). The gene defined by the ste73 mutation thus has two roles--inhibition of a-specific functions in α cells and promotion of sporulation in a/α cells.

One other mutant isolated by MacKay and Manney, strain VP1, also behaved in some respects like VC73. In particular, VP1 also exhibited the Bar phenotype (6) and mated a bit as a (28). One notable difference between VC73 and VP1 was that VP1 did not seem to exhibit the sporulation defect of VC73:

diploids formed between VP1 and MATa strains sporulated. Later, George Sprague made us feel a bit more "comfortable" by finding that sporulation by such strains was less efficient than by MATα strains (30). VP1 thus appears to carry a leakier mutation than VC73.

There were at the time two other ste mutations that were isolated in α strains and that mapped to MATα (20), what we now call ste1-2 and ste1-5. These mutants were quite different from VC73 and VP1: strains carrying ste1-2 or ste1-5 did not exhibit any phenotypes of a cells, neither mating as a nor exhibiting Bar activity. Furthermore, ste1-2/MATa and ste1-5/MATa sporulated as well as did MATα/MATa strains. The phenotypes of mutations at MATα thus fell into two categories, those like ste73 and those like ste1-5. It was satisfying when Jeff Strathern observed, as described next, that MATα contains two complementation groups, α1 and α2, that correspond to ste1- and ste73-type mutations, respectively (32, 35).

Genetic analysis of mutants that are defective in mating is fraught with potential complications because isolating diploids requires selecting for prototrophs from a mixture of parents. To form such diploids, we count on the mutation being leaky to allow some mating, and watch out for reversion of the mutation or isolation of suppressor mutations (which as we shall see has proved immensely informative). Fortunately, it is easy to determine whether the diploids obtained by selection have been formed due to leakiness or due to a genetic change: one simply sees whether the original mutation segregates appropriately after sporulation of the diploid. Strathern formed diploids by pairwise rare matings among VC73, VP1, and the ste1 mutants and observed that some diploids exhibited a non-mating phenotype, whereas others mated as α. The latter type were formed by matings between VC73 or VP1 with ste1-2 or ste1-5 strains and thus indicated the existence of two complementation groups. Critical to this conclusion was the observation that these diploids sporulated (again by virtue of carrying CSP1) to yield the parental ste mutations. We have renamed the mutations in VC73 and VP1 matα2-1 and matα2-4, respectively.

Behavior of $\alpha1^-\alpha2^-$ double mutants. Dissecting asci has some similarities to constructing phage recombinants by toothpick tests--it is a repetitive manual activity during which the practitioner has the opportunity to meditate over the plates or the agar slab. There is thus time to ponder about what might be wrong with the cross, about whether the desired recombinant will be viable or whether it will have some unusual properties. Perhaps it was in the course of dissecting the asci from an $\alpha1^-/\alpha2^-$ (matα1-5/matα2-1) diploid that Jeff Strathern realized something quite striking about one of the segregants from this cross. Of course wildtype recombinants ($MAT\alpha^+$) should be formed. What would be the phenotype of the reciprocal recombinant, defective in both MATα1 and MATα2? Our limited analysis of the parent $\alpha1^-$ or $\alpha2^-$ single mutants led to the view that α1 was a positive regulator of α functions and that α2 was a negative regulator of a functions. The matα1 matα2 double mutant thus would not express the α behaviors (because it lacked α1) but would exhibit a behaviors (because it lacked α2). In other words, the $\alpha1^-/\alpha2^-$ diploid should produce segregants that mate as a. Indeed, such segregants were obtained at a rather high frequency, approximately 1% (35). MATα recombinants were found at a comparable frequency. It was immediately apparent that the segregants that mated efficiently as a were not switches to MATa because when they were mated to an α strain, the resultant diploid did not have the properties of a MATa/MATα cell. Rather the diploids mated as α and did not sporulate. Furthermore, it was possible to cross the putative matα1 matα2 double mutants to MATα and identify recombinant segregants that contained the original matα1 and matα2 mutations. Finding that matα1 matα2 mutants have the mating phenotype of a cells was very exciting in that it indicated that our views of α1 and α2 had predictive value. Likewise, the analysis showed that the a mating type locus need not code for analogues to α1 and α2: a-specific functions are simply constitutive in a cells. These results furthermore explained why no one had isolated mutations at MATa that were defective in mating: the only mutation of MATa that was known, called a* (now mata1), mates fine as an a but (like the matα1 matα2 double mutant) does not form a sporulating diploid upon mating with α cells (13). Finally, the mating of $\alpha1^-\alpha2^-$ cells as a provided a way of understanding a set of observations concerning rare matings between ho cells of like mating type (35). When MATa ho cells carrying complementary auxotrophic mutations are mixed, prototrophs are obtained at a frequency of ca. 10^{-7} (24). These prototrophs are MATa/MATα diploids formed as result of one parent switching from MATa to MATα by genetic rearrangement. Matings between α strains yield a different result: although MATa/MATα prototrophs occur at a frequency of ca. 10^{-7}, the predominant class (approximately 10 times more frequent) is not MATa/MATα. These prototrophs mate as α and do not sporulate. Subsequent analysis indicates that these prototrophs are formed by matings between MATα cells that mate as α and MATα cells that transiently have a $Mat\alpha1^-$ $Mat\alpha2^-$ phenotype or that have lost MATα and thus that mate as a.

The role of α2 in determining properties of a/α cells. We have seen above that α2 is important in determining the properties of haploid α and a cells. In α cells, it turns off a-specific functions. In a cells, its absence allows these functions to be expressed. How does α2 promote the properties of a/α diploids, in particular, inhibition of mating and activation of sporulation? Mating: α2 inhibits mating in a/α cells in at least two ways: First, α2 is inferred to act as in haploid α cells to inhibit expression of a-specific genes (28, 35). Secondly, α2 acts in conjunction with a1 to turn off synthesis of α1 product (18, 22). Consequently, α-specific genes necessary for mating whose expression requires the α1 product (such as the STE3 gene; 31, 29) remain unexpressed. Sporulation: We attribute the induction of sporulation in a/α cells also to result from α2 and a1 acting as a negative regulator (26). (Another negative regulatory action of α2 and a1, on HO gene expression, is described below.) Sporulation can occur in a/a and α/α cells (as well as in mata1/MATα and in MATa/matα2 cells, as noted earlier) when the product of the RME gene is absent (13, 26). The RME gene product thus can be viewed as a negative regulator of sporulation, which is expressed in a and in α cells but which is not expressed in a/α cells due to repression by α2 and a1.

WHAT MATα2 HAS TOLD US ABOUT THE STRUCTURE OF MAT AND ABOUT MATING TYPE INTERCONVERSION

Mutations in MATα2 have acted as an experimental foil to matα1 mutations in genetic studies of the structure of the mating type locus, in our studies of unlinked suppressors of mat mutations, and in mating type interconversion. More often than not, we were somewhat surprised by the results that were obtained.

Structure of MAT. In thinking about the structure of MAT, we originally considered two general types of structures: in one, the information for both a and α is present and differentially expressed in a or α cells; in the other, the a and α alleles are non-homologous blocks of DNA, analogous to the imm434 substitution of lambdoid phages (8, 30). According to the latter view, it should not be possible to rescue the wildtype allele of matα1 mutations by recombination in matα1/MATa diploids. This prediction was confirmed for both matα1-2 and matα1-5 (30). Surprisingly, the wildtype alleles of both matα2-1 and matα2-4 were rescued from MATa. Thus, the MAT alleles share regions of homology and non-homology. Parallel physical analysis of MATa and MATα confirmed these views and allowed us to explain our results in a specific manner. MATα and MATa contain nonhomologous sequences (termed Yα and Ya) of 747 and 642 base pairs respectively (10, 21, 36, 1). MATα produces two divergent transcripts that begin within the Yα region but which extend into flanking homologous regions, into the X region to the left and into the Z region to the right (18, 22). These two transcripts contain coding regions which were shown to correspond to α2 and α1 by complementation tests between the "classical" α1 and α2 mutations of our analysis and mutations produced in vitro (38). The nucleotide sequence of MATα shows that α1 product should be 175 amino acids in length, 78% of which is coded within Yα. In contrast, α2 is 210 amino acids in length and is coded primarily (88%) within the X region (1). Our genetic results are thus explained by proposing that matα1-2 and matα1-5 are located within the Yα region, whereas the matα2 mutations are within the X region.

A suppressor of matα mutations. In selecting for mating between matα1-5 and MATa strains, Jim Hicks found that some diploids segregated an unlinked suppressor of the matα1-5 mutation (5). At this point, we did not know whether this suppressor, ssp515, was an informational suppressor (for example, affecting translation) or a physiological suppressor (for example, a function that is negatively regulated by α1). Soon thereafter, Jeff Strathern showed that ssp515 also suppressed matα1-2 and thus was not allele specific (32). More remarkably, he found that ssp515 was also able to suppress matα2 mutations. Thus, ssp515 suppressed mutations in two very different functions of MATα. Our understanding of ssp515 was precipitated by Jasper Rine, who found that ssp515 also suppressed the sporulation defects of α/α and a/a diploids (25, 27). Thus, ssp515 could suppress mutations of both MATα and MATa. Our analysis of mating type interconversion had led us to propose the existence of silent mating type locus information in the yeast genome, at loci termed HMa (now called HMLα) and HMα (now called HMRa) (8, 7). Using Occam's Razor, we had argued that the silent information was not expressed because it was not adjacent to an active promoter (or other essential controlling site) and that the mating type locus contained such a playback site and thus expressed the cassette located at this position. Rine proposed that the situation was a bit different, namely that the cassettes at HML and HMR were kept silent by the action of a negative regulator (now called Sir) (4, 27). ssp515 (sir1-1 of ref. 27) thus allows expression of both an α cassette at HMLα and an a cassette at HMRa to provide α1, α2 and a1 products.

Other mutations like ssp515 were subsequently found (mar1, 15; cmt, 3); a systematic analysis of sir mutations reveals four SIR genes (25).

Mating type interconversion. α2 has played a significant role in developing our understanding of mating type interconversion. After Jim Hicks found that matα1 mutations could be healed by mating type interconversion (7), it was of interest to know whether matα2 mutations could likewise be healed. Performing this analysis not only revealed that matα2 mutations were healable but also revealed important information of how mating type interconversion is controlled. To appreciate the matα2 healing studies, it is necessary to digress a bit. Mating type interconversion occurs efficiently in a and α haploids but not in a/α diploids. The absence of switching in a/α cells does not result from diploidy per se, because MATa/MATa and MATα/MATα diploids exhibit mating type interconversion (9, 14, 37). The inhibition of mating type interconversion in a/α cells does not require α1 product--matα1/MATa cells that carry HO do not undergo mating type interconversion (7). It is therefore possible to produce HO matα1 strains by sporulation of a matα1/MATa ho/HO diploid formed by mating between ho matα1 and HO MATa strains. The HO matα1 segregants can then be analyzed and were observed to switch, first to MATa and then to MATα. Mating between MATa HO and matα2 ho strains, in contrast, does not yield stable matα2/MATa diploids: such strains switch mating types until stable MATa/MATα diploids are formed (32, 34). These studies show (i) that the α2 product is necessary to inhibit mating type interconversion and (ii) that matα2 mutations can be healed in this process. The latter observation also provides information on the transposition event itself. Because mating type interconversion results in a substitution of one nonhomologous block of information for another (Ya to Yα and the reverse), the Y region must be transposed from HML or HMR to the mating type locus. Because the cassettes at HML and HMR are flanked by sequences that are also found at MAT (X and Z), one question that arises is whether any of the X and Z regions are also transposed to MAT. As discussed above, mutations matα2-1 and matα2-4 are inferred to be located in the X region (30). Hence mating type interconversion can result in movement of at least the part of the X region that contains the wildtype alleles of these mutations. Because our healing studies of matα2 mutations relied on the production of MATα/MATa diploids, we cannot answer the more refined question of whether all cassette transpositions result in movement of the same part of the X region in each switching event. The production of ochre mutations in MATα2 (23) should allow this question to be answered.

Why does mating type interconversion not occur in a/α cells? Studies of switching in mata1/MATα cells show that the a1 product is required in addition to the α2 product to block mating type interconversion (34, 16). One simple way for such inhibition to occur would be if synthesis of the HO gene product were blocked in a/α cells. To test this idea, Rob Jensen and George Sprague cloned the HO gene and assayed production of its RNA in a/α cells. The simple result was obtained: HO RNA is not produced (12). Furthermore, inhibition of synthesis requires both a1 and α2. These studies thus show a negative feedback control of mating type interconversion by the mating type locus: in a or α cells, the HO gene is expressed, which results in mating type switching. Mating between siblings of opposite mating type yields the final product of mating type interconversion, a stable a/α cell, which is stable because a1 and α2 turn off synthesis of HO gene product. Speculations on how a1 and α2 exert this negative control are discussed below.

α2 product may also provide an explanation for another aspect of mating type interconversion. Cells carrying HO can exhibit mating type interconversion only after they have become "experienced", that is, budded once (33). These experienced cells then exhibit "directionality" of switching in that they switch preferentially to the opposite mating type. Thus experienced α cells do not chose to transpose a cassette from HML or HMR at random. Rather, HO MATα cells that are HMLα HMRa transpose the a cassette in more then 75% of cell divisions. Recent work indicates that α cells preferentially mobilize the cassette at HMR and that a cells mobilize the cassette at HML (17). How is directionality brought about? A reasonable explanation is that a product of MAT directs this process. We know that α1 is not involved, because matα1 mutants exhibit normal directionality. A plausible explanation is that α2 product is responsible, for example, by binding to HML to prevent mobilization of the cassette at this locus or by binding to HMR to stimulate mobilization of the cassette at this position. Similar arguments can be made for directionality in a cells being mediated by a1.

Functional antagonism in matα2 mutants. As described above, we have proposed that α2 is a negative regulator of a-specific genes (35). The various defects of matα2 mutants are thus attributed to such cells expressing both α-specific genes (due to the presence of α1) as well as a-specific genes (due to the absence of α2). To ascertain whether this view is correct and to provide some insights into cellular physiological processes, we have examined two types of defects of α2 mutants: defects in α-factor secretion and in mating.

Neither α2 nor α1 mutants secrete α-factor (20, 28). In the case of α1 mutants, we anticipate that this defect results from lack of α-factor production itself: α1 may be necessary to activate expression of the α-factor gene (MAFα) itself or to activate expression of genes necessary for α-factor synthesis or secretion. In contrast, we explain the inability of α2 mutants to secrete α-factor to result from production of the α-factor degradation function (Bar). Thus, α2 mutants produce α-factor but they simultaneously degrade it. This hypothesis has been confirmed by isolating derivatives of α2 mutants that now secrete α-factor (28). These mutants are defective in the BAR1 gene. This finding has several implications. First, it confirms that α2 mutants do indeed produce α-factor and that the only function of MATα necessary for α-factor synthesis is α1. Secondly, they identify a gene, BAR1, which is a candidate for negative regulation by the α2 product. It will now be possible to clone the BAR1 gene by complementation and to determine by Northern analysis whether BAR1 is controlled in this manner. Finally, the behavior of α2 mutants provided a way for us to clone the α-factor gene.

Lacking mutants defective in the α-factor structural gene itself, we were not able to clone the gene by complementation. We argued, however, that a high copy number plasmid carrying MAFα might allow an α2 mutant to produce more α-factor than could be degraded by Bar. Janet Kurjan screened for such a plasmid and by nucleotide sequencing showed that it codes for α-factor (19). Remarkably, the MAFα gene carried on this plasmid appears to encode an α-factor precursor of 165 amino acids that contains four tandem copies of mature α-factor. The first half is hypothesized to contain information for targetting α-factor to the secretion pathway and processing locales; the second half contains the four α-factor segments, each preceded by a six or eight amino acid "spacer" segment (variations of the sequence Lys-Arg-Glu-Ala-Asp-Ala-Glu-Ala) that contains proteolytic processing sites.

Functional antagonism in α2 mutants is manifested also in defective mating. We noted earlier that α1⁻ α2⁻ mutants mate efficiently as a. Thus the inability of α2 mutants to mate as a results from expression of α1 product itself or from expression of some function(s) under its control. To test the latter possibility, we have determined what mutations will allow α2 mutants to mate as a. We find that inactivation of the α-specific gene STE3 allows α2 mutants to mate as a (31, 35). These results are of course consistent with α1 being a positive regulator of the STE3 gene. This contention has been confirmed (29).

By analogous reasoning, we have attempted to determine whether we can relieve functional antagonism in α2 mutants to allow efficient mating as α (31). Our efforts so far are negative: α2 mutants defective in two a-specific genes (STE2 and STE6) are still unable to mate as α. Additional mutations may be necessary to turn the trick.

Molecular mechanism of α2 action in a and in a/α cells. We have seen above that MATα2 is a regulatory gene with two roles in determining yeast cell type. In both a and a/α cells, we believe that it acts as a negative regulator to control expression of genes such as BAR1 and certain a-specific genes as well. In a/α cells α2 has an additional role exerted in conjunction with the a1 product of MATa.

α2 and a1 are necessary for a wide variety of actions--inhibition of mating and mating type interconversion. Why are both a1 and α2 required? One possibility is that a1 and α2 products associate to form a regulatory protein with distinctive specificity, for example, binding to target sites on genes such as α1 and HO that it regulates. Another possibility is that a1 and α2 interact cooperatively at the target site. For example, a1 might bind near the HO gene to allow binding of α2, which thereby blocks expression of the HO gene.

It should be apparent that studies of the α2 gene have provided information on and inroads into many different aspects of cell type determination, mating physiology, and mating type interconversion. It is hoped that future studies of α2 will shed light on the molecular mechanisms of action of regulatory proteins.

ACKNOWLEDGEMENTS

I would like to thank Alexander Johnson for comments on the manuscript, Virginia Staubach for superb preparation of the manuscript, and the National Institutes of Health, the American Cancer Society, and the Damon Runyon-Walter Winchell Cancer Fund for support of our work.

LITERATURE CITED

1. Astell, C.R., Ahlstrom-Jonasson, L., Smith, M., Tatchell, K., Nasmyth, K.A. and Hall, B.D. 1981. The sequence of the DNAs coding for the mating-type loci of Saccharomyces cerevisiae. Cell 27:15-23.

2. Blair, L.C. Personal communication.

3. Haber, J.E. and George, J.P. 1979. A mutation that permits the expression of normally silent copies of mating-type information in Saccharomyces cerevisiae. Genetics 93:13-35.

4. Herskowitz, I., Strathern, J.N., Hicks, J.B. and Rine, J. 1977. Mating type interconversion and its relationship to development in higher eukaryotes, p. 193-202. In: G. Wilcox, J. Abelson, and C.F. Fox (eds.), Molecular Approaches to Eucaryotic Genetic Systems. Academic Press, New York.

5. Hicks, J.B. 1975. Interconversion of mating types in yeast. Ph.D. thesis, University of Oregon.

6. Hicks, J.B. and Herskowitz, I. 1976. Evidence for a new diffusible element of mating pheromones in yeast. Nature 260:246-248.

7. Hicks, J.B. and Herskowitz, I. 1977. Interconversion of yeast mating types. II. Restoration of mating ability to sterile mutants in homothallic and heterothallic strains. Genetics 85:373-393.

8. Hicks, J.B., Strathern, J.N. and Herskowitz, I. 1977. The cassette model of mating-type interconversion, p. 457-462. In: A.I. Bukhari, J.A. Shapiro and S.L. Adhya (eds.), DNA Insertion Elements, Plasmids and Episomes. Cold Spring Harbor Laboratory, Cold Spring Harbor, N.Y.

9. Hicks, J.B., Strathern, J.N. and Herskowitz, I. 1979. Interconversion of yeast mating types. III. Action of the homothallism (HO) gene in cells homozygous for the mating type locus. Genetics 85:395-405.

10. Hicks, J., Strathern, J. and Klar, A.J.S. 1979. Transposable mating type genes in Saccharomyces cerevisiae. Nature 282:478-483.

11. Hopper, A.K. and Hall, B.D. 1975. Mating type and sporulation in yeast. I. Mutations which alter mating-type control over sporulation. Genetics 80:41-59.

12. Jensen, R., Sprague, Jr., G.F. and Herskowitz, I. 1982. Feedback control of mating type interconversion by the yeast mating type locus: Control of HO gene expression. In preparation.

13. Kassir, Y. and Simchen, G. 1976. Regulation of mating and meiosis in yeast by the mating type region. Genetics 82:187-206.

14. Klar, A.J.S. and Fogel, S. 1977. The action of homothallism genes in Saccharomyces diploids during vegetative growth and the equivalence of hma and HMα loci functions. Genetics 85:407-416.

15. Klar, A.J.S., Fogel, S. and MacLeod, K. 1979. MAR1, a regulator of the HMa and HMα loci in Saccharomyces cerevisiae. Genetics 93:37-50.

16. Klar, A.J.S., Fogel, S. and Radin, D.N. 1979. Switching of a mating-type a mutant allele in budding yeast Saccharomyces cerevisiae. Genetics 92:759-776.

17. Klar, A.J.S., Hicks, J.B. and Strathern, J.N. 1982. Directionality of yeast mating-type interconversion. Cell 28:551-561.

18. Klar, A.J.S., Strathern, J.N., Broach, J.R. and Hicks, J.B. 1981. Regulation of transcription in expressed and unexpressed mating type cassettes of yeast. Nature 289:239-244.

19. Kurjan, J. and Herskowitz, I. 1982. Structure of a yeast pheromone gene (MAFα): A putative precursor contains four tandem copies of mature α-factor. Submitted.

20. MacKay, V. and Manney, T.R. 1974. Mutations affecting sexual conjugation and related processes in Saccharomyces cerevisiae. I. Isolation and phenotypic characterization of nonmating mutants. Genetics 76:255-271. II. Genetic Analysis of nonmating mutants. Genetics 76:273-288.

21. Nasmyth, K.A. and Tatchell, K. 1980. The structure of transposable yeast mating type loci. Cell 19:753-764.

22. Nasmyth, K.A., Tatchell, K., Hall, B.D., Astell, C. and Smith, M. 1981. A position effect in the control of transcription at yeast mating type loci. Nature 289:244-250.

23. Porter, S. and Smith, M. Personal communication.

24. Rabin, M. 1970. Mating type mutations obtained from "rare mating" of cells of like mating type. M.S. thesis, University of Washington.

25. Rine, J.D. 1979. Regulation and transposition of cryptic mating type genes in Saccharomyces cerevisiae. Ph.D. thesis, University of Oregon.

26. Rine, J., Sprague, Jr., G.F. and Herskowitz, I. 1981. rme1 mutation of Saccharomyces cerevisiae: Map position and bypass of mating type locus control of sporulation. Molec. Cell. Biol. 1:958-960.

27. Rine, J., Strathern, J.N., Hicks, J.B. and Herskowitz, I. 1979. A suppressor of mating type locus mutations in Saccharomyces cerevisiae: evidence for and identification of cryptic mating type loci. Genetics 93:877-901.

28. Sprague, Jr., G.F. and Herskowitz, I. 1981. Control of yeast cell type by the mating type locus. I. Identification and control of expression of the a-specific gene, BAR1. J. Mol. Biol. 153:305-321.

29. Sprague, Jr., G.F., Jensen, R. and Herskowitz, I. In preparation.

30. Sprague, Jr., G.F., Rine, J. and Herskowitz, I. 1981. Homology and non-homology at the yeast mating type locus. Nature 289:250-252.

31. Sprague, Jr., G.F., Rine, J. and Herskowitz, I. 1981. Control of yeast cell type by the mating type locus. II. Genetic interactions between MATα and unlinked α-specific STE genes. J. Mol. Biol. 153:323-335.

32. Strathern, J.N. 1977. Regulation of cell type in Saccharomyces cerevisiae. Ph.D. thesis, University of Oregon.

33. Strathern, J.N. and Herskowitz, I. 1979. Asymmetry and directionality in production of new cell types during clonal growth: the switching pattern of homothallic yeast. Cell 17:371-381.

34. Strathern, J.N., Blair, L.C. and Herksowitz, I. 1979. Healing of mat mutations and control of mating type interconversion by the mating type locus in Saccharomyces cerevisiae. Proc. Natl. Acad. Sci. USA 76:3425-3429.

35. Strathern, J.N., Hicks, J.B. and Herskowitz, I. 1981. Control of cell type in yeast by the mating type locus: the α1-α2 hypothesis. J. Mol. Biol. 147:357-372.

36. Strathern, J.N., Spatola, E., McGill, C. and Hicks, J.B. 1980. The structure and organization of transposable mating type cassettes in Saccharomyces yeasts. Proc. Natl. Acad. Sci. USA 77:2839-2843.

37. Takano, I., Oshima, T., Harashima, S. and Oshima, Y. 1977. Tetraploid formation through the conversion of the mating-type alleles by the action of homothallic genes in the diploid cells of Saccharomyces yeasts. J. Ferment. Technol. 55:1-12.

38. Tatchell, K., Nasmyth, K., Hall, B.D., Astell, C.R. and Smith, M. 1981. In vitro mutation analysis of the mating-type locus in yeast. Cell 27:25-35.

39. Tkacz, J.S. and MacKay, V.L. 1979. Sexual conjugation in yeast. Cell surface changes in response to the action of mating hormones. J. Cell. Biol. 80:326-333.

40. Tohoyama, H. and Yanagishima, N. 1982. Control of the production of the sexual agglutination substances by the mating type locus in Saccharomyces cerevisiae: Simultaneous expression of specific genes for a and α agglutination substances in matα2 mutant. Submitted.

25. MECHANISM OF HOMOTHALLIC SWITCHING OF YEAST MATING TYPE GENES*

James Haber, Anne Comeau, Pei-Syan Liu, Dave Rogers, Sue Stewart, Michael Resnick+ and Barabera Weiffenbach

Department of Biology and Rosenstiel Center
Brandeis University, Waltham, Massachusetts 02254

Rec. Adv. Yeast Mol. Biol. (1982) 1:332-347

Homothallic switching of yeast mating type (MAT) genes can best be understood as a highly directional form of intrachromosomal mitotic gene conversion. We have addressed two aspects of this problem: 1) how switching is initiated, and 2) how the extent of pairing of homologous sequences shared by MAT and the donor loci, HML and HMR, affect switching. Switching is apparently initiated by a double-strand cut in MAT at or near a site defined by a series of deletions and point mutations. Two MAT-inc mutations were shown to be single base pair changes in the sequence CGCAAC that lies in the Z region of homology immediately adjacent to a- or α-specific (Y) sequences. The double-strand break is envisioned as promoting the conversion of MAT by a DNA repair mechanism, in which an intact, homologous sequence is used as a template to repair (switch) MAT. When HML and HMR are also "activated" in haploid HO strains carrying the cmt mutation, most switching events appear to be lethal; however, the transiently-viable products of switching can be rescued by mating with a Cmt+ strain. A high proportion of the rescued HO cmt cells have lost all of chromosome III or contain deletions joining MAT with HML or HMR.

We have also found similar directionality in gene conversions of the lys2 locus, where γ-rays were used to create double strand breaks in one parental chromosome before zygote formation.

Finally, we have explored how the alteration of homologous sequences at MAT and the two donors affects switching. Plasmids containing MAT or HML sequences have been integrated at different sites. Tandem duplications of MAT switch efficiently, a usually replacing the duplication by a single MAT locus of opposite mating type. The presence of an inc mutation in the right MAT region of the duplication prevents such replacements, while the same inc mutation in the left MAT site does not. The extent of pairing of HMLα with the duplication can be increased by constructing MAT/HML sequences at MAT.

INTRODUCTION

Under the control of the homothallism gene, HO, yeast mating type genes can switch from MAT**a** to MATα or vice versa as often as every cell division (11). MAT switching involves the transposition of **a**- or α-specific sequences from one of two unexpressed copies of mating type information, HMLα or HMR**a** (12,13,24). HML, HMR, and MAT all share extensive regions of homology

*Support was obtained from National Institutes of Health grant GM20056.

+ NIEHS, Research Triangle Park, NC

(segments W,X, and Z), flanking completely non-homologous **a** and α sequences (Ya and Yα) (1,22,27). A variety of experiments have suggested that transposition of mating type genes appears to involve a highly site-specific mitotic recombination event. Several observations suggest that switching requires direct physical pairing of homologous sequences shared by the MAT locus with either HML or HMR. For example, Klar et al. (15) have presented evidence that a heteroduplex structure is formed between the donor and MAT sequences during switching. In addition, although MAT switching is normally not reciprocal (that is, MAT is converted without affecting HML or HMR), reciprocal recombination events, joining MAT to HML or to HMR, occur in 1 - 2% of all switching events (9). Furthermore, the efficiency of switching was greatly impaired when the donor locus was translocated to another chromosome (10) so that pairing of MAT**a** and HMLα would be inter- instead of intra- chromosomal. In contrast, the efficiency of switching is only slightly impaired then a MAT locus was inserted in either orientation at another locus, leu2, on the opposite side of the centromere of the same chromosome III where MAT is located (Davidow and Haber, unpublished). Other experiments have suggested a similarity between MAT switching and other intrachromosomal mitotic gene conversion events (14,18). For example, MAT conversions are prevented by the rad52 mutation (19,29), as are intrachromosomal gene conversions between duplications of his4 heteroalleles (14). A more extensive review can be found by Haber (4).

Recently we have been carrying out a wide variety of experiments to substantiate how MAT switching occurs. The experiments summarized below address several questions: 1) how is MAT switching initiated; 2) what other predictions of a gene conversion model can be tested experimentally; 3) how similar is MAT switching to other mitotic gene conversion events? Our data show that it may be useful to consider mating type switching as a highly specialized mitotic gene conversion event dependent on the site-specific activation of the MAT locus.

RESULTS AND DISCUSSION

Cis-acting Deletions and Point Mutations that Prevent Switching.

Both MAT**a** and MATα cells carrying rad52 die when they attempt to switch mating type (19,29). Lethality occurs because of the formation of a double-strand chromosome break apparently within the MAT locus that cannot be repaired in rad52 strains (29). Lethality is prevented by mutations that prevent efficient MAT switching. The fact that HO rad52 cells die provides a strong selection for mutations that prevent mating type conversions and therefore allow cells to survive. Such mutations arise at very high frequency in HO rad52 MATα strains where the efficiency of switching is reduced by the swi1 mutation (29). All of these mutations proved also to be "sterile" matα1 mutations. We have now analyzed 9 of these mutations by restriction endonuclease mapping and found that all of them were deletions ranging in size from 50 to 800 bp (Figure 1). The deletions all removed a HhaI site that lies just at the Yα/Z border. The largest of the deletions we have isolated was not "sterile", but **a**-like in its mating behavior and proved to have lost not only part of the MATα1 cistron but part of the MATα2 cistron as well.

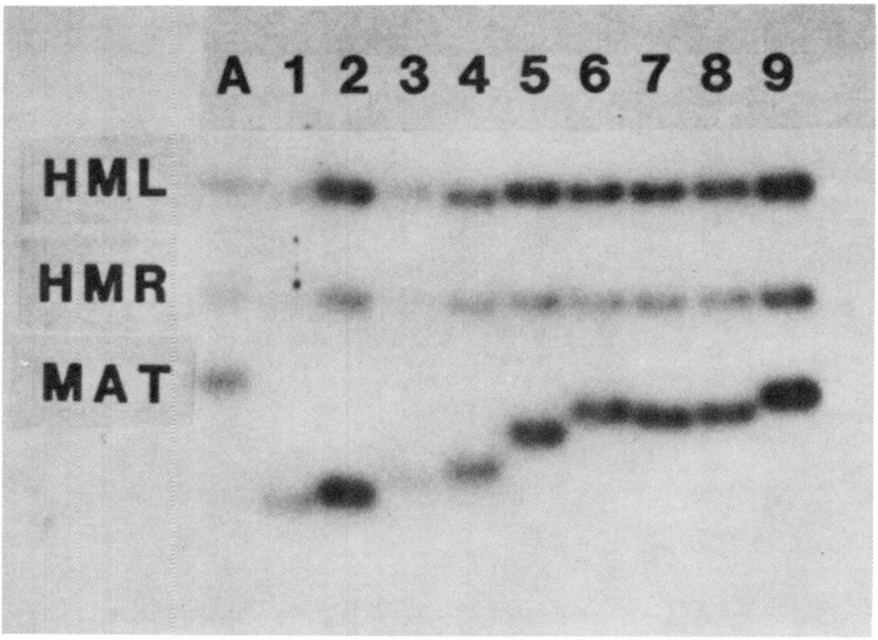

Figure 1. A. Southern blot of HindIII-digested DNA showing deletions in the MAT region in non-switching, matα1 derivatives of an HO rad52 swi1 MATα strain. The DNA was hybridized with a labelled probe containing the HindIII MATa fragment, which also hybridizes to HMLα and HMRa. The largest of the deletions (BWd1) was also mutant in the matα2 cistron.

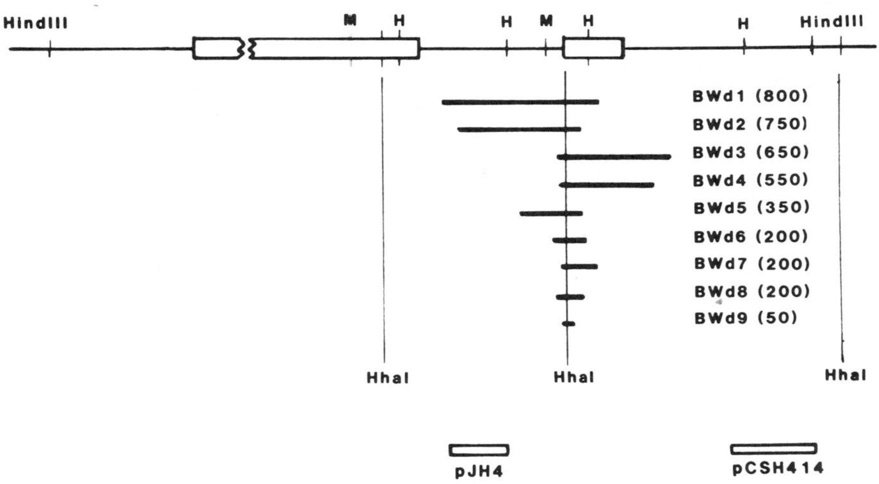

Figure 1. B. Restriction site analysis of the deletions within MATα. In addition to HindIII and HhaI sites, MspI (M) and HaeIII (H) sites were also mapped. Restriction digests were probed with both plasmids pJH4 and pCSH414, which contain the MATα sequences, as shown. All of the deletions eliminated the HhaI site at the Y/Z border.

```
Yα  CTTCG|CGCAACAGTA   Z        Ya   CTTTC|CGCAACAGTA   Z
    GAAGC|GCGTTGTCAT                 GAAAG|GCGTTGTCAT
          ↓                                ↓
          A  MATα-inc (3-7)                T  MATa-inc (4-28)
          T                                A
```

Figure 2. Single base pair substitutions in two inc mutations that reduce the efficiency of swithing by at 100 fold. Shown is the region spanning the Y/Z border. The nucleotide sequences of the two mutant strains were compared with the wild type sequences of MATα and MATa (1). More than 1 kb of additional sequences were also determined for the MATa (4-28) mutation without finding any other base changes (Weiffenbach, Rogers, Haber, Zoller, Russell, and Smith, manuscript submitted).

In conjunction with Mark Zoller, David Russell and Michael Smith, at the University of British Columbia, we have also cloned and sequenced two "inc" mutations that reduce the efficency of switching by at least 100-fold. Both MATα-inc (3-7) and MATa-inc (4-28) originally isolated by Oshima and Takano (25) have proven to be single base pair changes within 7 bp of the Y/Z border (Figure 2). We are now in the process of cloning and analyzing a number of other inc mutations. Our preliminary characterization of both MATα-inc (5-22) and of the MATα-inc variant from S. diastaticus (28) show that both have lost the HhaI site at the Y/Z border. On the other hand, a base pair variant 11 bp from the Y/Z border in a MATa strain is not significantly altered in the efficiency of MAT switching.

These results argue that the there is a specific sequence near the Y/Z border that is essential for the initiation of MAT conversion, probably by the formation of a double-strand break. The position of the two inc mutations we have sequenced argues that this sequence includes CGCAAC.

A Gene Conversion Model for Mating Type Switching.

As we have indicated, a large body of data argues in favor of an intra-chromosomal mitotic gene conversion model for MAT switching. A version of a gene conversion model is illustrated in Figure 3. Switching requires pairing of MAT and either HML or HMR in two different homologous regions flanking unique **a** or α sequences in the Y region. The switching event is initiated by a site-specific DNA cleavage near the Y/Z border of MAT (A). Strand invasion of the broken DNA into the intact, homologous Z region then promotes DNA replication, which in turn begins to displace an intact strand (containing a non-homologous Y region) to form a heteroduplex with MAT DNA (B). Further DNA replication allows strand displacement to continue until a stable heteroduplex is formed in X (C). Finally this recombination structure is resolved, in essentially the manner described by Meselson and Radding (21). This generally occurs without isomerization of the DNA, so that conversion of MAT is not accompanied by reciprocal recombination of flanking sequences (D); however, reciprocal recombinations, leading to the deletion of sequences between MAT and the donor could also occur (E).

The model is quite similar to Resnick's (26) model for double-strand break repair by recombination of homologous sequences and to that of Orr-Weaver et al. (23) for the homologous integration of plasmids containing double-strand breaks. In essence, this model argues that MAT switching occurs by a DNA repair event not fundamentally different from what might occur at another locus. What would make MAT switching special would be the highly efficient, genetically directed, site-specific cleavage of DNA to initiate this event. There might also be genetically controlled facilitation of homologous pairing of MAT and HML or HMR. This model requires no special replication of the donor sequences before transposition; consequently, it can also account for "illegal transpositions" (7) where MAT sequences are used to replace HML or HMR.

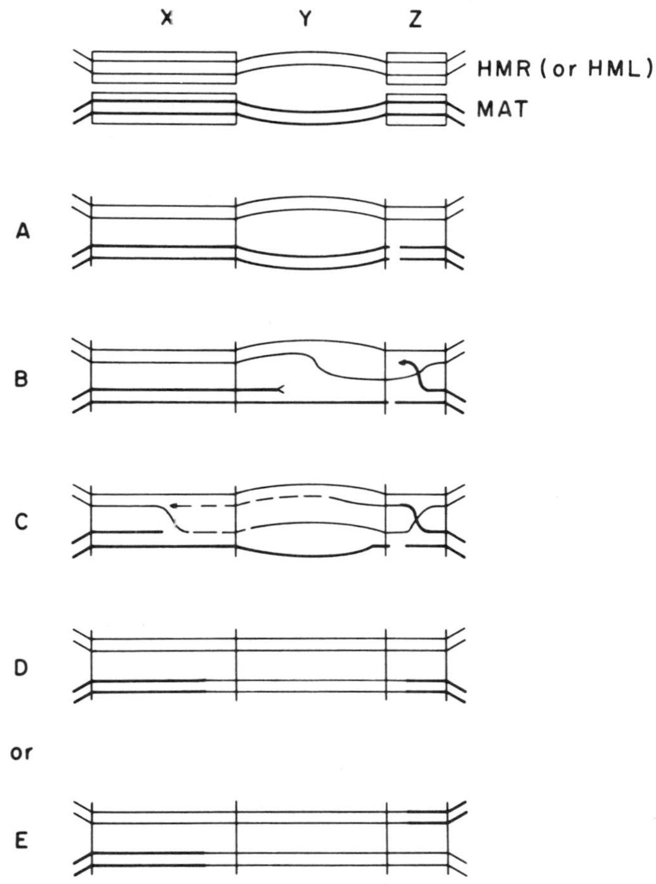

Figure 3. A gene conversion model for the switching of yeast mating type genes. Switching of MAT requires homologous pairing between the W,X, and Z regions shared by MAT and the donor locus (HML or HMR). The **a** and α-specific sequences (Ya or Yα) are completely non-homologous. MAT switching is initiated by a site-specific double-strand DNA cut in the Z region (A). A single strand from the MAT Z region then may invade and displace a strand of the donor locus (B). The initiation of DNA replication of the invading MAT strand, using the donor locus as a template will continue to diplace the strand of donor DNA, allowing it to form a stable heteoduplex with both the Z and W homologous regions (C). Preferential mis-match repair, favoring the incoming strand would result in the conversion of MAT without affecting the donor locus (D). The heteroduplex structure shown in (C) contains a cross-over structure that may isomerize during the switching event. In that case, switching would be accompanied by a reciprocal recombination event joining MAT with HML or HMR.

Role of Double-Strand Breaks in Other Mitotic Gene Conversions.

We have asked whether double-strand breaks might also cause directionality in the gene conversion of other loci in yeast. As a first step in studying this question, we have examined the gene conversion of heteroalleles of the *lys2* locus, using γ-rays to create double-strand breaks at or near one of the two alleles. Two heterothallic haploid strains of opposite mating type and carrying different *lys2* alleles were grown to stationary phase. One or the other parent was then irradiated with 5 krad γ-rays and allowed to stand in H_2O for 30 min to allow single-strand breaks to be repaired. The irradiated and unirradiated parents were then mated and zygotes were plated on medium lacking lysine to select for Lys^+ colonies. To determine what kind of recombination event had produced the Lys^+ allele, we then analyzed each diploid to determine the genotype of the unconverted recessive *lys2* allele remaining in the diploid. We recovered the recessive allele by selecting Lys^- mitotic recombinants homozygous for the remaining allele, using the α-amino adipic acid selection described by Chattoo and Sherman (2). Irradiation of one parent increased the frequency of Lys^+ colonies 20-50 fold. When neither parent was irradiated, the frequency of converting *lys2-1* (34/59) was nearly equal to the frequency of converting *lys2-2* (25/59). In contrast, when the *lys2-2* parent was irradiated, approximately 85% (97/114) of the Lys^+ diploids still contained *lys2-1* (that is, had converted *lys2-2*). Conversely, about 80% (73/89) of the Lys^+ diploids arising after irradiating the *lys2-1* parent had converted the *lys2-1* allele. Similar results have been obtained using a third *lys2* allele.

These results suggest that the directionality of switching seen at the *MAT* locus is not strikingly different from mitotic gene conversion at other loci that are initiated by a double-strand break. It should be pointed out that the breaks initiated by γ-irradiation are in no way site specific as the break initiated at *MAT* appears to be; nor is it certain that all of the increase in recombination stimulated by γ-rays results from double-strand break repair. Nevertheless, it is clear that γ-rays do not simply stimulate recombination in a general way, but promote recombination on the damaged chromosome. Again, these data are consistent with the model for recombination associated with double-strand break repair proposed by Resnick (26). These data are supported by independent experiments of Linda Friedman (personal communication) who has studied the effects of x-rays on mitotic gene conversions at the *his1* locus and reached the same conclusions: about 80% of the conversions are in the direction of the irradiated chromosome.

Changes in Switching When HML and HMR Are Expressed.

The normal directionality of *MAT* switching apparently depends on the fact that the *MAT* locus is expressed while *HML* and *HMR* are not. Klar et al. (16) have shown that when the normally silent donor loci are "turned on" by mutations such as *mar1*, both *HML* and *HMR* can also be efficiently converted. We have been carrying out similar experiments with the *cmt* mutation (6), which is allelic to *mar2*. We obtained meiotic segregants of genotype *HO cmt hml**a** MATα hmr**a*** (that is, where both *hml**a*** and *hmr**a*** carried recessive mutations in the **a**1 cistron). Thus, even though both donor loci are transcribed in these strains, the cells were initially α-mating. In equivalent

Cmt^+ derivatives, homothallic switching was very efficient and MATα was replaced by mat**a**. In cmt strains, however, switching was apparently much less efficient, as the cells had an (α > **a**) phenotype. This phenotype could either reflect an actual decrease in the efficieny of converting MATα or could indicate that either hml**a** or hmr**a** were switching to HMLα or HMRα. When these (α > **a**) strains were subcloned we discovered that there were surprisingly few **a**-mating colonies. Of 15 different HO cmt hml**a** MATα hmr**a** colonies that we have analyzed, fewer than 1% of the subclones had the genotype hml**a** mat**a** hmr**a**. Nearly all of the remaining subclones were again (α > **a**). From analogous experiments with other phenotypically slow-switching strains (3,5) we would have expected at least 10 - 20 % of the subclones to be **a**-mating. In addition, when we observed the growth of individual cells under the microscope, we found more than 50 % of the cells failed to grow into colonies. This raised the possiblity that a large fraction of the **a**-mating cells in these colonies were not haploid-viable, in much the same way that the **a**-mating cells in an HO rad52 swil MATα strain were not viable (29).

To show that switching of HO cmt hml**a** MATα hmr**a** strains led to transiently viable **a**-mating cells, we have carried out several types of experiments. First, we have rescued these **a**-maters by crossing them with a Cmt^+ HMLα MATα-inc HMR**a** strain carrying the recessive markers his4 and thr4, on the left and right arms of chromosome III, respectively. Approximately 10 % of the zygotes formed were His^- Thr^- and appeared to be instances where the entire chromosome III from the HO cmt strain had been lost. Another 30% of the matings yielded His^+ Thr^- diploids which, when sporulated and dissected, yielded only 2 viable spores per tetrad. These properties are characteristic of diploids carrying partial deletions of one homologue of chromosome III. Such partial deletions have been observed previously; for example, in the conversions of mat**a**/MATα-inc diploids, about 1 - 2% of the diploids became His^+ Thr^- by deletions joining MAT with either HML or HMR (8). We have confirmed that similar events can account for the partial deletion strains rescued from the HO cmt (α > **a**) colonies. Three such independent colonies have so far been analyzed by Southern blots; all three proved to contain HindIII restriction fragments characteristic of a MAT/HML**a** fusion. The diploids also did not have either an HML**a** or a MAT**a** fragment. Thus, all three diploids apparently carried a deficiency circular chromosome III joining MAT and HML. It should also be pointed out that the MAT/HML structures we have recovered on the "repaired" chromosome had switched to a functional **a** allele derived by postzygotic transposition from HMR**a** on the other homologue.

The remaining 60 % of the matings yielded His^+ Thr^+ diploids that Southern blot analysis confirmed had two intact chromosome III homologues wher **a**-specific sequences were found at HML, MAT, and HMR. These seem to be cases where an intact chromosome III was recovered by mating. These results are again quite similar to those obtained with the HO rad52 swil MATα strains, where transiently viable **a**-maters could be rescued and repaired to an intact chromosome III by mating with a Rad^+ MATα-inc strain (29, see also 20). These results suggest that a very large fraction of the **a**-mating cells in the HO cmt strains are transiently viable and may contain lesions , (chromosome breaks?) that can be repaired in Cmt^+ diploids.

In addition to the fact that most **a**-mating progeny were not viable, we also found that the α sequences did not readily transpose. In 8 out of the 9 (α > **a**) colonies whose DNA we have analyzed on Southern blots, using both **a**- and α- specific probes, the colonies showed predominantly the same position of **a** and α sequences expected from the genotype of the initial HO cmt hml**a** MATα hmr**a** spore. An **a**-specific probe hybridized to HML and HMR restriction fragments and the α-specific probe hybridized to the MAT fragment. There was a lower, but detectable, amount of hybridization of the **a**-specific probe to MAT, as well, but virtually no hybridization of the α-specific probe to HML or HMR. These observations suggested that MATα sequences did not frequently transpose to HML or HMR (at least not forming viable cells). The ninth (α > **a**) strain proved to have the genotype HO cmt HMLα mat**a** hmr**a**. This strain was derived from a spore that initially had an hml**a** MATα hmr**a** genotype. It, too, gave rise to transiently-viable **a**-mating cells and did not readily transpose α sequences back to MAT or HMR.

These observations suggest that most attempts to switch mating type in an HO cmt strain are lethal. This may be explained by the possiblity that, in a cmt strain, double strand chromosome breaks are generated at HML, MAT, and HMR, and that it is not possible to "repair" these breaks by switching events at all three loci within one cell divison. This would generate transiently-viable, **a**-mating strains containing a broken chromosome III and which could be rescued by mating with a Cmt$^+$ strain. Alternatively, lethality may arise by the formation of reciprocal recombination events joining MAT with HML or HMR. These structures may, however, only form after mating with a Cmt$^+$ strain. Although it is possible that growth of these cells exerts a strong selection for mutations that have lost the ability to switch efficiently, our results do not support this idea. For example, by mating (α > **a**) colonies with an HO Cmt$^+$ hml**a** mat**a** hmr**a** strain we have shown both that MATα is efficiently replaced by mat**a** and that HMLα acts as an efficient donor. Thus, is more likely that the effects of cmt that we observe reflect the failure to restore an undeleted, intact chromosome III after switching is initiated.

Switching of Tandem Duplications of MAT Genes.

We have been able to test several aspects of the gene conversion model by examining the switching of tandem duplications of MAT genes. Such tandem duplications are readily constructed by the integration of recombinant DNA plasmids containing MAT genes by homologous recombination. For example, we have constructed a pBR322 plasmid containing the 3.5 kb EcoR1 - HindIII frag ment containing MAT**a** as well as the 1.1 kb HindIII URA3 fragment. We have obtained transformants of strain DBY745 (MATα HMLα HMR**a** ura3 leu2) with plasmids integrated at all four sites homologous to the plasmid. Of 21 transformants analyzed, 16 were at MAT, 1 was at HML, 3 were at HMR, and 1 was at URA3. The transformants at MAT were of four types. Seven had the structure: MAT**a**-URA3-pBR322-MATα (abbreviated [**a**-URA3-α]) and 7 had [α-URA3-**a**]. All of these were non-mating haploids that expressed both MAT**a** and MATα. The remaining two integrants had become homozygous for MAT**a** or MATα; one was [**a**-URA3-**a**]; and the last was [α-URA3-α]. Thus, two of the 16 integration events between MATα on chromosome III and MAT**a** on the plasmid apparently involved a gene conversion event similar to MAT switching. The Southern blots confirming some of these structures have already been published

(8). The tandem duplications were quite stable, reverting to Ura$^-$ at frequencies of less than 10^{-3}.

We have examined the homothallic switching of both [**a**-URA3-**a**] and [α-URA3-α] duplications. For example, the heterothallic transformant DR210 carrying the [**a**-URA3-**a**] duplication was crossed with a homothallic ura3 MATα-inc strain and the resulting diploids were then sporulated. All of the homothallic segregants carrying the [**a**-URA3-**a**] duplication switched efficiently to form non-mating Ura$^+$ diploids. When these diploids were sporulated and dissected, we found that more than 90% of the tetrads contained two Ura$^+$ and two Ura$^-$ segregants, all of which were non-mating and sporulated well. The remaining tetrads gave four Ura$^+$ spores. It appeared that during the switching of tandem duplications of MAT, the duplication (and therefore URA3) was frequently lost, yielding diploids formed by mating an [**a**-URA3-**a**] cell with its Ura$^-$ MATα offspring. We substantiated this conclusion in several ways. First, from the diploids that segregated 2+:2- for URA3, sporulated colonies were mated with heterothallic ura3 strains of both mating types. The matings were carried out so that the germinating spores did not grow for many generations before mating. The complementation tests showed strong linkage between URA3 and **a**-mating spores. Thus, switching of [MAT**a**-URA3-pBR322 -MAT**a**] yielded Ura$^-$ MATα cells. We confirmed this conclusion by Southern blot analysis of DNA both from the Ura$^+$ diploids and from subsequent heterothallic derivatives.

When similar spore mating tests were carried out on diploids that yielded 4 Ura$^+$ spores, the spore mating test showed nearly equal linkage of URA3 with both **a**- and α-mating spores. These appeared to be cases where [MAT**a**-URA3-pBR322-MAT**a**] switched to [MATα-URA3-pBR322-MATα]. Again, we confirmed this conclusion by Southern blot and further genetic analysis.

It was striking that almost none of these switching events yielded Ura$^+$ non-mating haploid derivatives. As we had previously constructed both [**a**-URA3-α] and [α-URA3-**a**] duplications, we knew that these would be non-mating, but unable to sporulate. Either the switching process never resulted in the conversion of one half of the duplication, or else the formation of these duplications did not turn off the homothallic switching process before subsequent switching events had occurred.

Only rarely will all the cells of a colony be derived from a single switching event, because neither switching nor mating are 100 % efficient. To look for less frequent outcomes of switching the tandem duplications and to estimate the frequency of the different types, we subcloned the non-mating, homothallic Ura$^+$ segregants and repeated spore mating tests on at least 20 subclones per colony. The results of these analyses are summarized in Table 1A (for the switching of [**a**-URA3-**a**] cells) and in Table 1B (for the switching of [α-URA3-α] cells). These results showed that in most colonies there were more than one outcome. The predominant switching event in nearly every colony was the simultaneous excision of the duplication and its replacement by a single MAT locus of opposite mating type. About 20% of the time, switching did not lead to the excision of the duplication; rather, both copies switched to the opposite mating type. There was a small proportion of Ura$^-$ MAT**a**/MATα diploids that may have arisen by subsequent switching of a Ura$^-$ cell that did not mate with a Ura$^+$ cell of the original genotype.

Table 1. Genotypes of Subclones from Colonies arising from Homothallic Switching of Different Tandem Duplications of Mating Type Genes

Duplication	Number of subclones having genotype indicated[a]			
A. MATa-URA3-pBR322-MATa	[a-URA3-a] / α	[a-URA3-a] / [α-URA3-α]	[a-URA3-α] or [α-URA3-a]	a / α
13 colonies subcloned	187	66	1	2
B. MATα-URA3-pBR322-MATα	[α-URA3-α] / a	[α-URA3-α] / [a-URA3-a]	[a-URA3-α] or [α-URA3-a]	a / α
10 colonies subcloned	178	16	0	5
C. MAT/HMLα-URA3-pBR322-HML/MATα	[α-URA3-α] / a	[α-URA3-α] / [a-URA3-a]	[a-URA3-α] or [α-URA3-a]	a / α
8 colonies subcloned	145	46	39	11
D. MAT/HMLa-URA3-pBR322-HML/MATa	[a-URA3-a] / α	[a-URA3-a] / [α-URA3-α]	[a-URA3-α] or [α-URA3-a]	a / α
11 colonies subcloned	153	103	241	6
E. MATα-URA3-pBR322-(mata-inc)	[(a-inc)-URA3-(a-inc)]		[α-URA3-(a-inc)] / [a-URA3-(a-inc)]	a-inc
10 colonies subcloned	338		9	2
F. (mata-inc)-URA3-pBR322-MATα	[(a-inc)-URA3-(a-inc)]		[(a-inc)-URA3-α] / [(a-inc)-URA3-a)]	a-inc
10 colonies subcloned	283		27	31
		[(a-inc)-URA3-α] / a		
		20		

[a] HO strains carrying the duplication also carried HMLα and HMRa. Subclones were analyzed by sproulation and spore mating tests (see text) to determine the types of diploid or haploid cell types that had arisen.

These results argued that the switching of a tandem duplication of MAT genes is an efficient process, resulting primarily in the loss of the tandem duplication and its replacement with a single MAT locus of opposite mating type. These experiments did not show whether the loss of the duplication occurred simultaneously with switching to the opposite mating type, or whether the tandem duplication first "looped out" to form a single MAT locus of the original mating type which was subsequently switched. The fact that there were so few Ura$^-$ diploids argued against this possibility, but a more direct demonstration was needed. By following cell pedigrees as first described by Hicks and Herskowitz (11) we could show that, in fact, the loss of URA3 occurred during the same cell division as switching to the opposite mating type. Homothallic diploids of genotype [α-URA3-α]/**a** were dissected on agar plates adjacent to a source of α-factor. Germinating α-factor-resistant cells (presumably [α-URA3-α] cells were followed and cell lineages separated until the four-cell stage. In 11/18 lineages, two of the four cells became sensitive to α-factor and stopped dividing. As in other pedigree studies, it was always the original spore cell and its second daughter that switched mating type. At this stage all four cells were micromanipulated away from α-factor and allowed to grow into colonies. In every case, the two cells that remained α-factor resistant gave rise to Ura$^+$ colonies. In 8 of the 11 cases, the two α-factor sensitive cells gave rise to Ura$^+$ non-mating colonies. In the other 3 cases, the pairs of sensitive cells grew into Ura$^+$ colonies in which URA3 could be shown (by subsequent spore mating tests) to be linked to **a**-mating spores. Thus, in 8 of 11 cases, the tandem duplication was lost at the same cell division as the switch in cell mating type. Had the duplication been lost in the cell division prior to that in which switching occurred, both the α-factor-sensitive and the α-factor-resistant cells should have been Ura$^-$. In 3/11 cases, switching occurred without loss of the tandem duplication.

A Test of the Gene Conversion Model Using Tandem Duplications.

Switching of a tandem duplication of MAT genes generally leads to the loss of the tandem duplication and its replacment by a single MAT gene of opposite mating type. One interpretation of such a switching event is that a single donor locus (HMLα or HMR**a**) must pair with the tandem duplication, and that pairing with the outside homologous segments (W and X of the left MAT and Z of the right MAT) is somehow favcred over pairing with only one or the other MAT gene (Figure 4). This could be because there are sequences to the left of W or to the right of Z that were not contained within the cloned EcoRI - HindIII restriction fragment used to construct the MAT duplication. In that case, these sequences would only be found flanking the entire duplication. Another possibility is that the simultaneous activation of both MAT loci might lead to the actual excision of the sequences between the two sites of double-strand cleavage at the Y/Z borders. Thus, essentially only a single MAT locus (with a cut at the Y/Z) boundary would be left to pair with the donor locus).

To try to distinguish between these and other possible interpretations, we have extended these studies to examine the effect of including inc mutations at one or the other MAT gene in the duplication. We have constructed a mat**a**l-inc locus that carries the inc (4-28) mutation that we have shown is a single base pair change in Z, six bp from the Ya/Z border. This locus can neither switch nor provide functional **a** information. By transformation we have constructed two tandem duplications: [MATα-URA3-pBR322-(mata1-inc)] and [(mata1-inc)-URA3-pBR322-MATα]. In heterothallic strains both constructions have an α-mating, Ura^+ phenotype. By subsequent crosses with homothallic strains we have obtained homothallic segregants of these duplications. The analysis of the switching of these duplications was complicated by an unexpected result. Although these α-mating spores should would be expected to use HMRa as the donor (17), most of the switching events resulted in the formation of [(mata1-inc)-URA3-pBR322-(mata1-inc)] duplications, where (mata1-inc) acted as the donor to switch MATα. Nevertheless, by examining subclones from these **a**-mating, Ura^+ colonies, we found non-mating subclones that showed how the position of the mata1-inc mutation within the tandem duplication affected switching (Table 1, E and F).

In the case of the [α-URA3-(**a**1-inc)] strains, all of these non-mating segregants proved to be diploids where URA3 was linked to both α- and **a**-mating spores (that is, diploids of genotype [α-URA3-(**a**1-inc)] / [**a**-URA3-(**a**1-inc)] (Table 1E). In contrast, about half the non-mating derivatives of the [(**a**1-inc)-URA3-α] duplication proved to be ones where the tandem duplication had switched to a single MAT**a** (Table 1F). Thus, although the analysis of switching is complicated by unfavorable competition between HMR**a** and mata1-inc as donors, it seems clear that the [(mata-inc)-URA3-α] duplication can efficiently excise the tandem duplication, while the reverse orientation [α-URA3-(**a**1-inc)] cannot. These results argue that the two MAT sequences in the tandem duplication are not treated identically. A "cartoon" of how this might be explained is shown in Figure 4. If HMR**a** tends to pair with the outermost homologous sequences of the tandem duplication, an inc mutation in the left MAT region would be part of a substitution loop to be replaced by gene conversion. In this location, the mutation might be largely "invisible". When the inc mutation is in the right

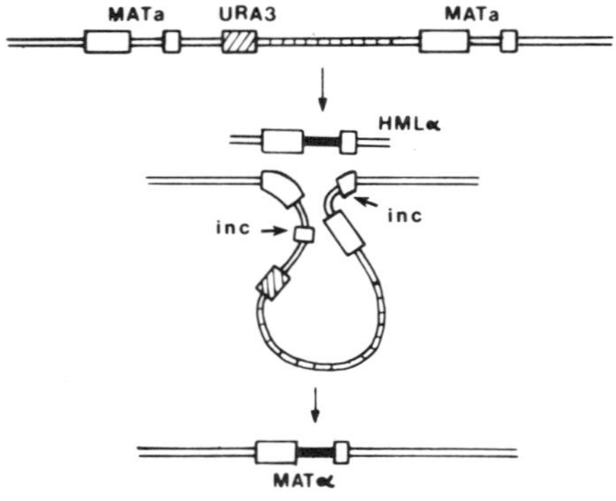

Figure 4. Switching of a tandem duplication of MAT sequences. A tandem duplication, constructed by integrating a MAT**a**-URA3-pBR322 plasmid at MAT**a**, switches to a single MATα sequence more than 80 % of the time. The presence of an inc mutation in the short Z region of homology shared by the right-hand MAT and HML prevents excision of the duplication, while the presence of the same inc mutation in the left MAT sequence (where it might not directly pair with the donor) does not prevent the replacement of the duplication by a single MAT.

MAT locus, the initiation of a switching event that would remove the tandem duplication would be prevented, though the other MAT region would still be able to switch. These results are not compatible with the notion that both Z regions must be cut in order to switch the tandem duplication to a single MAT locus.

Effect of the Extent of Homology between Donor and Recipient on Switching.

We have also constructed other tandem duplications that have allowed us to examine more directly the effect of homologous pairing on MAT switching. For these studies we have integrated an HMLα-containing URA3-pBR322 plasmid at MAT. The integration created a [(MAT/HMLα)-URA3-pBR322-(HML/MATα)] structure which we found switched efficiently in homothallic strains to give non-mating Ura$^+$ colonies. An analysis of subclones from 8 such colonies showed that the types of switching events, using HMR**a** as the donor, were essentially identical to those obtained for the [MATα-URA3-pBR322-MATα] duplication: most of the switching events resulted in the loss of the duplication and conversion to MAT**a** (Table 1C).

It should be pointed out that the HMLα-URA3-pBR322 plasmid contains an ars sequence that allows it to replicate autonomously. It was therefore significant that we did not recover cases where switching led to the reformation of the plasmid. Even when spores were germinated and grown of medium lacking uracil, there were no cases where URA3 behaved as if it were on a plasmid.

We have also recovered [(MAT/HML**a**)-URA3-pBR322-(HML/MAT**a**)] duplications. In this case we presume that the duplication will switch using HMLα as the donor. Instead of sharing 1400 bp and 320 bp of homology with MAT**a**, as is normally found, HMLα shares much more extensive regions of homology with both the MAT/HML**a** and HML/MAT**a** hybrid loci of the tandem duplication. The "Z" region shared between HMLα and MAT/HML**a** is 2300 bp, while the "W and X" region shared by HMLα and HML/MAT**a** is 1600 bp. When these duplications underwent homothallic switching we found a distinct different pattern of switching (Table 1D). Although switching was efficient, leading to non-mating colonies, the majority of the switching events led to Ura$^+$ non-mating haploid cells in which only one or other hybrid MAT locus had switched. These results would be expected if pairing with only one half of the duplication was now a favored state. We are also investigating the possiblity that some switching events have led to the formation of haploid-inviable circular chromosomes, resulting from reciprocal recombination joining HML and the MAT/HML locus. Thus the extent of pairing between MAT and a donor locus can significantly affect the type of transposition event. We have drawn similar conclusions about the affect of longer regions of homology affecting switching from our studies of transpositions of a tandem duplication of MAT sequences integrated at HMR to replace a single MAT sequence (8).

LITERATURE CITED

1) Astell, C.R., Ahlstrom-Jonasson, L., Smith, M., Tatchell, K., Nasmyth, K.A. and Hall, B.D. 1981. The sequence of the DNAs coding for the mating type loci of Saccharomyces cerevisiae. Cell 27:15-23.

2) Chattoo, B.B. and Sherman, F. 1979. Selection of lys2 mutants of the yeast Saccharomyces cerevisiae by the utilization of α - aminoadipate. Genetics 93: 51-65.

3) Davidow, L.S. and Haber, J.E. 1981. Relation between the efficiency of homothallic switching of yeast mating type genes and the distribution of cell types. Mol. Cell. Biol. 1:1120-1124.

4) Haber, J.E. 1982. Mating type genes of Saccharomyces cerevisiae. In: J. Shaprio (ed.), Mobile Genetic Elements. Academic Press. In press.

5) Haber, J.E. and Garvik, B. 1977. A new gene affecting the efficiency of mating type interconversions in homothallic strains of Saccharomyces cerevisiae. Genetics 87:33-50.

6) Haber, J.E. and George, J.P. 1979. A mutation that permits the expression of normally silent copies of mating-type information in Saccharomyces cerevisiae. Genetics 92:13-35.

7) Haber, J.E., Mascioli, D.W. and Rogers, D.T. 1980. Illegal transposition of mating type genes in yeast. Cell 20:519-528.

8) Haber, J.E. and Rogers, D.T. 1982. Transposition of a tandem duplication of yeast maing-type genes. Nature 296:768-773.

9) Haber, J.E., Rogers, D.T. and McCusker, J. 1980. Homothallic conversions of yeast mating-type genes occur by intrachromosomal recombination. Cell 22:277-289.

10) Haber, J.E., Rowe, J. and Rogers, D.T. 1981. Transposition of yeast mating type genes from two translocations of the left arm of chromosome III. Mol. Cell Biol. 1:1106-1123.

11) Hicks, J.B. and Herskowitz, I. 1976. Interconversion of yeast mating types. I. Direct observations of the action of homothallism (HO) gene. Genetics 83:245-258.

12) Hicks, J.B., Strathern, J.N. and Herskowitz, I. 1977. The cassette model of mating-type interconversion, p. 457-462. In: A.I. Bukhari, J.A. Shapiro and S.L. Adhya (ed.), DNA Insertion Elements, Plasmids and Episomes. Cold Spring Harbor Laboratory, Cold Spring Harbor, N.Y.

13) Hicks, J.B., Strathern, J.N. and Klar, A.J.S. 1979. Transposable mating type genes in Saccharomyces cerevisiae. Nature 282:478-483.

14) Jackson, J.A. and Fink, G.R. 1981. Gene conversion between duplicated genetic elements in yeast. Nature 292:306-311.

15) Klar, A.J.S., McIndoo, J., Strathern, J.N. and Hicks, J.B. 1980. Evidence for a physical interaction between the transposed and the substituted sequences during mating type gene transposition in yeast. Cell 22:291-298.

16) Klar, A.J.S., Strathern, J.N. and Hicks, J.B. 1981. A position-effect control for gene transposition: state of expression of yeast mating-type genes affects their ability to switch. Cell 25:517-524.

17) Klar, A.J.S., Hicks, J.B. and Strathern, J.N. 1982. Directionality of Yeast Mating-Type Interconversion. Cell 28:551-561.

18) Klein, H.L. and Petes, T.D. 1981. Intrachromosomal gene conversion in yeast. Nature 289:144-148.

19) Malone, R.E. and Esposito, R.E. 1980. The RAD52 gene as required for homothallic interconversion of mating types and spontaneous mitotic recombination in yeast. Proc. Nat. Acad. Sci. 77:503-507.

20) McCusker, J.H. and Haber, J.E. 1981. Evidence of chromosomal breaks near the mating type locus of S. cerevisiae that accompany MATα x MATα matings. Genetics 99:383-403.

21) Meselson, M. and Radding, C. 1975. A general model for genetic recombination. Proc. Natl. Acad. Sci. U.S. 72:358-361.

22) Nasmyth, K.A. and Tatchell, K. 1980. The structure of transposable yeast mating type loci. Cell 19:753-764.

23) Orr-Weaver, T.L., Szostak, J.W., and Rothstein, R.J. 1981. Yeast transformation: a model system for the study of recombination. Proc. Natl. Acad. Sci. USA 78:6354-6358.

24) Oshima, Y. and Takano, I. 1971. Mating types in Saccharomyces: their convertibility and homothallism. Genetics 67:327-335.

25) Oshima, T. and Takano, I. 1980. Mutants showing heterothallism from a homothallic strain of Saccharomyces cerevisiae. Genetics 94:841-857.

26) Resnick, M.A. 1976. Repair of double-strand breaks in DNA: a model involving recombination. J. Theor. Biol. 59:97-106.

27) Strathern, J., Spatola, E., McGill, C. and Hicks, J.B. 1980. Structure and organization of transposable mating type cassettes in Saccharomyces yeasts. Proc. Natl. Acad. Sci. USA 77:2839-2843.

28) Takano, I., Kusumi, T. and Oshima, Y. 1973. An α mating-type allele insensitive to the mutagenic action of the homothallic gene system in Saccharomyces diastaticus. Molec. Gen. Genet. 126:19-28.

29) Weiffenbach, B. and Haber, J.E. 1981. Homothallic mating type switching generates lethal chromosome breaks in rad52 strains of Saccharomyces cerevisiae. Mol. Cell. Biol. 1:522-534.

AUTHOR INDEX

Other Noyes Publications

GENETIC ENGINEERING AND NEW POLLUTION CONTROL TECHNOLOGIES

by

James B. Johnston and Susan G. Robinson
Institute for Environmental Studies
University of Illinois at Urbana-Champaign

Pollution Technology Review No. 106
Biotechnology Review No. 3

This book documents the basis for the use of genetic technology to develop new pollution controls in the area of waste treatment; it describes the current state of the art and recommends a future research approach. The book is based on information emerging from expert panel discussions covering pollution problems, gene manipulations, and natural limits to biodegradation.

Spectacular recent advances in gene manipulation provide excellent opportunities to design new organisms with defined degradative capacities; however problems arise in bringing the organisms into contact with a polluted environment in suitable fashion. How to release "engineered" microorganisms to the environment, safely, is a fundamental concern, and a public review policy is suggested.

The study provides background information for scientists, engineers, research managers, and other decision-makers for the evaluation of opportunities for improving biological waste treatment. Policy issues, research problems, and scientific resources that bear on pollution abatement through genetic engineering are identified. The book is intended to promote an interdisciplinary approach to a promising new genetic technology.

A condensed table of contents listing **chapter titles and selected subtitles** is given below.

ISBN 0-8155-0973-1 (1984) **131 pages**

Other Noyes Publications

ENZYME TECHNOLOGY
Preparation, Purification Stabilization, Immobilization
Recent Advances

Edited by S. Torrey

Biotechnology Review No. 2
Chemical Technology Review No. 222

Enzymes are the most efficient known catalysts for many industrial processes. This book reviews 155 recent processes involving enzyme technology—their preparation, purification, stabilization, and immobilization.

Modern enzyme technology is playing an increasing role in industrial processing, and is expected to become even more important as genetic engineering techniques advance. The world market for industrial enzymes is estimated at up to $400 million/year. Enzymes, and the microorganisms which produce them, are being used extensively in the food industry, pharmaceutical manufacture, and in medicine.

One of the most important areas of current research is immobilization. Enzymes bound to supports have the advantages of reusability, ease of separation from substrate, greater stability, and the potential for continuous reaction in columns or beds. One chapter of the book deals exclusively with immobilization processes.

Other chapters cover purification and stabilization, and therapeutic uses such as for anticoagulants to prevent thrombosis. Preparations of various oxidoreductases, hydrolases, transferases, isomerases, and lyases are described; and enzyme uses as reagents for immunoassays are also included.

The condensed table of contents listed below gives **chapter titles and selected subtitles.** Parenthetic numbers indicate the number of processes per topic.

ISBN 0-8155-0956-1 (1983)

308 pages

Other Noyes Publications

FERMENTATION AND BIOCHEMICAL ENGINEERING HANDBOOK
Principles, Process Design, and Equipment

Edited by

Henry C. Vogel
Davy McKee Corporation

This well-rounded, up-to-date handbook of fermentation and biochemical engineering presents the latest techniques for the commercial production of chemicals and pharmaceuticals via fermentation. Emphasis is given to unit operations—fermentation, separation, purification, and recovery. Principles, process design, and equipment are detailed.

The practical aspects of development, design, and operation are stressed, and some theory is included to provide the necessary insight for a particular operation. Problems addressed are the collection of pilot data, choice of scale-up parameters, selection of the right piece of equipment, pinpointing of likely trouble spots, and methods of troubleshooting.

The text, written from a practical and operating viewpoint, will assist development, design, engineering and production personnel in the fermentation industry. All of the contributors were selected based on their industrial background and orientation.

An introductory chapter on microbiology provides excellent background for those either entering or seeking to review the field.

Since much of fermentation involves antibiotic production, several chapters are specifically oriented to sterile production.

The book is illustrated with over 200 figures, photographs and schematic diagrams. **Chapter titles and contributors** are listed below.

ISBN 0-8155-0950-2 (1983) 439 pages

Other Noyes Publications

ENZYMATIC HYDROLYSIS OF CELLULOSE
Theory and Applications

by

C.R. Wilke, B. Maiorella, A. Sciamanna, K. Tangnu, D. Wiley, and H. Wong

Lawrence Berkeley Laboratory
and Department of Chemical Engineering
University of California, Berkeley

Chemical Technology Review No. 218

This book reviews the theory and applications of enzymatic hydrolysis of cellulosic biomass, a renewable raw material for chemicals and energy. The subject has been of great interest during recent energy crises; but it may possibly be of greater interest as genetic engineering techniques are applied to the conversion of cellulose to sugars.

In either case, this excellent and thorough study fills a gap in the available literature on the subject. The state of the art and potential industrial processes are detailed. A section is also included on high productivity fermentation systems for the production of ethanol.

The condensed table of contents given below lists **chapter titles and selected subtitles.**

ISBN 0-8155-0945-6(1983) **164 pages**